Wireless Blockchain

Wireless Blockchain

Principles, Technologies and Applications

Edited by

Bin Cao
Beijing University of Posts and Telecommunications
Beijing, China

Lei Zhang
University of Glasgow
Glasgow, UK

Mugen Peng
Beijing University of Posts and Telecommunications
Beijing, China

Muhammad Ali Imran
University of Glasgow
Glasgow, UK

IEEE PRESS

WILEY

Published by John Wiley & Sons Ltd., Chichester, United Kingdom.
Published simultaneously in Canada.

For general information on our other products and services or for technical support, please contact our Customer Care Department within the United States at (800) 762-2974, outside the United States at (317) 572-3993 or fax (317) 572-4002.

Wiley also publishes its books in a variety of electronic formats. Some content that appears in print may not be available in electronic formats. For more information about Wiley products, visit our web site at www.wiley.com.

Library of Congress Cataloging-in-Publication Data

Names: Cao, Bin, editor. | Zhang, Lei, editor. | Peng, Mugen, editor. |
 Imran, Muhammad Ali, editor.
Title: Wireless blockchain : principles, technologies and applications /
 Bin Cao, Beijing University of Posts and Telecommunications, Beijing,
 China, Lei Zhang, University of Glasgow, Glasgow, UK, Mugen Peng,
 Beijing University of Posts and Telecommunications, Beijing, China,
 Muhammad Ali Imran, University of Glasgow, Glasgow, UK.
Description: Chichester, United Kingdom ; Hoboken : Wiley-IEEE Press,
 [2022] | Includes bibliographical references and index.
Identifiers: LCCN 2021034990 (print) | LCCN 2021034991 (ebook) | ISBN
 9781119790808 (cloth) | ISBN 9781119790815 (adobe pdf) | ISBN
 9781119790822 (epub)
Subjects: LCSH: Blockchains (Databases) | Wireless communication
 systems–Industrial applications. | Personal communication service
 systems.
Classification: LCC QA76.9.B56 W57 2022 (print) | LCC QA76.9.B56 (ebook)
 | DDC 005.74–dc23
LC record available at https://lccn.loc.gov/2021034990
LC ebook record available at https://lccn.loc.gov/2021034991

Cover Design: Wiley
Cover Image: © phive/Shutterstock

Set in 9.5/12.5pt STIXTwoText by Straive, Chennai, India
Printed and bound by CPI Group (UK) Ltd, Croydon, CR0 4YY

C9781119790808_141021

Contents

List of Contributors

Nima Afraz
CONNECT Center, Trinity College
Dublin
Ireland

and

School of Computer Science
University College Dublin
Dublin
Ireland

Hamed Ahmadi
Department of Electronic Engineering
University of York
York
UK

Sandro Amofa
University of Electronic Science and
Technology of China
Chengdu
China

John G. Breslin
National University of Ireland
Galway
Ireland

Bin Cao
State Key Laboratory of Networking and
Switching Technology, Beijing University
of Posts and Telecommunications
Beijing
China

Volkan Dedeoglu
Data61, CSIRO
Brisbane
Australia

Zhi Ding
Department of Electrical and Computer
Engineering, University of California
Davis, CA
USA

Junyi Dong
James Watt School of Engineering
University of Glasgow
Glasgow
UK

Ali Dorri
School of Computer Science, QUT
Brisbane
Australia

Jaafar M. H. Elmirghani
School of Electronic and Electrical
Engineering, University of Leeds
Leeds
UK

Chenglin Feng
College of Science and Engineering
University of Glasgow
Glasgow
UK

Jianbin Gao
University of Electronic Science and
Technology of China
Chengdu
China

Xiqi Gao
National Mobile Communications
Research Laboratory, Southeast University
Nanjing
China

and

Purple Mountain Laboratories
Nanjing, Jiangsu
China

Dongyan Huang
College of Information and
Communications, Guilin University of
Electronic Technology
Guilin
China

Muhammad Ali Imran
James Watt School of Engineering
University of Glasgow
Glasgow
UK

Raja Jurdak
School of Computer Science, QUT
Brisbane
Australia

and

Data61, CSIRO
Brisbane
Australia

Hong Kang
James Watt School of Engineering
University of Glasgow
Glasgow
UK

Salil S. Kanhere
School of Computer Science and
Engineering, UNSW
Sydney
Australia

Samuel Karumba
School of Computer Science and
Engineering, UNSW
Sydney
Australia

Paulo Valente Klaine
James Watt School of Engineering
University of Glasgow
Glasgow
UK

Yuwei Le
National Mobile Communications
Research Laboratory, Southeast University
Nanjing
China

Wenyu Li
College of Science and Engineering
University of Glasgow
Glasgow
UK

Xintong Ling
National Mobile Communications
Research Laboratory, Southeast University
Nanjing
China

and

Purple Mountain Laboratories
Nanjing, Jiangsu
China

Weikang Liu
State Key Laboratory of Networking and
Switching Technology, Beijing University
of Posts and Telecommunications
Beijing
China

Kumudu Munasighe
Faculty of Science and Technology
University of Canberra
Canberra
Australia

Asuquo A. Okon
Faculty of Science and Technology
University of Canberra
Canberra
Australia

Mugen Peng
State Key Laboratory of Networking and
Switching Technology, Beijing University
of Posts and Telecommunications
Beijing
China

Marco Ruffini
CONNECT Center, Trinity College
Dublin
Ireland

Olusegun S. Sholiyi
National Space Research and Development
Agency, Obasanjo Space Centre
Abuja
Nigeria

Yao Sun
James Watt School of Engineering
College of Science and Engineering
University of Glasgow
Glasgow
UK

Subhasis Thakur
National University of Ireland
Galway
Ireland

Jiaheng Wang
National Mobile Communications
Research Laboratory, Southeast University
Nanjing
China

and

Purple Mountain Laboratories
Nanjing, Jiangsu
China

Taotao Wang
College of Electronics and Information
Engineering, Shenzhen University
Shenzhen
China

Qi Xia
University of Electronic Science and
Technology of China
Chengdu
China

Hao Xu
James Watt School of Engineering
University of Glasgow
Glasgow
UK

Bowen Yang
James Watt School of Engineering
College of Science and Engineering
University of Glasgow
Glasgow
UK

Lei Zhang
James Watt School of Engineering
College of Science and Engineering
University of Glasgow
Glasgow
UK

Shengli Zhang
College of Electronics and Information
Engineering, Shenzhen University
Shenzhen
China

Zaixin Zhang
James Watt School of Engineering
University of Glasgow
Glasgow
UK

Preface

Originally proposed as the backbone technology of Bitcoin, Ethereum, and many other cryptocurrencies, blockchain has become a revolutionary decentralized data management framework that establishes consensuses and agreements in trustless and distributed environments. Thus, in addition to its soaring popularity in the finance sector, blockchain has attracted much attention from many other major industrial sectors ranging from supply chain, transportation, entertainment, retail, healthcare, information management to financial services, etc.

Essentially, blockchain is built on a physical network that relies on the communications, computing, and caching, which serves the basis for blockchain functions such as incentive mechanism or consensus. As such, blockchain systems can be depicted as a two-tier architecture: an infrastructure layer and a blockchain layer. The infrastructure layer is the underlying entity responsible for maintaining the P2P network, building connection through wired/wireless communication, and computing and storing data. On the other hand, the top layer is the blockchain that is responsible for trust and security functions based on the underlying exchange of information. More specifically, blockchain features several key components that are summarized as transactions, blocks, and the chain of blocks. Transactions contain the information requested by the client and need to be recorded by the public ledger; blocks securely record a number of transactions or other useful information; using a consensus mechanism, blocks are linked orderly to constitute a chain of blocks, which indicates logical relation among the blocks to construct the blockchain.

As a core function of the blockchain, the consensus mechanism (CM, also referred to as consensus algorithm or consensus protocol) works in the blockchain layer in order to ensure a clear sequence of transactions and the integrity and consistency of the blockchain across geographically distributed nodes. The CM largely determines the blockchain system performance in terms of security level (fault tolerance level), transaction throughput, delay, and node scalability. Depending on application scenarios and performance requirements, different CMs can be used. In a permissionless public chain, nodes are allowed to join/leave the network without permission and authentication. Therefore, proof-based algorithms (PoX), such as proof-of-work (PoW), proof-of-stake (PoS), and their variants, are commonly used in many public blockchain applications (e.g. Bitcoin and Ethereum). PoX algorithms are designed with excellent node scalability performance through node competition; however, they could be very resource demanding. Also, these CMs have other limitations such as long transaction confirmation latency and low throughput. Unlike public chains, private and consortium blockchains prefer to adopt lighter protocols such as Raft and practical Byzantine fault tolerance (PBFT) to reduce computational power demand and improve the transaction throughput. A well-known example of PBFT implementation is the Hyper-Ledger Fabric, part of HyperLedger business blockchain frameworks. However, such CMs may require heavy communication resources.

Today, most state-of-the-art blockchains are primarily designed in stable wired communication networks running in advanced devices with sufficient communication resource provision. Hence, the blockchain performance degradation caused by communication is negligible. Nevertheless, this is not the case for the highly dynamic wireless connected digital society that is mainly composed of massive wireless devices encompassing finance, supply chain, healthcare, transportation, and energy. Especially through the upcoming 5G network, the majority of valuable information exchange may be through a wireless medium. Thus, it is critically important to answer one question, how much communication resource is needed to run a blockchain network (i.e. communication for blockchain).

From another equally important aspect when combining blockchain with communication (especially wireless communication), many works have focused on how to use blockchain to improve the communication network performance (i.e. blockchain for communication). This integration between wireless networks and blockchain allows the network to monitor and manage communication resource utilization in a more efficient manner, reducing its administration costs and improving the speed of communication resource trading. In addition, because it is the blockchain's inherit transparency, it can also record real-time spectrum utilization and massively improve spectrum efficiency by dynamically allocating spectrum bands according to the dynamic demands of devices. Moreover, it can also provide the necessary incentive for spectrum and resource sharing between devices, fully enabling new technologies and services that are bound to emerge. The resource coordination and optimization between resource requesters and providers can be automatically completed through smart contracts, thus improving the efficiency of resource optimization. Furthermore, with future wireless networks shifting toward decentralized solutions, with thousands of mobile cells deployed by operators and billions of devices communicating with each other, fixed spectrum allocation and operator-controlled resource sharing algorithms will not be scalable nor effective in future networks. As such, by designing a communications network coupled with blockchain as its underlying infrastructure from the beginning, the networks can be more scalable and provide better and more efficient solutions in terms of spectrum sharing and resource optimization, for example.

The book falls under a broad category of security and communication network and their transformation and development, which itself is a very hot topic for research these years. The book is written in such a way that it offers a wide range of benefits to the scientific community: while beginners can learn about blockchain technologies, experienced researchers and scientists can understand the extensive theoretical design and architecture development of blockchain, and industrial experts can learn about various perspectives of application-driven blockchains to facilitate different vertical sectors. Therefore, this feature topic can attract graduate/undergraduate level students, as well as researchers and leading experts from both academia and industry. In particular, some blockchain-enabled use cases included in the book are suitable for audiences from healthcare, computer, telecommunication, network, and automation societies.

In Chapter 1, the authors provide an overview of blockchain radio access network (B-RAN), which is a decentralized and secure wireless access paradigm. It leverages the principle of blockchain to integrate multiple trustless networks into a larger shared network and benefits multiple parties from positive network effects. The authors start from the block generation process and develop an analytical model to characterize B-RAN behaviors. By defining the work flow of B-RAN and introducing an original queuing model based on a time-homogeneous Markov chain, the steady state of B-RAN is characterized and the average service latency is derived. The authors then use the probability of a successful attack to define the safety property of B-RAN and evaluate potential factors that influence its security. Based on the modeling and analysis, the authors uncover an inherent trade-off relationship between security and latency and develop an in-depth understanding regarding the achievable performance of B-RAN. Finally, the authors verify the efficiency of the model through an innovative B-RAN prototype.

Chapter 2 theoretically and experimentally analyses different consensus algorithms in blockchains. The chapter firstly analyses the PoW consensus algorithm. The authors employ reinforcement learning (RL) to dynamically learn a mining strategy with the performance approaching that of the optimal mining strategy. Because the mining Markov decision process (MDP) problem has a non-linear objective function (rather than linear functions of standard MDP problems), the authors design a new multi-dimensional RL algorithm to solve the problem. Experimental results indicate that, without knowing the parameter values of the mining MDP model, the proposed multi-dimensional RL mining algorithm can still achieve optimal performance over time-varying blockchain networks. Moreover, the chapter analyzes the Raft consensus algorithm that is usually adopted in consortium/private blockchains. The authors investigate the performance of Raft in networks with non-negligible packet loss rate. They propose a simple but accurate analytical model to analyze the distributed network split probability. The authors conclude the chapter by providing simulation results to validate the analysis.

Chapter 3 describes a PBFT-based blockchain system, which makes it possible to break the communication complexity bottleneck of traditional PoW- or BFT-based systems. The authors discuss a double-layer PBFT-based consensus mechanism, which re-distributes nodes into two layers in groups. The analysis shows that this double-layer PBFT significantly reduces communication complexity. The authors then prove that the complexity is optimal when the nodes are evenly distributed in each group in the second layer. Further, the security threshold is analyzed based on faulty probability-determined (FPD) and the faulty number-determined (FND) models in the chapter. Finally, the chapter provides a practical protocol for the proposed double-layer PBFT system with a review of how PBFT is developed.

In Chapter 4, the authors start by introducing the basic concepts of blockchain and illustrating why a consensus mechanism plays an indispensable role in a blockchain-enabled Internet of Things (IoT) system. Then, the authors discuss the main ideas of two famous consensus mechanisms, PoW and PoS, and list their limitations in IoT. After that, the authors introduce PBFT and direct acyclic graph (DAG)-based consensus mechanisms as an effective solution. Next, several classic scenarios of blockchain applications in the IoT are introduced. Finally, the chapter is concluded with the discussion of potential issues and challenges of blockchain in IoT to be addressed in the future.

Chapter 5 addresses the issues associated with centralized marketplaces in 5G networks. The authors firstly study how a distributed alternative based on blockchain and smart contract technology could replace the costly and inefficient third-party-based trust intermediaries. Next, the authors propose a smart contract based on a sealed-bid double auction to allow resource providers and enterprise users to trade resources on a distributed marketplace. In addition, the authors explain the implementation of this marketplace application on HyperLedger Fabric permissioned blockchain while deploying the network using a pragmatic scenario over a public, commercial cloud. Finally, the authors evaluated the distributed marketplace's performance under different transaction loads.

In Chapter 6, the authors describe an integrated blockchain and software-defined network (SDN) architecture for multi-operator support in 6G networks. They present a unified SDN and blockchain architecture with enhanced spectrum management features for enabling seamless user roaming capabilities between mobile network operators (MNOs). The authors employ the smart contract feature of blockchain to enable the creation of business and technical agreements between MNOs for intelligent and efficient management of spectrum assets (i.e. the radio access network). The study shows that by integrating blockchain and SDN, the foundation for creating trusted interactions in a trustless environment can be established, and users can experience no disruption in service with very minimal delay as they traverse between operators.

Chapter 7 investigates and discusses the integration of blockchain and mobile edge computing (MEC). The authors firstly provide an overview of the MEC, which sinks computing power to the edge of networks and integrates mobile access networks and Internet services in 5G and beyond. Next, the authors introduce the typical framework for blockchain-enabled MEC and MEC-based blockchain, respectively. The authors further show that blockchain can be employed to ensure the reliability and irreversibility of data in MEC systems, and in turn, MEC can also solve the major challenge in the development of blockchain in IoT applications.

Chapter 8 establishes an analytical model for PoW-based blockchain-enabled wireless IoT systems by modeling their spatial and temporal characteristics as Poisson point processes (PPP). The authors derive the distribution of signal-to-interference-plus-noise ratio (SINR), blockchain transaction successful rate, as well as its overall throughput. Based on this performance analysis, the authors design an algorithm to determine the optimal full function node deployment for blockchain systems under the criterion of maximizing transaction throughput. In addition, the security performance of the proposed system is analyzed in the chapter considering three different types of malicious attacks. The chapter ends with a series of numerical results to validate the accuracy of the theoretical analysis and optimal node deployment algorithm.

In Chapter 9, the authors examine the factors governing successful deployment of blockchain-based distributed energy trading (DET) applications and their technical challenges. The chapter walks through the fundamentals of "citizen-utilities," primarily assessing its impact on efforts to manage distributed generation, storage, and consumption on the consumer side of the distribution network, while intelligently coordinating DET without relying on trusted third parties. Additionally, the chapter highlights some of the open research challenges including scalability, interoperability, and privacy that hinder the mainstream adoption of "citizen-utilities" in the energy sector. Then, to address these research challenges, the authors propose a scalable citizen-utility that supports interoperability and a Privacy-preserving Data Clearing House (PDCH), which is a blockchain-based data management tool for preserving on-ledger and off-ledger transactions data privacy. The chapter is finished with outlines of future research directions of PDCH.

In Chapter 10, the authors introduce a blockchain-enabled COVID-19 contact tracing solution named BeepTrace. This novel technology inherits the advantages of digital contract tracing (DCT) and blockchain, ensuring the privacy of users and eliminating the concerns about the third-party trust while protecting the population's health. Then, based on different sensing technologies, i.e. Bluetooth and GPS, the authors categorize BeepTrace into BeepTrace-active mode and BeepTrace-passive mode, respectively. In addition, the authors summarize and compare the two BeepTrace modes and indicate their working principles and privacy preservation mechanisms in detail. After that, the authors demonstrate a preliminary approach of BeepTrace to prove the feasibility of the scheme. At last, further development prospects of BeepTrace or other decentralized contact tracing applications are discussed, and potential challenges are pointed out.

Chapter 11 looks at the infusion of blockchain technology into medical data sharing. The chapter provides an overview of medical data sharing and defines the challenges in this filed. The authors revisit some already established angles of blockchain medical data sharing in order to properly contextualize it and to highlight new perspectives on the logical outworking of blockchain-enabled sharing arrangements. Then, the authors present three cases that are especially suited to blockchain medical data sharing. They also present an architecture to support each paradigm presented and analyze medical data sharing to highlight privacy and security benefits to data owners. Finally, the authors highlight some new and emerging services that can benefit from the security, privacy, data control, granular data access, and trust blockchain medical data sharing infuses into healthcare.

In Chapter 12, the authors propose a blockchain-based decentralized content vetting for social networks. The authors use Bitcoin as the underlying blockchain model and develop an unidirectional channel model to execute the vetting procedure. In this vetting procedure, all users get a chance to vote for and against a content. Content with sufficient positive votes is considered as vetted content. The authors then optimize the offline channel network topology to reduce computation overhead because of using blockchains. At last, the authors prove the efficiency of the vetting procedure with experiments using simulations of content propagation in social network.

Bin Cao, Lei Zhang, Mugen Peng, Muhammad Ali Imran
August 2021

Abbreviations

3GPP	third-Generation Partnership Project
4G	fourth generation
5G	fifth generation
6G	sixth generation
AAS	Authentication Agent Server
ABAC	Attribute-Based Access Control
ABIs	Application Binary Interfaces
ABM	Adaptive Blockchain Module
Abstract	Abortable Byzantine faulT toleRant stAte maChine replicaTion
ACC	Access Control Contract
AI	artificial intelligence
API	application programming interface
APs	access points
BaaS	Blockchain as a Service
BAS	Blockchain Agent Server
BASA	Blockchain-assisted Secure Authentication
BFT	Byzantine fault tolerance
BLE	Bluetooth low energy
BMap	Bandwidth Map
BN	blockchain network
BPL	building penetration losses
BPM	Business Process Management
bps	bits per second
B-RAN	Blockchain radio access network
BTC	bitcoin
CA	Certification Authority
CAGR	Compound annual growth rate
CAPEX	Capital expenditure
CBRS	Citizens Broadband Radio Services
CDC	Center for Diseases Control
CDF	cumulative distribution function
CFT	crash fault tolerance
CM	consensus mechanism

CoAP	Constrained application protocol
COVID-19	Coronavirus Disease 2019
CPU	Central Processing Unit
DAG	direct acyclic graph
DAS	distributed antenna systems
DCT	digital contact tracing
DDoS	Distributed Denial of Service
DEPs	Distributed Energy Prosumers
DER	Distributed Energy Resources
DET	Distributed Energy Trading
DIS	Data integrity verification systems
DLT	Distributed ledger technology
DoS	Denial of Service
DPoS	Delegate Proof-of-Stake
DR	demand response
DS	Directory Service
DSM	Demand Side Management
DSO	distributed system operator
dTAM	data Tagging and Anonymization Module
DTLS	Datagram Transport Layer Security
E2E	end to end
ECO	Energy Company Obligation
EE	energy efficiency
EMR	electronic medical record
eNBs	eNodeBs
ESPs	Edge computing service providers
ESS	energy storage systems
EV	electric vehicles
EVN	Electric Vehicle Networks
FAPs	femtocell access points
FCC	Federal Communications Commission
FDI	false data injection
FeGW	Femtocell gateways
FiT	Feedin Tariff
FL	Federated learning
FND	faulty number determined
FNs	function nodes
FPD	faulty probability determined
FSC	food supply chain
FSCD	fast smart contract deployment
FTTH	Fiber-to-the-Home
G2V	grid-to-vehicle
Gb/s	gigabyte per second
GDPR	General Data Protection Regulation
Geth	go-Ethereum

GPS	Global Position System
GTP	GPRS tunneling protocol
HARB	Hypergraph-based Adaptive Consortium Blockchain
HARQ	Hybrid Automatic Repeat Request
HeNB	home eNB
HLF	Hyperledger fabric
HSS	home subscriber server
HTLC	hash time-locked contract
HVAC	heating, ventilation, cooling, and air conditioning
IaaS	Infrastructure as a Service
IBC	Identity-based Cryptography
IBS	Identity-based Signature
IDC	International Data Corporation
IDE	Integrated development environment
IFA	Dentifier for advertisers
IIoT	Industrial Internet of things
IMDs	Internet of things/mobile devices
IMEI	International mobile equipment identity
IMT	International Mobile Telecommunications
InPs	Infrastructure Providers
IoT	Internet of things
IoVs	Internet of vehicles
KGC	Key Generation Center
KPIs	Key performance indicators
LAN	Local area network
LRSig	Linkable Ring Signatures
LSA	Licensed shared access
LTE	long-term evolution
MAC	Media access control
MadIoT	Manipulation of demand via IoT
MBS	Macrocell base station
MCMC	Markov Chain Monte Carlo
MDP	Markov decision process
MEC	mobile edge computing
MIMO	multiple-input, multiple-output
MME	mobility management entity
MNOs	mobile network operators
MOCN	multi-operator core network
μOs	micro-operators
MSP	Membership Service Providers
MSP	multi-sided platform
MTT	maximum transaction throughput
MVNO	Mobile Virtual Network Operator
MW	megawatts
Naas	Network as a Service

NAT	nucleic acid testing
NFV	Network Function Virtualization
NGN	next-generation network
NHS	National Health Service
NPI	Non-pharmaceutical intervention
ns-3	Network simulator 3
OAMC	Object Attribute Management Contract
ODN	Optical Distribution Network
OFSwitch	open-flow switch
OPEX	Operating expenditure
OSN	Online Social Network
OTP	one time programmable
OTT	over-the-top
P2P	peer-to-peer
PaaS	Platform as a Service
PBFT	Practical Byzantine Fault Tolerance
PBN	public blockchain network
PCG	Prosumer Community Groups
PCRF	Policy Charging and Rules Function
PDCH	Privacy-preserving Data Clearing House
PDF	probability density function
P-GW	packet data network gateway
PHY	physical
PKI	Public key infrastructure
PMC	Policy Management Contract
PoD	proof-of-device
PONs	Passive optical networks
PoO	proof-of-object
PoS	proof of stake
PoW	proof of work
PPP	Poisson point processes
QoS	Quality of Service
RAN	Radio access network
RES	renewable energy sources
RL	reinforcement learning
RMG	relative mining gain
RSRP	reference signal received power
RSRQ	reference signal received quality
SaaS	software as a Service
SAMC	Subject Attribute Management Contract
SARS-CoV-2	Severe Acute Respiratory Syndrome Coronavirus 2
SBCs	single-board computers
SBSs	small base stations
SDN	software-defined network
SEMC	Smart Energy Management Controller

S-GW	Serving gateway
SHeNB	Serving HeNB
SINR	signal-to-interference-plus-noise ratio
SLA	service-level agreement
SM	supermassive
SPF	single point of failure
SPs	service providers
SR	spinning reserve
SUTs	System under tests
Tb/s	terabyte per second
TDP	transaction data packet
THeNB	target HeNB
TNs	transaction nodes
TPA	Third Party Auditor
TPS	transactions per second
TTI	transmission time interval
TTP	Trusted Third Party
TTT	time to trigger
UE	user equipment
UE RRC	UE radio resource control
URI	Uniform Resource Identifier
URL	Uniform Resource Locator
UUID	Universally Unique Identifier
V2G	vehicle-to-grid
V2V	vehicle-to-vehicle
vCPUs	Virtual Central Processing Units
VEN	Vehicular energy networks
VLC	visible light communications
VM	virtual machine
VNO	Virtual Network Operator
VPP	Virtual Power Plants
WAN	Wide Area Network
WHO	World Health Organization
ZKP	zero-knowledge-proofs

1

What is Blockchain Radio Access Network?*

Xintong Ling[1,2], Yuwei Le[1], Jiaheng Wang[1,2], Zhi Ding[3], and Xiqi Gao[1,2]

[1]*National Mobile Communications Research Laboratory, Southeast University, Nanjing, 210096, China*
[2]*Purple Mountain Laboratories, Nanjing, Jiangsu, 211111, China*
[3]*Department of Electrical and Computer Engineering, University of California, 95616, Davis, CA, USA*

1.1 Introduction

The past decade has witnessed tremendous growth in emerging wireless technologies geared toward diverse applications [1]. Radio access networks (RANs) are becoming more heterogeneous and highly complex. Without well-designed inter-operation, mobile network operators (MNOs) must rely on their independent infrastructures and spectra to deliver data, often leading to duplication, redundancy, and inefficiency. A huge number of currently deployed business or individual access points (APs) have not been coordinated in the existing architecture of RANs and are therefore under-utilized. Meanwhile, user equipments (UEs) are not granted to access to APs of operators other than their own, even though some of them may provide better link quality and economically sensible. The present state of rising traffic demands coupled with the under-utilization of existing spectra and infrastructures motivates the development of a novel network architecture to integrate multiple parties of service providers (SPs) and clients to transform the rigid network access paradigm that we face today.

Recently, blockchain has been recognized as a disruptive innovation shockwave [2–4]. Federal Communications Commission (FCC) has been suggested that blockchain may be integrated into wireless communications for the next-generation network (NGN) in the Mobile World Congress 2018. Along the same line, the new concept of blockchain radio access network (B-RAN) was formally proposed and defined in [5, 6]. In a nutshell, B-RAN is a decentralized and secure wireless access paradigm that leverages the principle of blockchain to integrate multiple trustless networks into a larger shared network and benefits multiple parties from positive network effects [6]. It is a new architecture that integrates both characteristics of wireless networks and distributed ledger technologies. As revealed in [5, 6], B-RAN can improve the overall throughput through simplified inter-operator

---

* X. Ling, Y. Le, J. Wang, Z. Ding, and X. Gao. Practical modeling and analysis of blockchain radio access network. *IEEE Transactions on Communications*, 69(2): 1021–1037, Feb. 2021.

cooperation in the network layer (rather than increasing the channel capacity in the physical layer). B-RAN can enhance the data delivery capability by connecting these RANs into a big network and leveraging the power of multi-sided platform (MSP). The positive network effect can help B-RAN recruit and attract more players, including network operators, spectral owners, infrastructure manufacturers, and service clients alike [6]. The subsequent expansion of such a shared network platform would make the network platform more valuable, thereby generating a positive feedback loop. In time, a vast number of individual APs can be organized into B-RAN and commodified to form a sizable and ubiquitous wireless network, which can largely improve the utility of spectra and infrastructures. In practice, rights, responsibilities, and obligations of each participant in B-RAN can be flexibly codified as smart contracts executed by blockchain.

Among the existing studies on leveraging blockchain in networks, most have focused on Internet of Things (IoT) [7–11], cloud/edge computing [12–15], wireless sensor networks [16], and consensus mechanisms [17–19]. Only a few considered the future integration of blockchain in wireless communications [20–26]. Weiss et al. [20] discussed several potentials of blockchain in spectrum management. Kuo et al. [21] summarized some critical issues when applying blockchain to wireless networks and pointed out the versatility of blockchain. Pascale et al. [22] adopted smart contracts as an enabler to achieve service level agreement (SLA) for access. Kotobi and Bilen [23] proposed a secure blockchain verification protocol associated with virtual currency to enable spectrum sharing. Le et al. [27] developed an early prototype to demonstrate the functionality of B-RAN.

Despite the growing number of papers and heightened interests to blockchain-based networking, works including fundamental analysis are rather limited. A number of critical difficulties remain unsolved. (i) Existing works have not assessed the impact of decentralization on RANs after introducing blockchain. Decentralization always comes with a price that should be characterized and quantified. (ii) Very few papers have noticed that service latency will be a crucial debacle for B-RAN as a price of decentralization [22]. Unfortunately, the length of such delay and its controllability are still open issues. (iii) Security is yet another critical aspect of blockchain-based protocols. In particular, alternative history attack, as an inherent risk of decentralized databases, is always possible and must be assessed. (iv) A proper model is urgently needed to exploit the characteristics of B-RAN (such as latency and security) and to further provide insights and guidelines for real-world implementations.

To address the aforementioned open issues, this chapter establishes a framework to concretely model and evaluate B-RAN. We start from the block generation process and develop an analytical model to characterize B-RAN behaviors. We shall evaluate the performance in terms of latency and security in order to present a more comprehensive view of B-RAN. We further verify the efficacy of our model through an innovative B-RAN prototype. The key contributions are summarized as follows:

- We define the workflow of B-RAN and introduce an original queuing model based on a time-homogeneous Markov chain, the first known analytical model for B-RAN.
- From the queuing model, we analytically characterize the steady state of B-RAN and further derive the average service latency.
- We use the probability of a successful attack to define the safety property of B-RAN and evaluate potential factors that influence the security.

Table 1.1 Important variables in the modeling and analysis

Symbols	Explanations	Symbols	Explanations
τ_k^c	Required service time of request k	t_k^a	Arrival epoch of request k
λ^a	Request arrival rate	$T^a = 1/\lambda^a$	Average inter-arrival time
λ^b	Block generation rate	$T^b = 1/\lambda^b$	Average block time
λ^c	Service rate	$T^c = 1/\lambda^c$	Average service time
N	Number of required confirmations	s	Number of access links
β	Relative mining rate of an attacker	$\rho = \dfrac{\lambda^a}{s\lambda^c}$	Traffic intensity
$\Phi = \{\lambda^a, \lambda^b, \lambda^c, s\}$		Basic configuration of B-RAN	

- Based on the modeling and analysis, we uncover an inherent trade-off relationship between security and latency, and develop an in-depth understanding regarding the achievable performance of B-RAN.
- Finally, we build a B-RAN prototype that can be used in comprehensive experiments to validate the accuracy of our analytical model and results.

We organize this chapter as follows. Section 1.2 presents the B-RAN framework and the prototype. Section 1.3 provides the mining model to describe the block generation process. In Section 1.4, we establish the B-RAN queuing model, with which we analyze and evaluate the B-RAN performance concerning latency and security in Sections 1.5 and 1.6, respectively. We demonstrate the latency-security trade-off in Section 1.7 and provide some in-depth insights into B-RAN. Section 1.8 concludes this chapter. Given the large number of symbols to be used, we summarize the important variables in Table 1.1.

1.2 What is B-RAN

1.2.1 B-RAN Framework

B-RAN offers a decentralized and secure wireless access paradigm for large-scale, heterogeneous, and trustworthy wireless networks [5, 6]. B-RAN unites multilateral inherently trustless network entities without any trusted middleman and manages network access, authentication, authorization, and accounting via direct interactions. As an open unified framework for diverse applications to achieve resource pooling and sharing across sectors, B-RAN presents an attractive solution for future 6G networking.

With the help of blockchain, B-RAN can form an expansive cooperative network including not only telecommunication giants but also small contract holders or individual MNOs to deliver excellent quality services at high spectrum efficiency. B-RAN can integrate multiple networks across SPs for diverse applications. As illustrated in Figure 1.1, B-RAN is self-organized by APs belonging to multiple SPs, massive UEs, and a blockchain maintained by miners. In B-RAN, a confederacy of SPs (organizations or individuals) act as a virtual SP (VSP) to provide public wireless access under shared control. These SPs in B-RAN allow

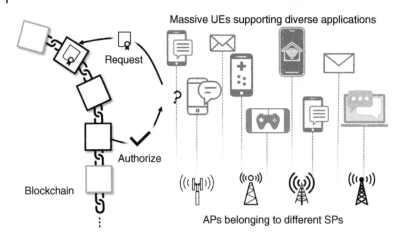

Figure 1.1 Conceptual illustration of self-organized B-RAN.

the greater pool of UEs to access their APs and networks by receiving payment or credit for reciprocal services. Blockchain acts as a public ledger in B-RAN for recording, confirming, and enforcing digital actions in smart contracts to protect the interests of all participants.

B-RAN is envisioned to be broadly inter-operative and to support diverse advanced wireless services and standards. In this chapter, we focus on the fundamental access approach for which the procedure is shown in Figure 1.2.

- In preparation for access, UEs and SPs should first enter an SLA containing the details including service types, compensation rates, among other terms. (For example, SPs can first publish their service quality and charge standard, and UEs select suitable SPs according to the expenditure and quality of service.) The service terms and fees will be explicitly recorded in a smart contract authorized by the digital signatures of both sides.
- In phase 1, the smart contract with the access request is committed to the mining network and is then verified by miners. The verified contracts are assembled into a new block, which is then added at the end of the chain.
- In phase 2, the block is accepted into the main chain after sufficient blocks as confirmations built on top of it.
- In phase 3, the request is waiting for service in the service queue.
- In phase 4, the access service is delivered according to the smart contract.

The above procedure can be viewed as a process of trust establishment between clients and SPs, similar to negotiating and signing monthly contracts between users and MNOs. Thus, in B-RAN, clients can obtain access services more conveniently through the above process instead of signing contracts with a specific MNO in advance. The service duration in B-RAN is flexible and can be as short as a few minutes or hours, which is different from typical long-term plans (e.g. monthly plans). UEs can prolong access services by renewing the contract earlier before the previous one expires in order to continue the connection status. Therefore, service latency in this context refers to the delay when a UE accesses an unknown network for the first time and can be viewed as the time establishing trust between two trustless parties, which is significantly different from the transmission delay in the physical layer.

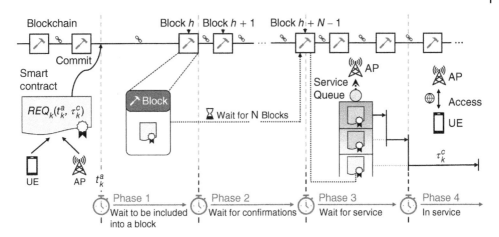

Figure 1.2 Four stages of the access workflow in B-RAN.

Mathematically, we can describe the request structure by $\text{REQ}_k\left(t_k^a, \tau_k^c\right)$ shown in Figure 1.2, where t_k^a and τ_k^c are the arrival epoch and the service duration of request k, respectively. Assume that the access requests are mutually independent and arrive as a Poisson process with rate λ^a. Equivalently, the inter-arrival time between two requests U^a follows exponential distribution with mean $T^a = 1/\lambda^a$. Based on well-known studies such as [28], the random service time τ_k^c is also expected to be exponential with mean $T^c = 1/\lambda^c$. Note that in this chapter, we consider a tract covered by multiple trustless SPs (organizations or individuals). Usually, the block size limit is much larger than the request rate of a single tract and thus can be ignored in this case.

In the context of B-RAN, we introduce the concept of "virtual link" instead of the physical channel, where one virtual link represents a tunnel providing access services to one client at a time.[1] Hence, the number of links means the maximum number of UEs that can receive access services simultaneously from the APs belonging to these SPs in the tract and reflects the access capability of a network. In the considered tract, we assume the maximum number of links to be s.

It is worth pointing out that the efficiency improvement of B-RAN roots in the network pooling principle, which requires a flexible offloading and sharing between subnetworks. Inspired by the Schengen Agreement, B-RAN adopts the mechanism that, if a client establishes trust with specific SPs via the procedure in Figure 1.2, the client may use resources pooled by SPs in B-RAN, e.g. a frequency band belonging to another SP. The miners can use some intelligent algorithms[2] to allocate and distribute the pooled resources for higher network efficiency [6]. As a result, mobile devices may access suitable APs belonging to the SPs, which likely provide higher quality coverage for the UEs in their current locations. The trading among SPs caused by roaming would be calculated and settled via blockchain

---

1 Usually, the physical channel may provide services to many clients via proper multiple access techniques. We use the concept of "virtual link" to avoid ambiguity.
2 Note that, in this chapter, the miners still follow the basic first-in-first-out principle to put the requests into the queue.

periodically. Based on the above mechanism, B-RAN can take advantage of pooling and sharing across subnetworks. Please refer to Figure 1.9 in this chapter and [6, 10] for more details and evidence for the pooling effect in B-RAN.

1.2.2 Consensus Mechanism

B-RAN, as a decentralized system, requires proper consensus mechanisms for consistency [18]. Proof-of-Work (PoW) has been widely used in practice and proven to be secure in cryptocurrencies such as Bitcoin. In PoW, network maintainers, also known as miners, need to obtain a hash value below a given target by repeatedly guessing a random variable named nonce. However, PoW-based consensus mechanisms consume a tremendous amount of energy, which is likely unbearable for energy-limited mobile devices.

Consequently, proof-of-device (PoD) is proposed for B-RAN as a low- cost alternative [5]. PoD utilizes the fact that wireless access usually depends on a hardware device associated with a unique identifier in order to elevate the cost of creating new identities. To be more specific, PoD can create new identifiers or rely on some existing ones, such as the international mobile equipment identity (IMEI) and identifier for advertisers (IFAs), in one time programmable (OTP) memory to distinguish different network entities and prevent identity fraud. Also, because of variations during manufacture, every device has multiple hardware-dependent features, which could constitute a unique RF fingerprinting for each device and can be identified from the transmitted RF signal [29, 30]. Forging an identity of a device is often costly in the real world, whereas creating multiple identities is almost costless in cryptocurrencies. Therefore, PoD can safeguard the security of B-RAN without expending immense computing power and is thus suitable for wireless networks. Notably, PoW, PoD, and other alternatives can be put in the same class since all of them are based on hash puzzles. We will further discuss and model the block generation process of a hash-based consensus mechanism in Section 1.3.

1.2.3 Implementation

In order to evaluate our established model, we will provide demonstrative experimental results from a home-built prototype throughout the whole article. We implement this version of B-RAN prototype on four single board computers (SBCs) and use a workstation with Intel Core CPU I7-8700K and 32GB RAM in order to provide sufficient computing power. Our prototype consists of a standard file system for data storage, a key-value database for file index, and the core modules written in Python. The prototype supports both PoW and PoD consensus mechanisms as two available options and can adopt an appropriate one according to the specific environment and requirements. We configure different SBCs as UEs and APs and set up the integrated development environment (IDE), wherein UEs propose access requests according to the input configurations, and APs provide services based on the workflow given in Section 1.2.1. During tests, the prototype can track running statistics and provide them as output results.

The B-RAN prototype is a hierarchical architecture with a number of modules and components [6]. For example, the fast smart contract deployment (FSCD) was proposed in [27] to accelerate the service deployment, and the hash time-locked contract (HTLC) is designed in [6] to enforce the contract and secure the trading process. This chapter is important by modeling and assessing B-RAN, and thus, we cannot include all the technical details here. Please see these citations for more details. Note that, in this chapter, the average service time T^c as time unit is set to unity without loss of generality, so time is measured as relative variables in terms of time unit T^c.

In Sections 1.5–1.7, we will assess the performance of B-RAN from different points of view and verify our model step by step through prototype verifications. Although these verifications focus on different aspects, all of them are obtained from the same B-RAN prototype described above.

1.3 Mining Model

1.3.1 Hash-Based Mining

In this section, we will present a general model for hash-based puzzles to describe the block generation process, also known as mining. Usually, block propagation in a network is much faster than block generation. We thus ignore the block spreading delay in the modeling and verify the assumption by using the real data of Bitcoin and Ethereum.

Generally, a hash puzzle can be modeled as a problem to find a suitable answer to satisfy the following conditions:

$$\text{Hash}(\text{HP} + \text{DP} + \text{TS} + \text{OF}) < \text{GT}. \tag{1.1}$$

Here, "HP" stands for the hash pointer to a previous block, "DP" means the data payload, "TS" is the current timestamp, "GT" represents a given target, and "OF" is the optional field depending on the specific type of the hash puzzle. For instance, in the PoW protocol, the optional field can be any random number, whereas in the PoD protocol, the optional field is given by the hardware identifier. In PoW, the range of the optional field is unlimited. Hence, a miner can guess many times to find a correct nonce, and hence, the number of trials is only restricted by the mining rigs. In PoD, the optional field is given by the tamper-proof identifier such that each device can perform the hash computation only once for each timestamp, thereby largely reducing the power consumption. The premise behind that is that the entities in real RANs cannot be effortlessly forged or created. The characteristics, e.g. security and power consumption, of different hash-based consensus mechanisms can be traded off by properly choosing the optional field.

1.3.2 Modeling of Hash Trials

For a general hash-based mining process, each hash trial can be regarded as an independent Bernoulli experiment with success probability p as the timestamp keeps changing. In a sequence of independent Bernoulli trials, the probability that the first block is generated after m failures is $(1 - p)^m p$. Let W^b be the number of failures preceding the first success. Then, W^b follows geometric distribution:

$$\Pr\left\{W^b = m\right\} = (1 - p)^m p, \quad m = 0, 1, \dots . \tag{1.2}$$

Easily, we can obtain

$$\Pr\left\{W^b \geq m\right\} = (1-p)^m, \quad m = 0, 1, \dots. \tag{1.3}$$

Note that W^b can be viewed as the waiting period before a block is successfully generated, and its distribution can be described by (1.2) and (1.3). The average number of successes in m independent trials is mp.

Hence, if m hash trials are conducted in a time interval of length τ, then the success rate defined as the mean number of successes per unit time is $\lambda^b = mp/\tau$. Now, let $p \to 0$ and $m \to \infty$ in the way that keeps λ^b constant. We can visualize an experiment with infinite hash trials performed within interval τ. Successive trials are infinitesimally close with vanishingly small probability of success, but the mean number of successes remains a non-zero constant $\lambda^b \tau$. By using the fact that the geometric distribution approaches the exponential distribution in the limit, we have

$$\lim_{m \to \infty} (1-p)^m \Big|_{p=\frac{\lambda^b \tau}{m}} = \lim_{m \to \infty} \left(1 - \frac{\lambda^b \tau}{m}\right)^m = \exp(-\lambda^b \tau). \tag{1.4}$$

Letting a random variable U^b be the continuous time before preceding the first success, we have

$$\Pr\left\{U^b > \tau\right\} = \Pr\left\{W^b > m\right\} = \exp(-\lambda^b \tau). \tag{1.5}$$

We name U^b as the block time in the context of blockchain. According to (1.5), the average block time $\mathbb{E}\left[U^b\right]$, denoted by T^b, is equal to $1/\lambda^b$, where λ^b is thus called as the mining rate representing how fast blocks generate.

Equations (1.4) and (1.5) imply that, if the number of hash computations of the whole network in a unit time tends to infinity, then the length of time between two successive blocks, i.e. U^b's, would follow the exponential distribution. Interestingly, most of mature PoW blockchain networks indeed perform a huge number of hash computations every moment. For example, the minimum hash rates of bitcoin and Ethereum during 2018 are 14 891 and 159TH/s.[3]. These are humongous numbers in real world and practically support the limiting condition that the number of trials tends to infinity. When massive hardware devices are participating in mining, the condition also holds for PoD. Remark that, even if the number of trials is finite and the exponential approximation no longer holds, the block times U^b's are still mutually independent and identically distributed because of the memoryless property of geometric distribution.

Now, we further prove that block generation forms a Poisson process. Let $B(n, t, t+h)$ denote the event that n blocks are generated in interval $(t, t+h)$. As $h \to 0$, the probability that at least one block is generated in $(t, t+h)$ is

$$\begin{aligned}
\Pr\{B(n \geq 1, t, t+h)\} &= 1 - \Pr\{B(n = 0, t, t+h)\} \\
&= 1 - \Pr\left\{U^b > h\right\} \\
&= 1 - \exp(-\lambda^b h) \\
&= \lambda^b h + o(h), \quad h \to 0.
\end{aligned}$$

---

3 Source: bitinfocharts.com/, accessed February 2021.

We note that $o(h)$ is an infinitesimal of higher order such that $\lim_{h \to 0} o(h)/h = 0$. Exactly, one block is generated in $(t, t+h)$, i.e. $B(n = 1, t, t+h)$, if and only if the first block occurs within $(t, t+h)$ and the second one occurs after the epoch $t + h$. Hence, the probability of $\Pr\{B(n = 1, t, t+h)\}$ is

$$\Pr\{B(n = 1, t, t+h)\}$$

$$= \int_0^h \Pr\{U_1^b = h - u\} \ \Pr\{U_2^b > u\} \, du$$

$$= \int_0^h \lambda^b \exp(-\lambda^b(h-u)) \exp(-\lambda^b u) du$$

$$= \lambda^b h \exp(-\lambda^b h) = \lambda^b h + o(h), \quad h \to 0,$$

where U_1^b and U_2^b represent the generation times of first and second blocks since t, respectively. As $h \to 0$, the probability of occurrence of two or more blocks in $(t, t+h)$ is

$$\Pr\{B(n \geq 2, t, t+h)\} = 1 - \exp(-\lambda^b h) - \lambda^b h \exp(-\lambda^b h) = o(h), \quad h \to 0.$$

In conclusion, the block generation can be modeled as a Poisson process with mining rate λ^b. Since the negligible propagation delay and infinity hash rate are assumed in the modeling, we shall carefully verify the validation of the Poisson model by real data.

In **Prototype Verification A**, we illustrate the validity of the Poisson model in Figure 1.3 by empirical data. In our self-built B-RAN prototype described in Figure 1.3a, we deploy five miners with equal mining rates and set the block generation rate to $\lambda^b = 1/10$ or equivalently the average block time $T^b = 1/\lambda^b = 10$. We measure the block time for 10 000 blocks and plot the histogram to approximate the distribution of block time U^b. One can see that the Poisson model closely fits the data. Furthermore, we collected the empirical data from Bitcoin (51 036 blocks starting at height 530 114[4]) and Ethereum (2 078 000 blocks starting at height 5 806 289[5]) to demonstrate the validity of the model in practice. The collected Bitcoin and Ethereum data are also consistent with the model in a real network where propagation delay does exist. To be more specific, we calculate R^2 as a goodness-of-fit indicator

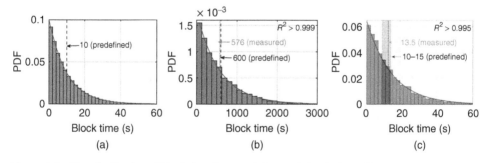

Figure 1.3 The distribution of block time from real data and simulations. (Please see footnotes 4 and 5 for sources.) (a) Self-built B-RAN prototype. (b) Bitcoin. (c) Ethereum.

4 Source: www.blockchain.com/en/btc/blocks/, accessed February 2021.
5 Source: etherscan.io/block/, accessed February 2021.

and obtain $R^2 > 0.999$ for Bitcoin and $R^2 > 0.995$ for Ethereum. The above results strongly support that the Poisson model perfectly fits the data, even in Ethereum where the block propagation is in the same order of magnitude as the block time. Therefore, even though the propagation delay is simply neglected, the Poisson model can still well characterize the block generation in practice.

1.3.3 Threat Model

In this study, we consider an adversary who mounts an alternative history attack by attempting to generate a longer fraudulent chain. The adversary has to generate new blocks in the same way as honest miners because the hash value can hardly be tampered. Hence, the fraudulent blocks are also generated as Poisson. Now, assume that the mining rate of the attacker is $\beta \lambda^b$, while the mining rate of honest miners is λ^b, i.e. the attacker controls $\frac{\beta}{1+\beta}$ fraction of hash power.[6] Both the attacker and the benign network are mining independently. From the additive property of Poisson processes, the sum generation rate of both benign and fraudulent block is $(\beta + 1)\lambda^b$. Note that, in an alternative history attack, the attacker will not publish the fraudulent chain until it creates a more extended branch. Consequently, the benign participants of B-RAN are unaware of the existence of the attacker and can only observe the blocks generated by honest miners with the generation rate λ^b until a fraud succeeds. We will provide a detailed procedure of the alternative history attack in Section 1.7.

1.4 B-RAN Queuing Model

According to the B-RAN framework in Section 1.2, we divide the service process of a valid request into four stages: (i) waiting to be included into a block, (ii) waiting for confirmations, (iii) waiting for service, and (iv) in service. Naturally, we can model the process using several queues in tandem based on the four phases. However, it is worth pointing out that, in the third stage, the requests in the same block arrive simultaneously, and the number of requests is related to the block generation time U^b. Hence, this queue is non-Markovian because some previous events (e.g. U^b) beside the current state of the queue may affect its future state. Usually, a non-Markovian queue is challenging to tackle. Therefore, we should carefully select the state space of B-RAN for further analysis.

Let i_n be the number of pending requests that already have n confirmations. A pending request is confirmed after receiving N confirmations. Then, $j = \sum_{n=N}^{+\infty} i_n$ denotes the number of confirmed requests that have not been served yet. In this approach, B-RAN can be fully identified by state $E(i_0, i_1, \ldots, i_{N-1}, j)$ belonging to the $(N + 1)$-dimensional state space $\mathbb{Z}_+^{N+1}$, where $\mathbb{Z}_+^{N+1}$ represents the set of all $(N + 1)$-tuples of non-negative integers. Formally, the queuing process of B-RAN is completely described by a vector stochastic process as:

$$\{X(t) \in \mathbb{Z}_+^{N+1}, t \geq 0\}.$$

---

6 Usually, an attacker can hardly amass more hash power than the sum of other honest miners, i.e. $\beta < 1$; Otherwise, the attacker already dominates the mining network.

B-RAN is said to be in state $E(i_0, i_1, \ldots, i_{N-1}, j)$ at time t if $X(t) = E(i_0, i_1, \ldots, i_{N-1}, j)$. Note that the way to establish a queuing model is not unique. We define B-RAN by using the queuing model $\{X(t), t \geq 0\}$ owing to two critical properties, as shown in Theorem 1.1.

Theorem 1.1 *The queuing model $\{X(t), t \geq 0\}$ is a continuous process with Markov property and time homogeneity or mathematically:*

$$(a) \; Pr\{X(t + h) = E | X(t) = E', X(u) \text{ for } 0 \leq u \leq t\} \tag{1.6}$$
$$= Pr\{X(t + h) = E | X(t) = E'\};$$

$$(b) \; Pr\{X(t + h) = E | X(t) = E'\} = Pr\{X(t) = E | X(0) = E'\}. \tag{1.7}$$

Proof: Recall that, as claimed in Section 1.2, requests arrive according to a Poisson process with rate λ^a, and the service times are exponentially distributed with mean $1/\lambda^c$, independently of each other. According to the mining model in Section 1.3, the block generation is a Poisson process with rate λ^b. In a nutshell, request inter-arrival times U^a, block times U^b, and service times U^c are exponential with mean $1/\lambda^a$, $1/\lambda^b$, and $1/\lambda^c$, respectively, and independent of each other.

The exponential service times U^c imply that if j UEs are in service, the rate at which service completions occur is $j\lambda^c$. To show this, suppose that $U_1^c, U_2^c, \ldots, U_j^c$ are the duration of j i.i.d. exponential simultaneously running time intervals with mean $1/\lambda^c$. Let $U_{min}^c = \min\{U_1^c, U_2^c, \ldots, U_j^c\}$ be the minimum service time. Observe that U_{min}^c will exceed u if and only if all U_j^c exceed u. Hence,

$$Pr\{U_{min}^c > u\} = Pr\{U_1^c > u\} \cdot Pr\{U_2^c > u\} \cdots Pr\{U_j^c > u\} = \exp(-j\lambda^c u).$$

Only those UEs that are in service can possibly leave. Hence, the service completion rate with j simultaneous services is $j\lambda^c$ for $0 \leq j \leq s$. Since at most s UEs can be in service simultaneously, obviously the service completion rate is at most $s\lambda^c$. Hence, we denote

$$\lambda_j^c = \min(j, s) \cdot \lambda^c, \tag{1.8}$$

to represent the service completion rate of B-RAN compactly. Similarly, the time until the next event (a request arrival, a block generation, or a service completion) is also exponentially distributed with rate $(\lambda^a + \lambda^b + \lambda_j^c)$. The probability that an event occurs in $(t, t + h)$ is $1 - \exp\left(-\left(\lambda^a + \lambda^b + \lambda_j^c\right)h\right)$, which tends to $\left(\lambda^a + \lambda^b + \lambda_j^c\right)h + o(h)$ as $h \to 0$. The length of time required for the event to occur and the type of the event are independent. There are several possible changes to a state. If a new request arrives, the state of B-RAN will switch from $E'\left(i_0, i_1, \ldots, i_{N-1}, j\right)$ to $E\left(i_0 + 1, i_1, \ldots, i_{N-1}, j\right)$. The probability that a new request arrives in $(t, t + h)$ is given by

$$Pr\{X(t + h) = E\left(i_0 + 1, i_1, \ldots, i_{N-1}, j\right) | X(t) = E'\left(i_0, i_1, \ldots, i_{N-1}, j\right)\} \tag{1.9}$$
$$= \frac{\lambda^a}{\lambda^a + \lambda^b + \lambda_j^c}\left(\left(\lambda^a + \lambda^b + \lambda_j^c\right)h + o(h)\right) = \lambda^a h + o(h), \; (h \to 0).$$

Similarly, if a new block is generated, all the existing blocks will get one more confirmation, and the state of B-RAN will move from $E'\left(i_0, i_1, \ldots, i_{N-1}, j\right)$ to $E\left(0, i_0, i_1, \ldots, i_{N-1} + j\right)$. The

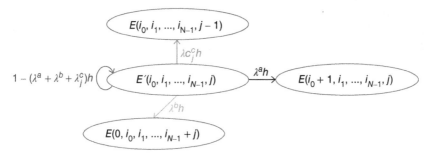

Figure 1.4 State transition graph of $E\left(i_0, i_1, \ldots, i_{N-1}, j\right)$. ($j \geq 1$.)

probability that a block is generated in $(t, t + h)$ is $\lambda^b h + o(h)$ as $h \to 0$. If an access service is ended ($j \geq 1$ at this instant), the state of B-RAN will switch from $E'\left(i_0, i_1, \ldots, i_{N-1}, j\right)$ to $E\left(i_0, i_1, \ldots, i_{N-1}, j - 1\right)$ with probability $\lambda_j^c h + o(h)$ as $h \to 0$. The probability that no event occurs in $(t, t + h)$ is $1 - (\lambda^a + \lambda^b + \lambda_j^c)h + o(h)$, and the probability that more than one event occurs in $(t, t + h)$ is $o(h)$ as $h \to 0$.

Now, we have obtained the transition probabilities $\Pr\left\{X(t + h) = E | X(t) = E'\right\}$ for any E and E' in the state space. Observe that the transition probabilities are irrelevant to the starting time t, which indicates that $\{X(t), \ t \geq 0\}$ is homogeneous in time. Moreover, the transition probabilities are independent of the states of previous moments, which implies the Markov property. Therefore, we have proven that $\{X(t), \ t \geq 0\}$ is a time-homogeneous Markov process. ∎

Although the queuing model established by Theorem 1.1 is a vector stochastic process with possibly high dimensions, we would like to emphasize that such a queuing model is more tractable than the original non-Markovian process. According to the proof of Theorem 1.1, we can directly obtain the transition probabilities $\Pr\left\{X(h) = E | X(0) = E'\right\}$ and characterize the queuing model $\{X(t), \ t \geq 0\}$ completely.

Figure 1.4 visualizes the transition relationships according to the proof of Theorem 1.1. One can see that the basic elements $\left\{\lambda^a, \lambda^b, \lambda^c, s\right\}$ and the number of confirmations N are enough to determine the transition probabilities and thus characterize the behaviors of B-RAN. Hence, we introduce a four-tuple $\Phi = \left\{\lambda^a, \lambda^b, \lambda^c, s\right\}$ as basic configurations to describe B-RAN. In Sections 1.5 and 1.6, we will analyze B-RAN from a deeper view in more dimensions.

1.5 Latency Analysis of B-RAN

1.5.1 Steady-State Analysis

Now, we have thus far modeled the B-RAN as a time-homogeneous Markov process with $(N + 1)$ dimensions. However, the dimensionality of the state space results in a complex probability transition graph and is difficult to analyze in general. In this section, we will analyze the access service latency of B-RAN by starting from a relatively simple one-confirmation case (i.e. only one confirmation is required to confirm a request). In other words, a request is confirmed, as long as it is assembled into a block. We must stress that the service latency here is significantly different from the physical-layer transmission delay.

In the one-confirmation case, the queuing model is presented by $\{X(t) = E(i,j), \ t \geq 0\}$, where we drop the subscript of i_0 for notational simplicity. State $E(i,j)$ means that i pending requests are waiting for assembling into a block and j confirmed requests are waiting for service. Define $w_{i,j}(t) = \Pr\{X(t) = E(i,j)\}$ as the probability of the queue in state $E(i,j)$ at time t. Now, let us investigate the transition probabilities during time h with the help of Theorem 1.1. By comparing the state of B-RAN at time $t + h$ with that at time t, for all $i = 1, 2, \ldots$ and $j = 0, 1, 2, \ldots$, we have

$$w_{i,j}(t + h) - w_{i,j}(t) = w_{i-1,j}(t)\lambda^a h + w_{i,j+1}(t)\lambda^c_{j+1} h - w_{i,j}(t)\left(\lambda^a + \lambda^b + \lambda^c_j\right) h,$$

where λ^c_j is defined by (1.8). By letting $h \to 0$, we get the following differential-difference equation:

$$\frac{d}{dt} w_{i,j}(t) = w_{i-1,j}(t)\lambda^a + w_{i,j+1}(t)\lambda^c_{j+1} - w_{i,j}(t)\left(\lambda^a + \lambda^b + \lambda^c_j\right).$$

Let $\{w_{i,j}\}$ be the steady-state distribution of B-RAN. The equilibrium condition $\frac{d}{dt} w_{i,j}(t) = 0$ yields

$$w_{i-1,j}\lambda^a + w_{i,j+1}\lambda^c_{j+1} - w_{i,j}\left(\lambda^a + \lambda^b + \lambda^c_j\right) = 0. \tag{1.10}$$

For the boundary cases ($i = 0$), we have

$$\left(\sum_{\ell=1}^{j} w_{\ell, j-\ell}\right)\lambda^b + w_{0,j+1}\lambda^c_{j+1} - w_{0,j}\left(\lambda^a + \lambda^c_j\right) = 0, \ \forall j = 0, 1, 2, \ldots \tag{1.11}$$

$$w_{0,1}\lambda^c_1 - w_{0,0}\lambda^a = 0. \tag{1.12}$$

The differential-difference equations (1.10)–(1.12) are known as the forward Kolmogorov equations [31]. We illustrate the state transition relationships with one confirmation in Figure 1.5.

In order to present the forward Kolmogorov equations in a compact form, we rearrange the two-dimension states $\{w_{i,j}\}$ by a particular order as shown in Figure 1.6, captured by the probability vector:

$$\mathbf{w} = \begin{bmatrix} w_{0,0} \ | \ w_{1,0} \ \ w_{0,1} \ | \ w_{2,0} \ \ w_{1,1} \ \ w_{0,2} \ | \ \cdots \end{bmatrix}^T.$$

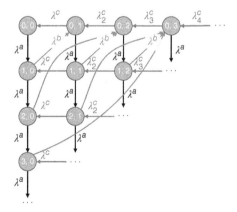

Figure 1.5 State space diagram in B-RAN.

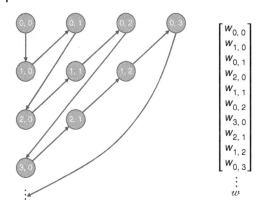

Figure 1.6 States $E(i,j)$ rearrangement into a row.

Now, the forward Kolmogorov equations can be rewritten in a matrix form:

$$\mathbf{Q}\mathbf{w} = \mathbf{0}, \tag{1.13}$$

where $\mathbf{Q}$ is the infinitesimal generator, or transition rate matrix, given by

$$\mathbf{Q} = \begin{bmatrix} -\lambda^a & & \lambda_1^c & & & & \cdots \\ \lambda^a & -(\lambda^a + \lambda^b) & & & \lambda_1^c & & \cdots \\ & \lambda^b & -(\lambda^a + \lambda_1^c) & & & \lambda_2^c & \cdots \\ & \lambda^a & & -(\lambda^a + \lambda^b) & & & \cdots \\ & & \lambda^a & & -(\lambda^a + \lambda^b + \lambda_1^c) & & \cdots \\ & & & \lambda^b & \lambda^b & -(\lambda^a + \lambda_2^c) & \cdots \\ \vdots & \vdots & \vdots & \vdots & \vdots & \vdots & \ddots \end{bmatrix} \tag{1.14}$$

The entry in $\mathbf{Q}$ equals to the corresponding transition rate given by $\frac{d}{dh}\Pr\{X(h) = E|X(0) = E'\}$ only depending on the B-RAN configuration tuple $\Phi = \{\lambda^a, \lambda^b, \lambda^c, s\}$. We can numerically solve the matrix equation by combining with the sum probability condition of $\mathbf{1}^T\mathbf{w} = 1$, i.e.

$$\begin{bmatrix} \mathbf{Q} \\ \mathbf{1}^T \end{bmatrix}\mathbf{w} = \begin{bmatrix} \mathbf{0} \\ 1 \end{bmatrix}. \tag{1.15}$$

From (1.15), the steady-state distribution $\mathbf{w}(\Phi)$ can be expressed as an implicit function of Φ. Note that the waiting space of B-RAN has no maximum limit. The number of states, i.e. the dimensions of the vector $\mathbf{w}$, should be infinite. In numerical calculations, we can use the solution with large enough but finite dimensions to approximate the infinite-dimension one. However, in practice, the number of UEs in a tract cannot be infinite, either. The aggressive load λ^a is required to be less than λ^c by the stable condition.

We can obtain the steady-state distribution of B-RAN via (1.15). In **Prototype Verification B**, we use our self-built prototype to measure the sojourn time of each state and estimate the probability of a state. The results show that analytical steady distributions are highly consistent with experimental outcomes, thereby validating our established queuing model. We illustrate the steady-state distributions of B-RAN with different T^b and different traffic intensities $\rho = \frac{\lambda^a}{s\lambda^c}$ in Figure 1.7. The low, medium, and high traffic

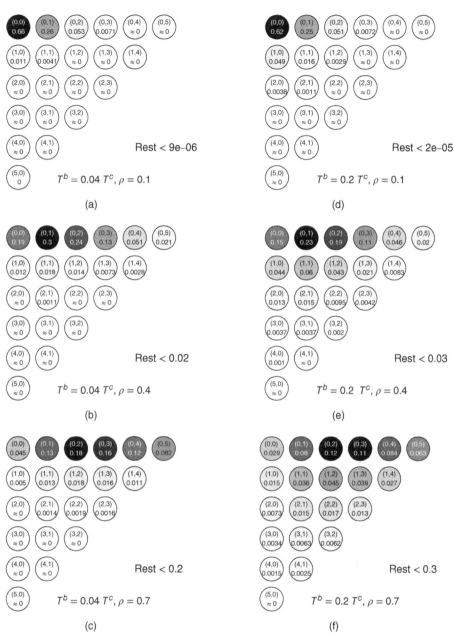

Figure 1.7 The distribution of steady states under different traffic intensities and block time with $s = 4$ links. (a) $T^b = 0.04T^c$ under low traffic intensity $\rho = 0.1$. (b) $T^b = 0.04T^c$ under medium traffic intensity $\rho = 0.4$. (c) $T^b = 0.04T^c$ under high traffic intensity $\rho = 0.7$. (d) $T^b = 0.2T^c$ under low traffic intensity $\rho = 0.1$. (e) $T^b = 0.2T^c$ under medium traffic intensity $\rho = 0.4$. (f) $T^b = 0.2T^c$ under high traffic intensity $\rho = 0.7$.

intensities $\rho = \frac{\lambda^a}{s\lambda^c}$ are set to 0.1, 0.4, and 0.7, respectively. Each node represents $E(i,j)$ in the figures. The results show that, on the one hand, under a higher traffic intensity, there are more confirmed requests waiting for service. On the other hand, when the block time T^b becomes larger, more requests wait to be assembled into a block.

1.5.2 Average Service Latency

To analyze the average latency in B-RAN, we consider the limiting distribution $\mathbf{w}(\Phi)$, from which we can obtain the average number of waiting requests $N(\Phi)$ in B-RAN:

$$\mathbb{E}\{N(\Phi)\} = \sum_{i,j}(i+j)\cdot w_{i,j}(\Phi). \tag{1.16}$$

We can apply Little's Law [28] as a bridge to connect the expected queue length and the average latency. The Little's Law states that, in a stable system, the average number of items is equal to the arrival rate multiplied by the average time an item spends in the system. Hence, the expected sojourn time $L_s(N,\Phi)$ is the sum of waiting time and service time (all the four stages of the workflow), and the expression of expected sojourn time in the one-confirmation case is given by

$$L_s(N=1,\Phi) = \mathbb{E}\{N(\Phi)\}/\lambda^a = T^a\sum_{i,j}(i+j)\,w_{i,j}(\Phi). \tag{1.17}$$

According to Section 1.2, the expected service time is T^c. Thus, the average latency $L(N,\Phi)$ is defined as the expected waiting time (the first three stages). In the one-confirmation case, $L(N,\Phi)$ can be written as

$$L(N=1,\Phi) = T^a\sum_{i,j}(i+j)\,w_{i,j}(\Phi) - T^c. \tag{1.18}$$

Equation (1.18) gives the average latency with one confirmation in terms of the limiting distribution in (1.13).

However, (1.18) only applies the one-confirmation special case. Our goal is to investigate the general N-confirmation problem. It is difficult to directly analyze a queue with $(N+1)$-dimensional state space, mainly due to the excessively large number variables in solving the equilibrium equations (1.13). Recall that we are more interested in the average system latency. A request, once assembled into a block, is required to wait for $N-1$ confirmations after the very first confirmation of assembling into a block. This waiting period is an additional stage compared to the one-confirmation case. The additional $N-1$ confirmations correspond to the extra waiting time of $N-1$ independent block time U_n^b. Hence, we obtain the average access service latency given by the following theorem.

Theorem 1.2 *[32] Given the B-RAN configurations $\Phi = \{\lambda^a, \lambda^b, \lambda^c, s\}$ and the number of confirmations N, the average access service latency is*

$$L(N,\Phi) = L(1,\Phi) + \mathbb{E}\left\{\sum_{n=2}^{N}U_n^b\right\} \tag{1.19}$$

$$= T^a\sum_{i,j}(i+j)\,w_{i,j}(\Phi) + T^b(N-1) - T^c.$$

Despite the implicit expression of $w_{i,j}(\Phi)$, Theorem 1.2 points out that the latency $L(N,\Phi)$ grows linearly with the number of confirmations N. Each extra confirmation leads to T^b longer waiting time on average and fewer confirmations will effectively reduce service latency. Note that we cannot simply conclude that latency is linear in T^b because T^b also influences the limiting distribution $w_{i,j}(\Phi)$, which in turn also affects the latency. For large N, however, we can conclude that latency is quasi-linear in T^b. Meanwhile, the more explicit relationships between the network parameters and the latency are revealed by the lower and upper bounds of service latency derived in [32].

Remark that the service latency is also different from the transaction confirmation latency in the context of blockchain. The average transaction confirmation latency is NT^b without considering the block size limit. However, from the above derivations, one can see that the service latency in B-RAN is much more complicated. The RAN capacity is another bottleneck limiting the network throughput and has a significant impact on service latency. As shown in the following prototype verification, under high traffic intensity, RANs may be congested and result in a lengthy delay.

In **Prototype Verification C**, we compare the analytical results of (1.19) with the outcomes from our B-RAN prototype under different settings of N and λ^b. Figure 1.8 shows both the average and the 95% confidence interval of latency under different traffic intensities. In the figure, service latency is presented relative to (r.t.) the service time T^c. First, the experimental results are highly consistent with the analysis of (1.19). Basically, higher traffic intensity will lead to longer service latency. Furthermore, the results show a strong linear relationship between latency $L(N,\Phi)$ and confirmation number N, for which the linear slope is close to T^b. This conclusion holds for different traffic intensities and is a simple corollary of (1.19): each extra confirmation leads to exactly T^b longer waiting time on average.

Moreover, we can assess the impact of s on the average latency by Figure 1.9. Given $N = 2$ and $\lambda^b = 25$, as the number of access links s increases, the average service latency $L(N,\Phi)$ becomes lower, especially for the high-intensity case. For asymptotically large s, the latency tends to a lower bound of $N\lambda^b$ [32]. As the number of access links s reflects the network scale, Figure 1.9 implies that B-RAN uniting more SPs and subnetworks is less likely to be congested under constant traffic $\rho = \frac{\lambda^a}{s\lambda^c}$. B-RAN, through inter-operative network sharing and offloading, can benefit from the pooling principle by delivering shorter and becomes

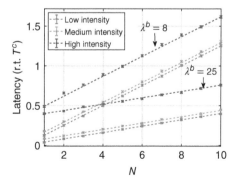

Figure 1.8 The analytical and experimental latency for different N and λ^b with $s = 4$.

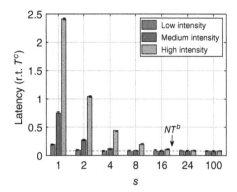

Figure 1.9 The impact of the maximal access channels s on service latency ($N = 2$ and $\lambda^b = 25$).

jointly more valuable and a longer bandwagon. This remarkable insight strongly motivates the future development of B-RAN. The pooling effect regarding network throughput is also illustrated in Section V-A in [6].

1.6 Security Considerations

1.6.1 Alternative History Attack

Since B-RAN is a distributed RAN enabled by blockchains, there can be a variety of vulnerabilities and security implications in blockchain systems. The alternative history attack, also referred to as "double spending attack," is a fundamental and inherent risk of distributed systems [3].

The alternative history attack implies that the history record in blockchain could be tampered by malicious miners, which essentially exposes the vulnerability of blockchain. In alternative history attacks, an attacker privately mines an alternative blockchain fork in which a fraudulent double spending event (e.g. an access service in B-RAN) is included. After the network accepts a benign chain, the attacker releases the fraudulent fork. If the fraudulent fork eventually becomes longer than the benign one, the attacker can successfully alter a confirmed history and spend the same coin (credit) twice, which would drive the blockchain system into a catastrophic inconsistent state. For wireless network, altering a confirmed chain in B-RAN may lead to interference issues when more than one UE attempts to access the same channel simultaneously. Hence, alternative history attack cannot be simply ignored, despite the differences between B-RAN and cryptocurrencies. In our security considerations, we focus mainly on the risk of alternative history attacks.

For blockchain to overcome alternative history attacks, miners are always guided to the "longest" locally known fork, that is, the one involving the highest amount of computational effort so far. The block generation process is coupled to some specific capability of miners, e.g. computational power in PoW and the number of available devices in PoD. Still, alternative history attacks are possible and only require adversaries to create a chain longer than the benign one. Clearly, the higher the number of confirmations, the lower the probability of a successful alternative history attack.

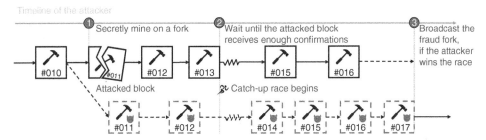

Figure 1.10 The process of a double spending attack.

1.6.2 Probability of a Successful Attack

Recall that the benign miners' and an attacker's mining rate are λ^b and $\beta\lambda^b$, respectively, as β represents the attacker's relative hash power. Rational benign miners always work on the longest chain for the mining rewards, while the attacker attempts to generate a fraudulent fork to revise the history for double spending or other purposes. Before initiating a fraud, the attacker will broadcast a regular event (e.g. paying for access) and wait for it to be assembled into a block. That block is a piece of confirmed history that the attacker attempts to revise. As shown in Figure 1.10, the attacker will mount an alternative history attack according to the following steps:

1. secretly mine on a fork excluding the attacked block;
2. wait until the attacked block receives enough confirmations (say, N blocks);
3. broadcast the respective blocks as soon as the fraudulent fork is longer than the benign fork.

Given the attacker's relative mining rate β, the probability of a successful alternative history attack, donated by $S(N, \beta)$, can reflect the safety properties of B-RAN, given by the following theorem.

Theorem 1.3 *[32] Given the attacker's relative mining rate β and the required number of confirmations N, the probability of a successful alternative history attack is*

$$S(N, \beta) = \begin{cases} 1 - \sum_{n=0}^{N} \binom{n+N-1}{n} \left(\frac{1}{1+\beta}\right)^N \left(\frac{\beta}{1+\beta}\right)^n \left(1 - \beta^{N-n+1}\right) & \text{if } \beta < 1 \\ 1 & \text{if } \beta \geq 1 \end{cases} \quad (1.20)$$

Theorem 1.3 intuitively indicates that a more powerful attacker with a larger β is more likely to revise history successfully, while more confirmations can effectively mitigate the risk of such malicious attacks. However, no matter how large N is, the alternative history attack is always possible if the attacker has a non-negative mining rate and sufficient luck.

The attacker's strategy also affects the probability of success. The dominant strategy of a powerful attacker with $\beta \geq 1$ is to never quit. Because $S(N, \beta) = 1$ for $\beta > 1$, the fraud would succeed eventually. This is known as the "51%" attack or Goldfinger attack [3]. For weaker hash power $\beta < 1$, although repeated trials can also increase the success probability, it costs a vast number of resources and could possibly lead to negative yields. If the fraudulent chain

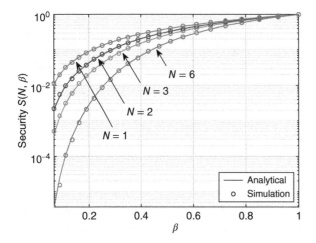

Figure 1.11 The analytical and experimental probability of success attack for different β and N.

is hundreds of blocks behind the benign chain, a rational attacker should give up and mount a new round of attack instead of sticking to the original one with mounting cost. $S(N, \beta)$, revealed by Theorem 1.3, is the maximum probability of an alternative history attack, which is achieved only if the attacker is never giving up. The impact of the attacker's strategy on the success probability was analyzed in [32] at length.

Furthermore, we would like to highlight the differences between Theorem 1.3 and the related results in [2, 3, 33, 34]. The original Bitcoin paper [2] claimed that, before the benign chain extends by N additional blocks, the number of fraudulent blocks generated by the attacker follows a Poisson distribution rather than the negative binomial distribution revealed in the proof of Theorem 1.3. In [3], the breakeven case was counted in computing the probability of successful attack, whereas in [3, 33], the attacker was assumed to pre-mine one block before initiating an attack.

In **Prototype Verification D**, we now show the security performance of B-RAN obtained from our home-built prototype. We estimate the successful attack probability from 2^{16} Monte-Carlo simulations according to the process in Figure 1.10 until the attacker eventually wins or gives up and present the simulation results in Figure 1.11. From the figure, one can clearly see that the experiment results firmly agree with our analytical probability of Theorem 1.3. As the relative mining rate β increases, the probability of a successful attack rises significantly. The chain is more secure if it is safeguarded by more confirmations, which suggests that the number of confirmations N can effectively reduce the risk.

1.7 Latency-Security Trade-off

From the above analysis, we discover that there exists an inherent relation between latency and security. This latency-security relationship and design trade-off capture a more complete picture of the achievable performance of B-RAN. We formally state this design trade-off as the following theorem.

Theorem 1.4 *The trade-off between latency and security is given by the piecewise-linear function connecting the points $(L(N, \Phi), S(N, \beta))$ for $N = 1, 2, \ldots$, where $L(N, \Phi)$ and $S(N, \beta)$ are given by (1.19) and (1.20), respectively.*

Given the network setting Φ and the attacker's relative hash power β, the performance of B-RAN can be thoroughly evaluated by the trade-off specified by Theorem 1.4. The latency-security trade-off can be used as a new performance metric to characterize B-RAN. Theorem 1.4 indicates that more confirmations N can effectively safeguard the security of B-RAN $S(N, \beta)$ but would unfortunately increase the access delay $L(N, \Phi)$. Conversely, fewer confirmations to verify a request can considerably reduce latency while suffering a higher risk from an adversary.

Figure 1.12 graphically illustrates t he trade-off curve $(L(N, \Phi), S(N, \beta))$. Essentially, the latency-security trade-off is between the convergence rate and confirmation error probability of a distributed system. Enhancing the security advantage comes at a price of longer access delay, whereas latency reduction comes at a price of security. Theorem 1.4 provides the achievable region of B-RAN for given network parameters Φ and the attacker's power β. The trade-off provides a more in-depth and complete view of B-RAN than just staring at latency or security.

Theorem 1.4 also makes another informative statement. If the network condition Φ becomes better, e.g. with less traffic load λ^a, then the entire trade-off curve will be shifted to the left along the axis of latency for the same security level. If the attacker becomes more powerful, e.g. a larger β, then the trade-off curve drops in the term of security at the same access delay.

Furthermore, Theorem 1.4 implies that latency and security can be balanced by adjusting the number of confirmations N. The confirmation number N is a positive integer that can be chosen to achieve the shortest latency under an acceptable security condition. Different from cryptocurrencies, wireless access is much more sensitive to latency. Therefore, we usually should set a relatively smaller N instead of the typical six confirmations suggested in Bitcoin [2]. The specific value of N should be determined by the latency requirement and also the security level.

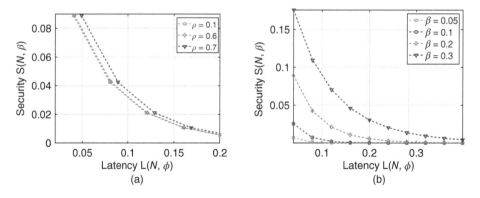

Figure 1.12 The latency-security trade-off ($s = 25$ and $T^b = 0.04$). (a) Under different traffic intensities, ρ and $\beta = 0.2$. (b) An attacker with different mining rates β and $\rho = 0.1$.

1.8 Conclusions and Future Works

In this chapter, we established an original framework to analytically characterize B-RAN properties and performances. Based on the established queuing model, we analyzed the system latency of B-RAN and evaluated the security level of B-RAN by considering the risk of alternative history attacks. We uncovered an inherent latency and security relationship. We proposed to assess the achievable performance of B-RAN according to the latency-security trade-off curve. We validated our analytical works by building an in-house B-RAN prototype. The results derived from the model will provide meaningful insights and guidelines for future B-RAN designs and implementations. As last, we would like to discuss several interesting insights for future research.

1.8.1 Network Effect and Congest Effect

The network scale can be a double-edged sword. Positive network effects coexist with the negative ones, also known as *congest effects*. For example, the scalability issue of blockchains can cause congest effect in B-RAN. Hence, B-RAN must be further strengthened to promote network effects and suppress congestion.

1.8.2 Chicken and Eggs

As an MSP, B-RAN also faces the start-up problem on how to attain a critical mass, i.e. the number of participants needed to spur network effects. This is known as the chicken-and-egg problem. How can B-RAN get enough clients interested before other sides join, and vice versa.

1.8.3 Decentralization and Centralization

Decentralization cannot outperform centralization in all aspects because each has its own merits and defects. Historically, traditional noncooperative RANs have evolved toward both centralized/cloud radio access network (C-RAN) for operating expenditure reduction and decentralized edge/fog networks. The evolution of RANs provides insights and guidelines for B-RAN to take advantage of both decentralization and centralization.

1.8.4 Beyond Bitcoin Blockchain

The recent advances on blockchain protocols, such as Conflux [35], Phantom [36], Prism [37], and Avalanche [38], aim to achieve great scalability, low latency, and high security via proper chain design. B-RAN is compatible with these underlying blockchain protocols, and it is possible to obtain better performance regarding both latency and security. The trade-off $(L(N, \Phi), S(N, \beta))$ derived in this chapter can be viewed as a benchmark for future B-RAN based on advanced blockchains.

References

1 Cisco Visual Networking Index. Cisco Visual Networking Index: Global Mobile Data Traffic Forecast Update, 2017–2022. *Tech. Rep.*, Feb. 2019.

2 S. Nakamoto. Bitcoin: A Peer-to-Peer Electronic Cash System. *Tech. Rep.*, Oct. 2008.

3 F. Tschorsch and B. Scheuermann. Bitcoin and beyond: a technical survey on decentralized digital currencies. *IEEE Communication Surveys and Tutorials*, 18 (3): 2084–2123, 2016.

4 J. Xie, H. Tang, T. Huang, F. R. Yu, R. Xie, J. Liu, and Y. Liu. A survey of blockchain technology applied to smart cities: research issues and challenges. *IEEE Communication Surveys and Tutorials*, 21 (3): 2794–2830, 2019.

5 X. Ling, J. Wang, T. Bouchoucha, B. C. Levy, and Z. Ding. Blockchain radio access network (B-RAN): towards decentralized secure radio access paradigm. *IEEE Access*, 7: 9714–9723, Jan. 2019.

6 X. Ling, J. Wang, Y. Le, Z. Ding, and X. Gao. Blockchain radio access network beyond 5G. *IEEE Wireless Communications*, 27 (6): 160–168, Dec. 2020.

7 K. Christidis and M. Devetsikiotis. Blockchains and smart contracts for the Internet of Things. *IEEE Access*, 4: 2292–2303, Jun. 2016.

8 B. Cao, Y. Li, L. Zhang, L. Zhang, S. Mumtaz, Z. Zhou, and M. Peng. When Internet of Things meets blockchain: challenges in distributed consensus. *IEEE Network*, 33 (6): 133–139, Jul. 2019.

9 Y. Sun, L. Zhang, G. Feng, B. Yang, B. Cao, and M. A. Imran. Blockchain-enabled wireless Internet of Things: performance analysis and optimal communication node deployment. *IEEE Internet of Things Journal*, 6 (3): 5791–5802, Jun. 2019.

10 X. Ling, Y. Le, J. Wang, and Z. Ding. Hash access: trustworthy grant-free IoT access enabled by blockchain radio access networks. *IEEE Network*, 34 (1): 54–61, Jan. 2020.

11 Z. Xiong, Y. Zhang, N. C. Luong, D. Niyato, P. Wang, and N. Guizani. The best of both worlds: a general architecture for data management in blockchain-enabled Internet-of-Things. *IEEE Network*, 34 (1): 166–173, Jan. 2020.

12 M. Liu, F. R. Yu, Y. Teng, V. C. M. Leung, and M. Song. Computation offloading and content caching in wireless blockchain networks with mobile edge computing. *IEEE Transactions on Vehicular Technology*, 67 (11): 11008–11021, Nov. 2018.

13 Z. Xiong, S. Feng, W. Wang, D. Niyato, P. Wang, and Z. Han. Cloud/Fog computing resource management and pricing for blockchain networks. *IEEE Internet of Things Journal*, 6 (3): 4585–4600, Jun. 2019.

14 R. Yang, F. R. Yu, P. Si, Z. Yang, and Y. Zhang. Integrated blockchain and edge computing systems: a survey, some research issues and challenges. *IEEE Communication Surveys and Tutorials*, 21 (2): 1508–1532, 2019.

15 Z. Xiong, J. Kang, D. Niyato, P. Wang, and V. Poor. Cloud/Edge computing service management in blockchain networks: multi-leader multi-follower game-based ADMM for pricing. *IEEE Transactions on Services Computing*, 13 (2): 356–367, 2020.

16 J. Yang, S. He, Y. Xu, L. Chen, and J. Ren. A trusted routing scheme using blockchain and reinforcement learning for wireless sensor networks. *Sensors*, 19 (4): 970, Feb. 2019.

17 Z. Liu, N. C. Luong, W. Wang, D. Niyato, P. Wang, Y. Liang, and D. I. Kim. A survey on blockchain: a game theoretical perspective. *IEEE Access*, 7: 47615–47643, Apr. 2019.

18 W. Wang, D. T. Hoang, P. Hu, Z. Xiong, D. Niyato, P. Wang, Y. Wen, and D. I. Kim. A survey on consensus mechanisms and mining strategy management in blockchain networks. *IEEE Access*, 7: 22328–22370, Jan. 2019.

19 X. Ling, Z. Gao, Y. Le, L. You, J. Wang, Z. Ding, and X. Gao. Satellite-aided consensus protocol for scalable blockchains. *Sensors*, 20 (19): 5616, Oct. 2020.

20 M. B. H. Weiss, K. Werbach, D. C. Sicker, and C. E. C. Bastidas. On the application of blockchains to spectrum management. *IEEE Transactions on Cognitive Communications and Networking*, 5 (2): 193–205, Apr. 2019.

21 P. Kuo, A. Mourad, and J. Ahn. Potential applicability of distributed ledger to wireless networking technologies. *IEEE Wireless Communications*, 25 (4): 4–6, Aug. 2018.

22 E. D. Pascale, J. McMenamy, I. Macaluso, and L. Doyle. Smart contract SLAs for dense small-cell-as-a-service. *arXiv preprint arXiv:1703.04502*, Mar. 2017.

23 K. Kotobi and S. G. Bilen. Secure blockchains for dynamic spectrum access a decentralized database in moving cognitive radio networks enhances security and user access. *IEEE Vehicular Technology Magazine*, 13 (1): 32–39, Mar. 2018.

24 J. Backman, S. Yrjölä, K. Valtanen, and O. Mämmelä. Blockchain network slice broker in 5G: slice leasing in factory of the future use case. In *Proceedings of Internet Things Business Models, Users, and Networks*, pages 1–8, Copenhagen, DK, Nov. 2017.

25 H. Xu, P. V. Klaine, O. Onireti, B. Cao, M. Imran, and L. Zhang. Blockchain-enabled resource management and sharing for 6G communications. *Digital Communications and Networks*, 6 (3): 261–269, Aug. 2020.

26 H. Xu, L. Zhang, E. Sun, and C. I. BE-RAN: blockchain-enabled RAN with decentralized identity management and privacy-preserving communication. *arXiv preprint arXiv:2101.10856*, Jan. 2021.

27 Y. Le, X. Ling, J. Wang, and Z. Ding. Prototype design and test of blockchain radio access network. In *Proceedings of IEEE International Conference on Communications Workshops (ICC'19)*, Shanghai, CN, May 2019.

28 R. B. Cooper. *Introduction to Gueueing Theory*. Macmillan, New York, 1972.

29 Q. Wu, C. Feres, D. Kuzmenko, D. Zhi, Z. Yu, and X. Liu. Deep learning based RF fingerprinting for device identification and wireless security. *Electronics Letters*, 54 (24): 1405–1407, Nov. 2018.

30 G. Li, J. Yu, Y. Xing, and A. Hu. Location-invariant physical layer identification approach for WiFi devices. *IEEE Access*, 7: 106974–106986, Aug. 2019.

31 L. Kleinrock. *Theory, Queueing Systems*, volume 1. Wiley-Interscience, New York, NY, USA, 1975.

32 X. Ling, Y. Le, J. Wang, Z. Ding, and X. Gao. Practical modeling and analysis of blockchain radio access network. *IEEE Transactions on Communications*, 69 (2): 1021–1037, Feb. 2021.

33 M. Rosenfeld. Analysis of hashrate-based double spending. *arXiv preprint arXiv:1402.2009*, Apr. 2014.

34 A. Gervais, G. O. Karame, K. Wüst, V. Glykantzis, H. Ritzdorf, and S. Capkun. On the security and performance of proof of work blockchains. In *Proceedings of ACM SIGSAC Conference on Computer and Communications Security (CCS'16)*, pages 3–16, New York, NY, USA, Oct. 2016.

35 C. Li, P. Li, D. Zhou, W. Xu, L. Fan, and A. Yao. Scaling Nakamoto consensus to thousands of transactions per second. *arXiv preprint arXiv:1805.03870*, Aug. 2018.

36 Y. Sompolinsky, S. Wyborski, and A. Zohar. PHANTOM and GHOSTDAG: a scalable generalization of Nakamoto consensus. *Cryptology ePrint Archive*, Report 2018/104, Jan. 2018. [Online]. Available: https://eprint.iacr.org/2018/104.

37 V. Bagaria, S. Kannan, D. Tse, G. Fanti, and P. Viswanath. Prism: deconstructing the blockchain to approach physical limits. In *Proceedings of ACM SIGSAC Conference on Computer Communications Security (CCS'19)*, pages 585–602, New York, USA, Nov. 2019.

38 T. Rocket, M. Yin, K. Sekniqi, R. Van Renesse, and E. G. Sirer. Scalable and probabilistic leaderless BFT consensus through metastability. *arXiv preprint arXiv:1906.08936*, Jun. 2019.

2

Consensus Algorithm Analysis in Blockchain: PoW and Raft

Taotao Wang[1], Dongyan Huang[2], and Shengli Zhang[1]

[1]*College of Electronics and Information Engineering, Shenzhen University, Shenzhen 518060, China*
[2]*College of Information and Communications, Guilin University of Electronic Technology, Guilin 541004, China*

2.1 Introduction

Blockchain technology, which was firstly coined in Bitcoin [1] by Satoshi Nakamoto in 2008, has received extensive attention recently. A blockchain is an encrypted, distributed database/transaction system where all the peers share information in a decentralized and secure manner. Because of its key characteristics as decentralization, immutability, anonymity, and auditability, blockchain becomes a promising technology for many kinds of asset transfers and point-to-point (P2P) transaction [2]. Recently, blockchain-based applications are springing up, covering numerous fields including financial services [3–5], Internet of Things (IoT) [6], reputation systems [7], and so on.

Because the essence of blockchain is a distributed system, consensus algorithms play a crucial role in maintaining the safety and efficiency of blockchains. A consensus algorithm for distributed systems is to figure out the coordination among multiple nodes, i.e. how to come to an agreement if there are multiple nodes. Several consensus algorithms have been proposed, e.g. proof-of-work (PoW) [1], proof-of-stake (PoS) [8, 9], delegate proof-of-stake (DPoS) [10], practical byzantine fault-tolerant (PBFT) [11], Paxos [12], and Raft [13]. Among them, PoW and PoS algorithms have a good support for safety, fault tolerance, and scalability of a blockchain. Thus, PoW and PoS are common choices of public blockchains, in which any one can join the network, and there are no trust relationships among the nodes. However, PoW and PoS have slow speed of transaction confirmation, which limits its applications to those requiring high confirmation speed. In a consortium/private blockchain network, all participants are whitelisted and bounded by strict contractual obligations to behave "correctly," and hence more efficient consensus algorithms such as PBFT and Raft are more appropriate choices. Consortium/private blockchains could be applied into many business applications. For example, Hyperledger [14] is developing business consortium blockchain frameworks. Ethereum has also provided tools for building consortium blockchains [15]. Raft algorithm is considered as a consensus algorithm for private blockchains [16], which is applied to more ad hoc networks such as the intranet. Furthermore, several hybrid consensus protocols have been proposed to improve

Wireless Blockchain: Principles, Technologies and Applications, First Edition.
Edited by Bin Cao, Lei Zhang, Mugen Peng and Muhammad Ali Imran.

consensus efficiency without undermining scalability. For example, Zilliqa [17] proposed PoW to elect directory service (DS) committee nodes, and then, the DS committee has to run PBFT consensus protocol on the transaction block. In this chapter, we investigate the PoW consensus algorithm for public blockchains and the Raft consensus algorithm for consortium/private blockchains.

The idea of PoW originated in [18] and is rediscovered and exploited in the implementation of Bitcoin. PoW provides strong probabilistic consensus guarantee with resilience against up to 1/2 malicious nodes [19, 20]. The successful operation of Bitcoin demonstrates the practicality of using PoW to achieve consensus. Subsequent to Bitcoin, many other cryptocurrencies, such as Litecoin [21] and Ethereum [22], also adopt the PoW consensus algorithm. Peers running the PoW consensus algorithm are miners who compete to solve a difficult cryptographic hash puzzle, called the PoW problem. The miner who successfully solves the PoW problem obtains the right to extend the blockchain with a block consisting of valid transactions. In doing so, the miner receives a reward in the form of a newly minted coin written into the added block. Solving the PoW problem for rewards is called mining, just like mining for precious metals.

Miners commit computation resources to solve the PoW problem. Previously, it was believed that the most profitable mining strategy is honest mining, wherein a miner will broadcast the newly added block as soon as it has solved the PoW problem. Let α be the ratio of a particular miner's computing power over the computing powers of all miners. This ratio is also the probability that the miner can solve the PoW problem before others in each round of an added block [23]. Over the long term, the rewards to a miner that executes the honest mining strategy are therefore α fraction of the total rewards issued by the Bitcoin network. This is reasonable since miners share the pie in proportion to their investments. Not known were whether there are other mining strategies more profitable than honest mining. Later, the authors of [24] developed a selfish mining strategy that can earn higher rewards than honest mining. A selfish miner does not broadcast its mined block immediately; it carries out a block-withholding attack by secretly linking its future mined blocks to the withheld mined block. If the selfish miner can mine two successive blocks before other miners do, it can broadcast its two blocks at the same time to override the block mined by others. Because Bitcoin has an inherent self-adjusting mechanism to ensure that on average only one block is added to the blockchain every 10 minutes [25], by invalidating the blocks of others (hence removing them from the blockchain), the selfish miner can increase its own profits. For example, with computing power ratio $\alpha = 1/4$, the rewards obtained by selfish mining can be up to 1/3 fraction of the total rewards [24]. Based on this observation, Nayak et al. [26] further proposed various selfish mining strategies with even higher rewards. Despite the many versions of selfish mining, the optimal (i.e. most profitable) mining strategy remained elusive until Sapirshtein et al. [27]. The authors of [27] formulated the mining problem as a general Markov decision process (MDP) with a large state-action space. The objective of the mining MDP, however, is not a linear function of the rewards as in standard MDPs. Thus, the mining MDP cannot be solved using a standard MDP solver. To solve the problem, Sapirshtein et al. [27] first transformed the mining MDP with the nonlinear objective to a family of MDPs with linear objectives and then employed a standard MDP solver over the family of MDPs to iteratively search for the optimal mining strategy.

The approach in [27] is model based in that various parameter values (e.g. α) must be known before the MDP can be set up. In real blockchain networks, the exact parameter values are not easy to obtain and may change over time, hindering the practical adoption of the solution. In the first part of this chapter, we propose a model-free approach that solves the mining MDP using machine learning tools. In particular, we solve the mining MDP using reinforcement learning (RL) without the need to know the parameter values in the mining MDP model. RL is a machine-learning paradigm, where agents learn successful strategies that yield the largest long-term reward from trial-and-error interactions with their environment [28, 29]. Q-learning is the most popular RL technique [30]. It can learn a good policy by updating a state-action value function without an operating model of the environment. RL has been successfully applied in many challenging tasks, e.g. playing video games [31] and Go [32] and controlling robotic movements [33]. The original RL algorithm cannot deal with the nonlinear objective function of our mining problem. In this part, we put forth a new multi-dimensional RL algorithm to tackle the problem. Experimental results indicate that our multi-dimensional RL mining algorithm can successfully find the optimal strategy. Importantly, it demonstrates robustness and adaptability to a changing environment (i.e. parameter values changing dynamically over time).

Compared with PBFT and Paxos, Raft algorithm has high efficiency and simplicity, and it has been widely adopted in the distributed systems. Raft is a leader-based algorithm, which uses leader election as an essential part for the consensus protocol. Ledger entries in Raft-based system flow in only one direction from the leader to other servers. Paxos and its improved algorithms do not take leader election as an essential part of the consensus protocol, which can balance load well among nodes because any node may commit commands [12, 34]. However, Paxos' architecture requires complex changes to support practical systems. Raft achieves the same safety performance as Paxos and is more convenient in engineering implementation and understanding. Raft consensus algorithm cannot tolerate malicious nodes and can tolerate up to 50% nodes of crash fault. For private blockchains, nodes are verified members. Hence, it is more important to solve the crash faults than Byzantine faults for private blockchains.

Network is called split when more than half of nodes are out of current leader's control. Node failures and communication interruption caused by packet loss are the main reasons of network split. If network split occurs, the blockchain network with Raft consensus algorithm would re-start a new leader election process. Meanwhile, the blockchain network stops accepting new transactions, i.e. the blockchain network becomes unavailable. Obviously, consensus efficiency of blockchain is degraded tremendously if network split occurs frequently. The existing works on blockchain mainly focus on designing algorithms or optimizing performance or safety certification, but a theoretical analysis of network split is lacking. The impact of packet loss rate on network split is rarely considered. In fact, packet loss rate plays an important role on network split.

In the second part of this chapter, we concentrate on analyzing the network split probability of the Raft algorithm. We provide a simple model that accounts for protocol details and allows for the performance computation of distributed networks. In particular, we derive the network performance in normal conditions, and we explore the parameters' (such as packet loss rate, period of election timeout, and size of network) impact on network split performance.

2.2 Mining Strategy Analysis for the PoW Consensus-Based Blockchain

In this part, we analyze the PoW consensus algorithm, specifically, different mining strategies. We propose an RL approach to achieve the optimal mining in the PoW-based blockchains.

2.2.1 Blockchain Preliminaries

Blockchain is a decentralized append-only ledger for digital assets. The data of blockchain is replicated and shared among all participants. Its past recorded data are tamper resistant and participants can only append new data to the tail end of the chain of blocks. The state of blockchain is changed according to transactions, and transactions are grouped into blocks that are appended to the blockchain. The header of the block encapsulates the hash of the preceding block, the hash of this block, the Merkle root[1] of all transactions contained in this block, and a number called nonce that is generated by PoW. Because each block must refer to its preceding block by placing the hash of its preceding block in its header, all the blocks form a chain of blocks arranged in chronological order. Figure 2.1 illustrates the data structure of blockchain.

2.2.2 Proof of Work and Mining

In this part, we focus on a Bitcoin-like blockchain that adopts the PoW consensus protocol to validate new blocks in a decentralized manner.[2] In each round, the PoW protocol selects

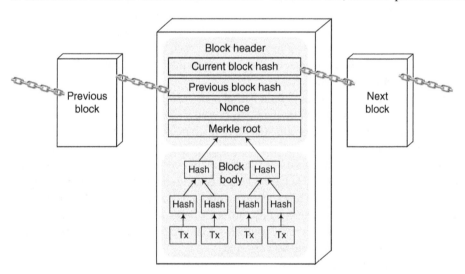

Figure 2.1 Data structure of blockchain.

---

1 The Merkle root of the transactions is the hash value of the Merkle tree whose leaves are the transactions [35].
2 There are also blockchains adopting other several consensus algorithms, such as proof of stake (PoS) and Byzantine fault tolerance (BFT) [36].

a leader that is responsible for packing transactions into a block and appends this block to the blockchain. To prevent adversaries from monopolizing the blockchain, the leader selection must be approximately random. Because Bitcoin-like blockchain is permissionless and anonymity is inherently designed as the goal, it must consider the Sybil attack where an adversary simply creates many participants with different identities to increase its probability of being selected as the leader. To address the above issues, the key idea behind PoW is that a participant will be randomly selected as the leader of each round with a probability in proportion to its computing power.

In particular, blockchain implements PoW using computational hash puzzles. To create a new block, the nonce placed into the header of the block must be a solution to the hash puzzle expressed by the following inequality:

$$\mathscr{H}(n, p, m) < D, \tag{2.1}$$

where the nonce n, the hash of the previous block p, and the Merkle root of all included transactions m are taken as the input of a cryptographic hash function $\mathscr{H}(\cdot)$, and the output of the hash function should fall below a target D that is small with respect to the whole range of the hash function outputs. The used hash function (e.g. SHA-256 hash is used for Bitcoin) satisfies the property of puzzle friendliness [37]: it is challenging to guess the nonce to fulfill (2.1) by a one-shot try. The only way to solve (2.1) is to try a large number of nonces one by one to check if (2.1) is fulfilled until one lucky nonce is found. Therefore, the probability of finding such a nonce is proportional to the computing power of the participant – the faster the hash function in (2.1) can be computed in each trial, the more nonces can be tried per unit time. Using the blockchain terminology, the process of computing hashes to find a nonce is called *mining*, and the participants involved are called *miners*.

2.2.3 Honest Mining Strategy

When a miner tries to append a new block to the latest legal block by placing the hash of the latest block in the header of the new block, we say that the miner mines on the latest block. The blockchain is maintained by miners in the following manner.

To encourage all miners to mine on, and maintain, the current blockchain, a *reward* is given as an incentive to the miner by placing a coin-mint transaction in its mined block that credits the miner with some new coins. If the block is verified and accepted by other peers in the blockchain network, the reward is effective and thus can be spent on the blockchain. When a miner has found an eligible nonce, it publishes his block to the whole blockchain network. Other miners will verify the nonce and transactions contained in that block. If the verification of the block is passed, other miners will mine on the block (implicitly accepting the block); otherwise, other miners discard the block and will continue to mine on the previous legal block.

If two miners publish two different legal blocks that refer to the same preceding block at the same time, the blockchain is then forked into two branches. This is called *forking* of blockchain. Forks are the manifestations of disagreement among peers on the blockchain structure. It can also compromise the integrity and security of the blockchain [38]. To resolve a fork, PoW prescribes that only the rewards of the blocks on the longest branch (called the main chain) are effective. Then, miners are incentivized to mine on the longest branch, i.e. miners always add new blocks after the last block on the longest main chain that is observed from their local perspectives. If the forked branches are of equal length, miners may mine subsequent blocks on either branch randomly. This is referred to as the rule of the longest chain extension. Eventually, one branch will predominate and the other branches are discarded by peers in the blockchain network.

The mining strategy adhering to the rule of the longest chain extension and publishing a block immediately after the block is mined is referred to as *honest mining* [23, 36, 39]. The miners that comply with honest mining are called honest miners. It was widely believed that the most profitable mining strategy for miners is honest mining; and that when all miners adopt honest mining, each miner is rewarded in proportion to its computing power [23, 36, 39]. As a result, any rational miner will not deviate from honest mining. This belief was later shown to be ill founded and that other mining strategies with higher profits are possible [24, 26, 27]. We will briefly discuss these mining strategies in Section 2.2.4. For a more concrete exposition, we will first present the mining model.

2.2.4 PoW Blockchain Mining Model

In this section, we present the MDP model for blockchain mining. Eyal and Sirer [24] first developed an MDP mining model and used the model to construct a selfish mining strategy with higher rewards than honest mining. Then, Nayak et al. [26] proposed even more profitable selfish mining strategies. Recently, Sapirshtein et al. [27] extended the MDP mining models of [24, 26] to a more general form. In this work, we adopt the mining model of Sapirshtein et al. [27].

Without loss of generality, we assume that the network is split into two mining pools: one is an adversary that controls a fraction α of the whole network's computing power and the other is the network of honest miners that controls a fraction $1 - \alpha$ of the computing power of the whole network.

Even if the adversary and an honest miner release their newly mined blocks to the network simultaneously, the blocks will not be received by all miners simultaneously because of propagation delays and network connectivity. We model the communication capability of the adversary using the parameter γ defined as the fraction of the honest miners that will first receive the block from the adversary when the adversary and one honest miner release their blocks approximately at a same time – more specifically, $\gamma(1 - \alpha)$ is the computing power of the honest network that will mine on the block of the adversary when the adversary and an honest miner release their blocks simultaneously.

As in [27], we model blockchain mining as a single-player MDP $M = \langle S, A, P, R \rangle$, where S is the state space, A is the action space, P is the transition probability matrix, and R is the reward matrix. Each transition is triggered by the event of a miner mining a new block, whether the block is mined by the adversary or one of the honest miners. The action taken by the adversary based on the previous state, together with the event, determines the next state to which the system evolves.

The objective of the adversary is to earn rewards higher than its computational power. To achieve this, the adversary will generally deviate from honest mining by building a private chain of blocks without releasing them the moment the blocks are mined; the adversary will release several blocks from its private chain at a time to undo the honest chain opportunistically.

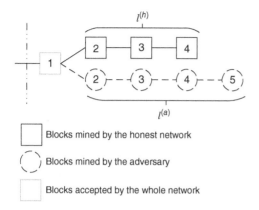

Blocks mined by the honest network

Blocks mined by the adversary

Blocks accepted by the whole network

Figure 2.2 An illustrating example of the state in the adopted MDP.

2.2.4.1 State

Each state in the state space is represented by a three-tuple form $(l^{(a)}, l^{(h)}, fork)$, where $l^{(a)}$ and $l^{(h)}$ are the lengths of the adversary's chain and the honest network's chain, respectively, after the latest fork (as illustrated in Figure 2.2). In general, *fork* can take three possible values (*irrelevant, relevant, active*). Their meanings will be explained later.

2.2.4.2 Action

The action space A includes four actions that can be executed by the adversary.

- *Adopt*: The adversary accepts the honest chain and mines on the last block of the honest chain. This action discards the $l^{(a)}$ blocks in the chain of the adversary and it renews the attack from the new starting point without a fork. This action is allowed by the MDP model for all $l^{(a)}$ and $l^{(h)}$.
- *Override*: The adversary publishes one block more than the honest chain (i.e. $l^{(h)} + 1$ blocks) to the whole network. This action overrides the conflicting blocks of the honest chain. This action is allowed when $l^{(a)} > l^{(h)}$.
- *Match*: The adversary publishes the same number of blocks as the honest chain (i.e. $l^{(h)}$ blocks) to the whole network. This action creates a fork deliberately and initiates an open mining competition between the two branches of the adversary and the honest network. This action is allowed when $l^{(a)} \geq l^{(h)}$ and *fork = relevant*.
- *Wait*: The adversary does not publish blocks and it just keeps mining on its own chain. This action is always feasible.

One remark about the actions of the MDP mining model is that some actions that can generally be performed are deliberately removed from the action-state space because these actions are not gainful for the adversary. For example, when $l^{(a)} < l^{(h)}$, the adversary can still release a certain number of its blocks. However, because releasing fewer blocks than the number of blocks on the honest chain will not increase its probability of mining the next block compared to mining it privately, these actions thus are excluded from the allowed actions.

We now explain the three values of the entry *fork* in the three-tuple state.

- *Relevant*: The value of relevant means that the latest block is mined by the honest network. Now, if *fork* = *relevant* and $l^{(a)} \geq l^{(h)}$, the action *match* is allowed. For example, if the previous state is $\left(l^{(a)}, l^{(h)} - 1, \bullet\right)$ and now the honest network successfully mines one block, the state then changes to $\left(l^{(a)}, l^{(h)}, relevant\right)$. If at this time, $l^{(a)} \geq l^{(h)}$, *match* is allowed. We remark that *match* here may be gainful for the adversary because $\gamma(1 - \alpha)$ computing power of the honest network would be dedicated to mining on the adversary chain because of the near-simultaneous releases of the latest block of the adversary chain and the latest block of the honest chain. In this state, as far as the public is concerned, there is no fork yet, as the $l^{(a)}$ mined blocks of the adversary are private and hidden from the public. However, if the adversary executes a match from this state, then a fork will be made known to the public and an active competition between the two branches will follow.
- *Irrelevant*: The value of *irrelevant* means that the latest block is mined by the adversary and the blocks published by the honest network have been already received by (the majority of) the honest network. Now, even if $l^{(a)} \geq l^{(h)}$, the action *match* is not allowed. For example, if the previous state is $\left(l^{(a)} - 1, l^{(h)}, \bullet\right)$ and now the adversary successfully mines a new block, the state changes to $\left(l^{(a)}, l^{(h)}, irrelevant\right)$. We emphasize that *match* is disallowed here even if $l^{(a)} \geq l^{(h)}$, not because it cannot be performed in the blockchain but rather *match* here is not gainful for the adversary. If *match* were to be performed here, no computing power of the honest network would shift to mining on the adversary chain because the miners in the honest network would have received the latest block of the honest chain first (well before the current transition triggered by the adversary mining a new block) and would have dedicated to mining on the honest chain already. Again, in this state, there is no fork as far as the public blockchain is concerned.
- *Active*: The value of *active* means that the adversary has executed the action *match* from the previous state, and the blockchain is now split into two branches. For example, if the previous state is $\left(l^{(a)}, l^{(h)}, relevant\right)$ with $l^{(a)} \geq l^{(h)}$ and the adversary executed the action *match*. If the new transition is triggered by the honest network mining a new block, then the state transitions to $\left(l^{(a)} - l^{(h)}, 1, active\right)$. In short, *active* means a fork is made known to the public and that an active competition between the two branches of the fork is ongoing.

2.2.4.3 Transition and Reward

After the execution of an action, the occurrence of each state transition is triggered by the creation of a new block (either by the adversary or by the honest network) and the corresponding transition probability is the probability of the block created by the adversary (α) or by the honest network ($1 - \alpha$). The initial state is $(1, 0, irrelevant)$ with probability α or $(0, 1, irrelevant)$ with probability $1 - \alpha$. Different actions performed by the adversary will have different effects on the state transitions. The specific description is as follows:

- The state transitions after the execution of action *adopt*: By executing the *adopt* action, the adversary accepts all the blocks on the branch mined by the honest network and mines on the latest block on the honest chain together with the honest network. An illustrating example of the state transitions after the execution of action *adopt* is given in Figure 2.3. As shown in Figure 2.3, with the probability of α, the adversary can successfully mine

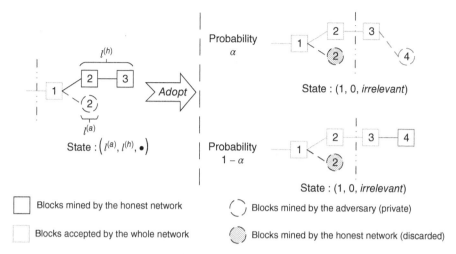

Figure 2.3 An illustrating example of the state transitions after the execution of action *adopt*.

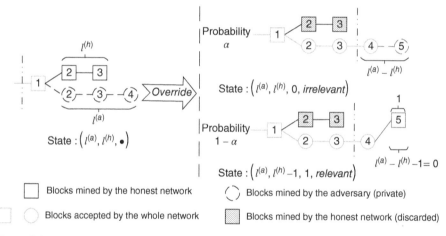

Figure 2.4 An illustrating example of the state transitions after the execution of action *override*.

the next block and then the state transits to $(1, 0, relevant)$; with the probability of $1 - \alpha$, the honest network can successfully mine the next block and then the state transits to $(0, 1, relevant)$.

- The state transition after the execution of action *override*: The adversary can only perform action *override* when the number of the blocks on its private branch is greater than the number of the blocks on the honest branch (i.e. when $l^{(a)} > l^{(h)}$). By performing *override*, the adversary publishes $l^{(h)} + 1$ blocks from its private branch to overwrite the latest $l^{(h)}$ blocks on the honest branch. After that, the branch of the adversary becomes the main chain and the whole network mines on the latest block of the adversary's branch. An illustrating example of the state transitions after the execution of action *override* is given in Figure 2.4. As shown in Figure 2.4, the adversary has the probability of α to successfully mine the next block and makes the state transit to $(l^{(a)} - l^{(h)}, 0, irrelevant)$; the honest network has the probability of $1 - \alpha$ to successfully mine the next block and makes the state transit to $\left(l^{(a)} - l^{(h)} - 1, 1, relevant\right)$.

- The state transition after the execution of action *match*: The *match* action can only be executed when *fork* = *relevant* and when the number of blocks on the private branch of the adversary is greater than or equal to the number of blocks on the public branch of the honest network (i.e. when $l^{(a)} \geq l^{(h)}$). After the adversary performs the *match* action, a fork will be formed on the blockchain that is observed by all the miners. After that, the adversary is still mining on its own branch; however, because of the fork, a γ fraction of the honest network will mine on the branch published by the adversary, and the other $1 - \gamma$ fraction of the honest network will mine on the branch published by the honest network. An illustrating example of the state transitions after the execution of action *match* is given in Figure 2.5. As shown in Figure 2.5, the next block may be published by the adversary on its own branch such that the state transits to $\left(l^{(a)} + 1, l^{(h)}, active\right)$ with the probability of α; the next block may be published by the honest network on the branch of the adversary such that the state transits to $\left(l^{(a)} - l^{(h)}, 1, relevant\right)$ with the probability of $\gamma\left(1 - \alpha\right)$; the next block may be published by the honest network on the

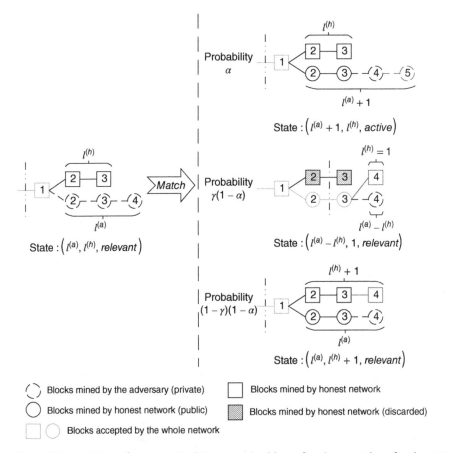

Figure 2.5 An illustrating example of the state transitions after the execution of action *match*.

branch of the honest network such that the state transits to $\left(l^{(a)}, l^{(h)} + 1, relevant\right)$ with the probability of $(1 - \gamma)(1 - \alpha)$. We must emphasize that after the execution of action *match*, among the $l^{(a)}$ blocks of the adversary, some of the blocks may be private while other blocks are public. Which parts of blocks are private/public are implied by the state implicitly. For example, suppose that the previous state is $\left(l^{(a)}, l^{(h)}, relevant\right)$ with $l^{(a)} > l^{(h)}$ (as illustrated in the left part of Figure 2.5) and the action *match* is performed. If the adversary subsequently mines a new block on its own branch, then the state changes to $\left(l^{(a)} + 1, l^{(h)}, active\right)$, where there are $l^{(a)} + 1 - l^{(h)}$ private blocks and $l^{(h)}$ public blocks among the $l^{(a)} + 1$ blocks owned by the adversary (as illustrated by the first case in the right part of Figure 2.5). If the honest miners mine a new block on the adversary's branch, the state changes to $\left(l^{(a)} - l^{(h)}, 1, relevant\right)$, where there are $l^{(a)} - l^{(h)}$ private blocks left for the adversary (as illustrated by the second case in the right part of Figure 2.5). If the honest miners mine a new block on the honest network's branch, the state changes to $l^{(a)} - l^{(h)}$, where there are $l^{(a)} - l^{(h)}$ private blocks and $l^{(h)}$ public blocks among the $l^{(a)}$ blocks owned by the adversary (as illustrated by the third case in the right part of Figure 2.5).

- The state transition triggered by action *wait*: The *wait* action means that the adversary does not perform any actions and continues to mine on its private branch. After the action *wait* is executed, if *fork* ≠ *active*, the adversary and the honest network mine on their own branches. An illustrating example of the state transitions after the execution of action *match* when *fork* ≠ *active* is given in Figure 2.6. As shown in Figure 2.6, when *fork* ≠ *active*, the next new block may be mined by the adversary on its own private branch

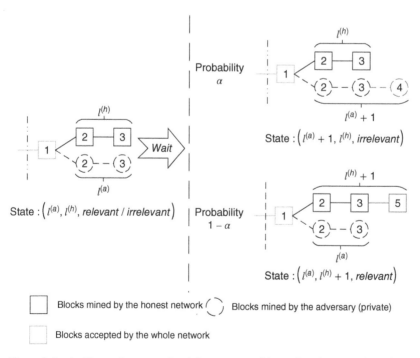

Figure 2.6 An illustrating example of the state transitions after the execution of action *wait* when *fork* ≠ *active*.

such that the state changes to $\left(l^{(a)} + 1, l^{(h)}, irrelevant\right)$ with the probability of α; or the next new block may be mined by the honest network on the public branch such that the state changes to $\left(l^{(a)}, l^{(h)} + 1, relevant\right)$ with the probability of $1 - \alpha$. After the action *wait* is executed, if *fork* = *active*, because of the fork that can be observed by the whole network, the mining behaviors of all miners are the same as that after the execution of the *match* action. An illustrating example of the state transitions after the execution of action *match* when *fork* = *active* is given in Figure 2.7. blackAs shown in Figure 2.7, when *fork* = *active*, the next new block may be mined by the adversary on its own branch such that the state changes to $\left(l^{(a)} + 1, l^{(h)}, active\right)$ with probability α; or the next new block may be mined by the honest network on the branch of the adversary such that the state changes to $\left(l^{(a)} - l^{(h)}, 1, relevant\right)$ with probability $\gamma\left(1 - \alpha\right)$; or the next new block may be mined by the honest network on the branch of the adversary such that the state changes to $\left(l^{(a)}, l^{(h)} + 1, relevant\right)$ with probability $\left(1 - \gamma\right)\left(1 - \alpha\right)$.

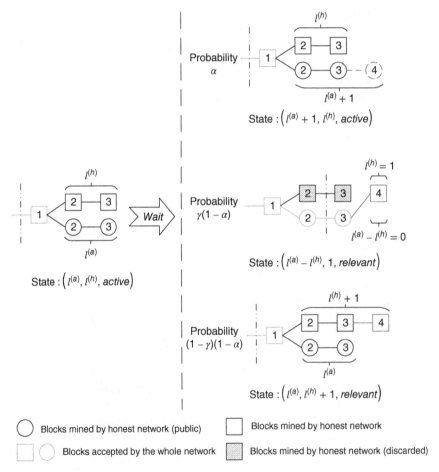

Figure 2.7 An illustrating example of the state transitions after the execution of action *wait* when *fork* = *active*.

Table 2.1 The state transitions and reward matrices of the MDP mining model.

Current state, action	Next state	Transition probability	Reward
$\left(l^{(a)}, l^{(h)}, \bullet\right), adopt$	$(1, 0, irrelevant)$	α	$\left(0, l^{(h)}\right)$
	$(0, 1, irrelevant)$	$1 - \alpha$	
$\left(l^{(a)}, l^{(h)}, \bullet\right), override$	$\left(l^{(a)} - l^{(h)}, 0, irrelevant\right)$	α	$\left(l^{(h)} + 1, 0\right)$
	$\left(l^{(a)} - l^{(h)} - 1, 1, relevant\right)$	$1 - \alpha$	
$\left(l^{(a)}, l^{(h)}, irrelevant\right), wait$	$\left(l^{(a)} + 1, l^{(h)}, irrelevant\right)$	α	$(0, 0)$
$\left(l^{(a)}, l^{(h)}, relevant\right), wait$	$\left(l^{(a)}, l^{(h)} + 1, relevant\right)$	$1 - \alpha$	$(0, 0)$
$\left(l^{(a)}, l^{(h)}, active\right), wait$	$\left(l^{(a)} + 1, l^{(h)}, active\right)$	α	$(0, 0)$
$\left(l^{(a)}, l^{(h)}, relevant\right), match$	$\left(l^{(a)} - l^{(h)}, 1, relevant\right)$	$\gamma(1 - \alpha)$	$\left(l^{(h)}, 0\right)$
	$\left(l^{(a)}, l^{(h)} + 1, relevant\right)$	$(1 - \gamma)(1 - \alpha)$	$(0, 0)$

The action *override* is allowed when $l^{(a)} > l^{(h)}$; the action *match* is allowed when $l^{(a)} \geq l^{(h)}$.

The reward is given as a tuple $\left(r^{(a)}, r^{(h)}\right)$, where $r^{(a)}$ denotes the number of blocks mined by the adversary and accepted by the whole network, and $r^{(h)}$ denotes the number of blocks mined by the honest network and accepted by the whole network. The state transitions and reward matrices are given in Table 2.1.

2.2.4.4 Objective Function

The objective of the adversary is to find the optimal mining strategy that can earn as much reward as possible. Because blockchain keeps adjusting the mining difficulty (i.e. the mining target on the right hand side (RHS) of inequality (2.1)) to ensure that on average one valid block is introduced to the overall blockchain per *valid block interval* (e.g. one block per 10 minutes for Bitcoin and per 10–20 seconds for Ethereum), the mining objective of the adversary is not to maximize its absolute cumulative reward but to maximize the ratio of its cumulative rewards over the cumulative rewards of the whole network (i.e. the cumulative rewards of the whole network advance by one reward per block interval – rewards of all miners/Time is fixed to 1 per block interval; then maximizing adversary rewards/Time is equivalent to maximizing the ratio of adversary rewards/Time to rewards of all miners/Time = adversary rewards/rewards of all miners). We emphasize that blocks mined by the adversary and the honest network that are discarded because of losing out in the competition are not considered as having been successfully introduced to the blockchain. Thus, the principle behind the strategy of the adversary is to maximize the number of blocks mined by the honest network that are later discarded while reducing its own discarded blocks.

As in [27], we define the following *relative mining gain (RMG)* as the objective function for blockchain mining:

$$RMG = E\left[\lim_{T \to \infty} \frac{\sum_{\tau=t}^{t+T-1} r_{\tau+1}^{(a)}}{\sum_{\tau=t}^{t+T-1} r_{\tau+1}^{(a)} + \sum_{\tau=t}^{t+T-1} r_{\tau+1}^{(h)}}\right], \tag{2.2}$$

where $\left(r_t^{(a)}, r_t^{(h)}\right)$ is the tuple of rewards issued in the block interval t, T is the size of the observing window. The objective of the adversary is to maximize this relative mining gain.

Under the above MDP mining model, we can now interpret honest mining, selfish mining [24], and lead stubborn mining [26] as examples of different mining strategies.

2.2.4.5 Honest Mining

For honest mining, miners will follow the rule of the longest chain extension. Thus, they will not maintain a private chain: when they have a new block, they will immediately publish it. The honest mining strategy can be written as

$$HM\left(l^{(a)}, l^{(h)}, \bullet\right) = \begin{cases} adopt & l^{(a)} < l^{(h)} \\ wait & l^{(a)} = l^{(h)} \\ override & l^{(a)} > l^{(h)} \end{cases}, \tag{2.3}$$

where we note that $l^{(a)}$, $l^{(h)}$ can only take a value of 0 or 1.

2.2.4.6 Selfish Mining

The main idea of selfish mining [24] is described as follows. If one block is found by the adversary, it does not publish it immediately and it keeps mining on its private chain. When the adversary already has one private block and then honest network publishes one block (immediately after an honest miner mines a new block), the adversary chooses to publish its block to match the honest network. This causes $\gamma(1 - \alpha)$ computing power of the honest network to mine on the adversary's chain. When the adversary already has some private blocks and then honest network catches up with only one block less than the adversary ($l^{(h)} = l^{(a)} - 1 \geq 1$), the adversary overrides the honest network's block by publishing all its blocks. The selfish mining strategy can be written as

$$SM\left(l^{(a)}, l^{(h)}, \bullet\right) = \begin{cases} adopt & l^{(a)} < l^{(h)} \\ match & l^{(a)} = l^{(h)} = 1 \\ override & l^{(h)} = l^{(a)} - 1 \geq 1 \\ wait & otherwise \end{cases}. \tag{2.4}$$

2.2.4.7 Lead Stubborn Mining

Lead stubborn mining [26] is different from selfish mining in the following way. A lead stubborn miner always publishes one block from its private chain to match with the honest network when the honest network mines a new block if $l^{(a)} \geq l^{(h)}$. The adversary never executes the action override. The lead stubborn mining can be written as

$$LSM\left(l^{(a)}, l^{(h)}, fork\right) = \begin{cases} adopt & l^{(a)} < l^{(h)}, \forall fork \\ match & otherwise \\ wait & l^{(a)} > l^{(h)}, fork = irrelevant \end{cases}. \tag{2.5}$$

It is shown that this lead stubborn mining can achieve higher profits than selfish mining [26].

2.2.4.8 Optimal Mining

Although there are many possible mining strategies that can obtain profits higher than honest mining, the optimal mining strategy is not obvious. Because the state-action space of the MDP is huge, it is not straightforward to derive the optimal mining strategy. The relative mining gain objective (2.2) is a nonlinear function of the rewards, and thus, the corresponding MDP cannot be solved using standard MDP solvers to give the optimal mining strategy. To solve this problem, Sapirshtein et al. [27] first transformed the MDP with the nonlinear objective to a family of MDPs with linear objectives and then employed a standard MDP solver combined with a numerical search over the family of MDPs to find the optimal mining strategy. As shown in [27], this solution indeed can find the optimal mining strategy. However, the solution of [27] is a model-based approach: it must know the parameters that characterize the MDP model exactly (i.e. the computing power distribution α and the communication capability γ). In real blockchain networks, these parameters are not easy to obtain and may change over time, hindering the use of the solution proposed in [27]. We propose a model-free approach that solves the MDP with the nonlinear objective using RL.

2.2.5 Mining Through RL

This section first provides preliminaries for RL and then presents a new RL algorithm that can derive the optimal mining strategy without knowing the parameters of the environment. We propose the new RL mining algorithm based on Q-learning, one popular algorithm from the RL family.

2.2.5.1 Preliminaries for Original Reinforcement Learning Algorithm

In RL, an agent interacts with an environment in a sequence of discrete time steps, $t = 0, 1, 2, \ldots$, as shown in Figure 2.8. At time t, the agent observes the state of the environment, s_t; it then takes an action, a_t. As a result of the state-action pair, (s_t, a_t), the agent receives a scalar reward r_{t+1}, and the environment moves to a new state s_{t+1} at time $t + 1$. Based on s_{t+1}, the agent then decides the next action a_{t+1}. The goal of the agent is

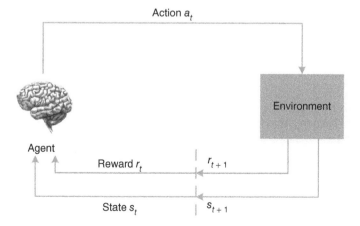

Figure 2.8 The agent-environment interaction process of RL algorithm.

to effect a series of rewards $\{r_t\}_{t=1,2,...}$ through its actions to maximize some performance criterion. For example, for Q-learning [30], the performance criterion to be maximized at time t is the discounted accumulated rewards going forward $R_t = \sum_{\tau=t}^{\infty} \lambda^{\tau-t} r_{\tau+1}$, where $\lambda \in (0, 1)$ is a discount factor for weighting future rewards [28]. In general, the agent takes actions according to some decision policy π. RL methods specify how the agent changes its policy as a result of its experiences. With sufficient experiences, the agent can learn an optimal decision policy π^* to maximize the long-term accumulated reward [28].

The desirability of state-action pair (s_t, a_t) under a decision policy decision π is captured by a Q function defined as $Q(s, a) = [R_t | s_t = s, a_t = a, \pi]$, i.e. the expected discounted accumulated reward going forward given the current state-action pair (s_t, a_t). The optimal decision policy π^* is one that maximizes the Q function. In Q-learning, the goal of the agent is to learn the optimal policy π^* through an online iterative process by observing the rewards while it takes action in successive time steps. In particular, the agent maintains the Q function, $Q(s, a)$, for all state-action pairs (s, a), in a tabular form.

Let $q(s, a)$ be the estimated action-value function during the iterative process. At time step t, given state s_t, the agent selects a greedy action $a_t = \arg\max_a q(s_t, a)$ based on its current Q function. This will cause the system to return a reward r_{t+1} and move to state s_{t+1}. The experience at time step t is captured by the quadruplet $e_t = (s_t, a_t, r_{t+1}, s_{t+1})$. At the end of time step t, experience e_t is used to update $q(s_t, a_t)$ for entry (s_t, a_t) as follows:

$$q(s_t, a_t) \leftarrow (1 - \beta) q(s_t, a_t) + \beta [r_{t+1} + \lambda \max q(s_{t+1}, a')], \tag{2.6}$$

where $\beta \in (0, 1]$ is a parameter that governs the learning rate. Q-learning learns from experiences gathered over time, $\{e_t\}_{t=0,1,...}$, through the iterative process in (2.6). Note that Q-learning is a model-free learning framework in that it tries to learn the optimal policy without having a model that describes the operating behavior of the environment beyond what can be observed through the experiences.

As a deviation from the above description, a caveat in Q-learning is that the so-called ε-greedy algorithm is often adopted in action selection. For the ε-greedy algorithm, the action $a_t = \arg\max_a q(s_t, a)$ is only chosen with probability $1 - \varepsilon$. With probability ε, a random action is chosen uniformly from the set of possible actions. This is to avoid the algorithm from zooming into a local optimal policy and to allow the agent to explore a wider spectrum of different actions in search of the optimal policy [28].

It has been shown that in a stationary environment that can be fully captured by an MDP, the Q-values will converge to optimality if the learning rate decays appropriately and each action in the state-action pair (s, a) is executed an infinite number of times in the process [28].

2.2.5.2 New Reinforcement Learning Algorithm for Mining

The original RL algorithm as presented above cannot be directly applied to maximize the mining objective function expressed in (2.2); there is one fundamental obstacle that must be overcome. The obstacle is the nonlinear combination of the rewards in the objective function. The original RL algorithm can only maximize an objective that is a linear function of the scalar rewards, e.g. the weighted sum of scalar rewards. To address this issue, we put forth a new algorithm that aims to optimize the original mining objective: the multi-dimensional RL algorithm.

We formulate the multi-dimensional RL algorithm as follows. At *mined block interval*[3] t ($t = 0, 1, 2, \ldots$), the state $s_t \in S$ takes a value from the state space S as defined in the MDP model of blockchain mining, and the action $a_t \in A$ is chosen from the action space A. The state transition occurs according to Table 2.1. The reward is the pair $\left(r_{t+1}^{(a)}, r_{t+1}^{(h)}\right)$ whose value is assigned according to Table 2.1. The experience at the end of mined block interval t is given by $e_t = \left(s_t, a_t, s_{t+1}, r_{t+1}^{(a)}, r_{t+1}^{(h)}\right)$. The objective of the multi-dimensional RL algorithm is to maximize the relative mining gain as expressed in (2.2).

For a state-action pair (s_t, a_t), instead of maintaining an action-value scalar $Q(s, a)$, the multi-dimensional RL algorithm maintains an action-value pair $\left(Q^{(a)}(s, a), Q^{(h)}(s, a)\right)$ corresponding to the Q function values of the adversary and the honest network, respectively. The Q functions defined by Q learning are the expected cumulative discounted rewards. Specifically, $Q^{(a)}(s, a)$ and $Q^{(h)}(s, a)$ are defined as

$$
\begin{aligned}
Q^{(a)}(s, a) &= E\left[\lim_{T \to \infty} \sum_{\tau=t}^{T} \lambda^{\tau-t} r_{\tau+1}^{(a)} \,\middle|\, S_t = s, a_t = a, \pi\right] \\
Q^{(h)}(s, a) &= E\left[\lim_{T \to \infty} \sum_{\tau=t}^{T} \lambda^{\tau-t} r_{\tau+1}^{(h)} \,\middle|\, S_t = s, a_t = a, \pi\right]
\end{aligned} \tag{2.7}
$$

Suppose that at mined block interval t, the Q functions in (2.7) are estimated to be $q^{(a)}(s, a)$, $q^{(h)}(s, a)$. For action selection, we still adopt the ε-greedy approach. To select the greedy action, we construct the following objective function:

$$
f(s, a) = \frac{q^{(a)}(s, a)}{q^{(a)}(s, a) + q^{(h)}(s, a)}. \tag{2.8}
$$

After taking action a_t, the state transitions to s_{t+1} and the reward pair $\left(r_{t+1}^{(a)}, r_{t+1}^{(h)}\right)$ are issued. With the experience $e_t = \left(s_t, a_t, s_{t+1}, r_{t+1}^{(a)}, r_{t+1}^{(h)}\right)$, the multi-dimensional RL algorithm updates the two Q functions as follows:

$$
\begin{aligned}
&q^{(a)}(s_t, a_t) \\
&\leftarrow (1 - \beta)\, q^{(a)}(s_t, a_t) + \beta \left[r_{t+1}^{(a)} + \lambda q^{(a)}\left(s_{t+1}, a'\right)\right] \\
&q^{(h)}(s_t, a_t) \\
&\leftarrow (1 - \beta)\, q^{(h)}(s_t, a_t) + \beta \left[r_{t+1}^{(h)} + \lambda q^{(h)}\left(s_{t+1}, a'\right)\right]
\end{aligned} \tag{2.9}
$$

where $a' = \arg\max_a f\left(s_{t+1}, a\right)$. Note that the update rule of (2.9) is very similar to the update rule of Q learning, except that the greedy action a' is chosen by maximizing the constructed objective function in (2.8) rather than maximizing the Q function itself as in Q learning. From the expressions in (2.7) and (2.8), we can verify that the adopted objective

---

3 A mined block interval is different from a valid block interval. A valid block interval separates two valid blocks that are ultimately adopted by the blockchain. The average duration of a valid block interval is a constant in many blockchain systems (e.g. 10 minutes in bitcoin). The average duration of the valid block interval is defined by the system designer and its constancy is maintained by adjusting the mining target. A mined block interval separates two mined blocks (by either the adversary of the honest network), regardless of whether the blocks become valid later or not. In the MDP model, each transition is triggered by the mining of a new block. Thus, the average duration of a mined block interval is the average time separated by two adjacent transitions. Because of the actions of the adversary, some of the mined blocks (by the adversary of the honest network) may be discarded later.

function in (2.8) is consistent with the relative mining gain objective function defined in (2.2), except the discount terms $\lambda^{\tau-t}$ used in the computation of the Q functions. The use of discount terms can ensure that the Q functions can converge to some bounded values; however, adding discount terms to the rewards will change the original mining objective. One simple way to ensure strict objective consistency is to set $\lambda = 1$. Although the setting of $\lambda = 1$ will result in unbounded values for the Q functions as the RL iteration gradually progresses to infinite time steps, this is not a big problem as long as the Q function values do not overflow during the execution of the algorithm. In practice, we can also set λ to be very close to one.

The RL algorithm expressed by the Q function updates in (2.9) is our multi-dimensional RL algorithm. We introduce one additional technical element to the ε-greedy action selection, as explained in the next paragraph.

As described above, when we select the action, we adopt the ε-greedy strategy that allows us to select the current best action $\left(a_t = \arg\max_a f\left(s_t, a\right)\right)$ with probability $1 - \varepsilon$ and to randomly select an action with probability ε. This random action selection is used to explore some unseen states and can avoid trapping at local optimal maximums. However, the tuning of parameter ε is not straightforward. A large ε reduces the possibility of trapping at local optimal maximums, but it also decreases the average reward as it wastes a fraction of the time to explore nonoptimal states. In our algorithm, we adopt the following strategy for dynamically tuning the parameter ε. Denote the number of times state was visited by $V\left(s_t\right)$. Then, the ε parameter used at state s_t for performing ε-greedy action selection is given by

$$\varepsilon\left(s_t\right) = \exp\left(-\frac{V\left(s_t\right)}{T_\varepsilon}\right), \tag{2.10}$$

where T_ε is a temperature parameter that governs how fast we gradually reduce the ε parameter. The pseudo-code of our multi-dimensional RL algorithm for blockchain mining is given in Algorithm 2.1.

2.2.6 Performance Evaluations

We have conducted simulations to investigate our proposed RL mining strategy. Following the simulation approach used in [24], we constructed a Bitcoin-like simulator that captures all the relevant PoW network details described in Section 2.2.4, except that the crypto puzzle solving processing was replaced by a Monte Carlo simulator that simulates the time required for block discovery without actually attempting to compute a hash function. We simulated 1000 miners mining at identical rates (i.e. they each can have one simulated hash test at each time step of the Monte Carlo simulation). A subset of the 1000 miners (1000α miners) forms an adversary pool running a malicious mining strategy that co-exists with honest mining adopted by the other $1000(1 - \alpha)$ miners. When co-existing with honest mining, the malicious mining strategy is one of the following mining strategies: (i) our RL mining strategy, (ii) the optimal mining strategy derived in [27], or (iii) the selfish mining strategy derived in [24]. Upon encountering two subchains of the same length, we divide the honest miners such that a fraction γ of them mine on the attacking pool's branch while the rest mine on the other branch. The performance metric used is the relative mining gain (RMG) computed over a window consisting of $T_w = 10^5$ time

---

Algorithm 2.1 Multi-dimensional RL Algorithm for Blockchain Mining.

Initialize $q^{(a)}(s, a) = 0, \forall s, \forall a$;
Initialize $q^{(h)}(s, a) = 0, \forall s, \forall a$;
Initialize $V(s) = 0, \forall s$;
Initialize $T_\varepsilon, \lambda, \beta$;
for $t = 0, 1, 2, \cdots$ **do**

    Receive s_t, r_t from the blockchain environment;
    Generate action $a_t = \text{SELECTACTION}(s_t)$;
    Input a_t to the blockchain environment;
    Observe $s_{t+1}, r_{t+1}^{(a)}, r_{t+1}^{(h)}$ from the blockchain environment;
    Compute $a' = \arg\max\limits_{a} \dfrac{q^{(a)}(s_{t+1}, a)}{q^{(a)}(s_{t+1}, a) + q^{(h)}(s_{t+1}, a)}$
    Update

$$q^{(a)}(s_t, a_t)$$
$$\leftarrow (1 - \beta) q^{(a)}(s_t, a_t) + \beta \left[r_{t+1}^{(a)} + \lambda q^{(a)}\left(s_{t+1}, a'\right) \right];$$
$$q^{(h)}(s_t, a_t)$$
$$\leftarrow (1 - \beta) q^{(h)}(s_t, a_t) + \beta \left[r_{t+1}^{(h)} + \lambda q^{(h)}\left(s_{t+1}, a'\right) \right];$$

end for
procedure SELECTACTION(s_t)
    Compute $\varepsilon(s_t) = \exp\left(-\dfrac{V(s_t)}{T_\varepsilon}\right)$;
    if random $< \varepsilon(s_t)$ **then**
        randomly select an action a_t from A;
    else
        $a_t = \arg\max\limits_{a \in A} \dfrac{q^{(a)}(s_t, a)}{q^{(a)}(s_t, a) + q^{(h)}(s_t, a)}$;
    end if return a_t
end procedure

---

steps: $\sum_{\tau=t}^{t+T_w-1} r_{\tau+1}^{(a)} / \left(\sum_{\tau=t}^{t+T_w-1} r_{\tau+1}^{(a)} + \sum_{\tau=t}^{t+T_w-1} r_{\tau+1}^{(h)} \right)$. The hyper-parameters used in the RL algorithm are set to as $\lambda = 0.999$, $\beta = 0.05$.

We first compare the performances of our RL mining, the optimal policy mining, and the selfish mining. Figures 2.9–2.11 plots the mining reward of the adversary vs. α for different values of γ and $\gamma \in \{0, 0.5, 1\}$. Note that the value of γ ranges in the interval $[0, 1]$. Therefore, $\gamma = 0, \gamma = 0.5$, and $\gamma = 1$ means a low, a median, and a high communication capability for the adversary, respectively. We just take these three values for γ to demonstrate that the adversary with our RL mining can dynamically adapt to the optimal mining when it has different communication capabilities. The relative mining gain of $\alpha / (1 - \alpha)$ is treated as a bound for the mining problem and it can only be achieved by optimal policy mining for $\gamma = 1$. To derive the optimal policy, we adopt the search algorithm proposed in [27] and set the search error to a very tiny number of 10^{-5}. As in [27], we truncate the MDP at $l^{(a)} = 100$ or $l^{(a)} = 100$ for both of optimal policy mining and RL mining. The temperature parameter T_ε is set to as $T_\varepsilon = 10^4$ and it is reset to $T_\varepsilon = 0$ after $t = 10^8$ time steps when convergence is

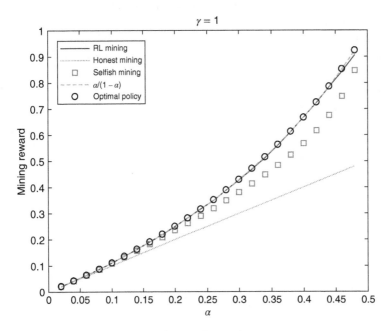

Figure 2.9　Achieved mining gain vs. α for $\gamma = 1$.

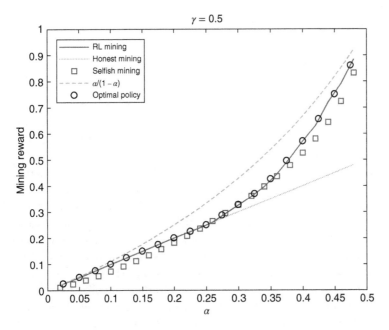

Figure 2.10　Achieved mining gain vs. α for $\gamma = 0.5$.

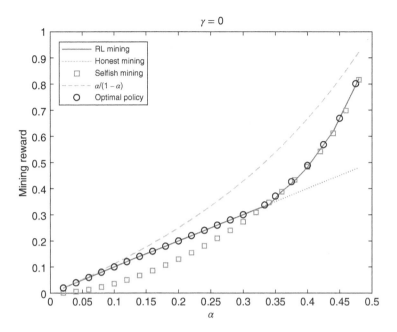

Figure 2.11 Achieved mining gain vs. α for $\gamma = 0$.

attained. All the results of RL mining are given after the algorithm has converged. From the results, we can see that the performance of our RL mining can converge to the performance of optimal policy mining without knowing the details about the environment model.

We next consider the impact of the temperature parameter T_ε on the convergence of RL mining. Figures 2.12–2.14 present the mining rewards obtained by RL mining with different T_ε over time for $\gamma \in \{0, 0.5, 1\}$, respectively (α is fixed to 0.45). blackIn fact, the parameter of T_ε determines the extent of the exploration performed by the RL algorithm in its learning process. A larger T_ε encourages more explorations, and eventually, the learning process can converge, although a larger T_ε needs more time to converge. The optimal value of T_ε can lead to the convergence of RL by enough explorations and does not waste learning time by having unnecessary explorations. How to tune to the optimal T_ε is an interesting research direction. In this work, we just investigate the impact of T_ε on our RL mining by simulations. From the simulation results in Figures 2.12–2.14, we can see that generally, RL mining with larger T_ε can have more explorations and can converge more closely to the optimal performance; however, RL mining with larger T_ε also have longer exploration phases that slow down the convergence process. Figure 2.15 presents the mining rewards of RL mining with different T_ε for different α (γ is fixed to 1). The mining reward results are given after $t = 10^7$ time steps and without resetting $T_\varepsilon = 0$. We see that for larger α, we need larger T_ε to ensure the convergence of RL mining, although it will slow down the convergence process. In practice, we can dynamically reduce the value of T_ε when we find that the mining gain has already converged.

Last, we investigate the mining performances of different mining strategies when the blockchain environment changes. The experimental results are given in Figures 2.16–2.18. The blockchain environment starts with parameter values of $(\alpha = 0.35, \gamma = 1)$ and the values of (α, γ) change sequentially in the experiment. The temperature parameter T_ε of RL

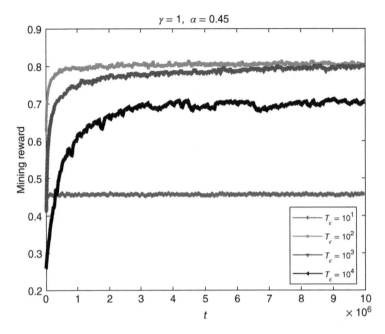

Figure 2.12 Achieved mining gain vs. time step for different T_ε and $\gamma = 1$, $\alpha = 0.45$.

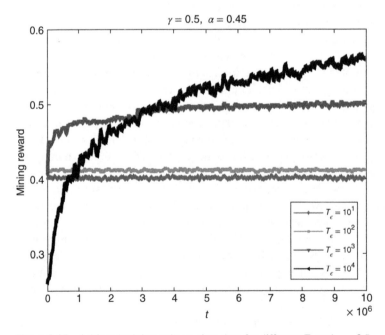

Figure 2.13 Achieved mining gain vs. time step for different T_ε and $\gamma = 0.5$, $\alpha = 0.45$.

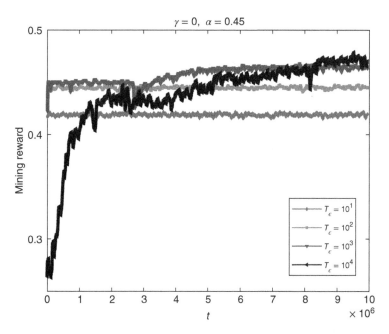

Figure 2.14 Achieved mining gain vs. time step for different T_ϵ and $\gamma = 0$, $\alpha = 0.45$.

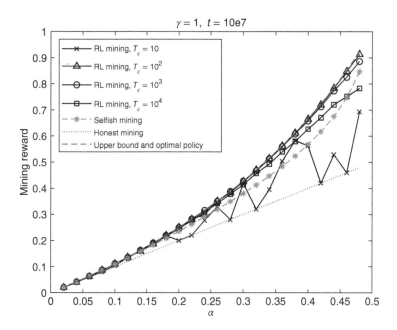

Figure 2.15 Achieved mining gain vs. the α for $\gamma = 1$ and different T_ϵ.

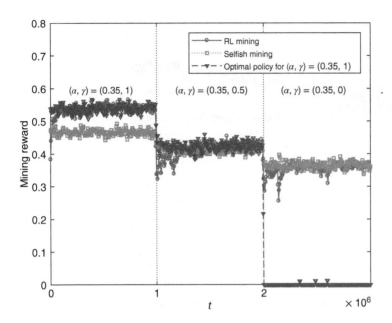

Figure 2.16 Achieved mining gain when the environment is changing and the values of (α, γ) change in the following order: (0.35, 1), (0.35, 0.5), and (0.35, 0).

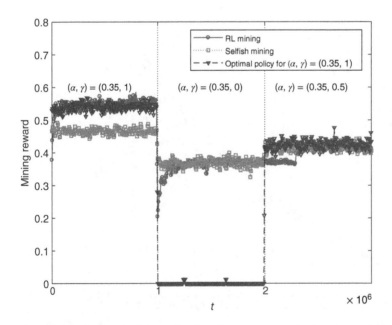

Figure 2.17 Achieved mining gain when the environment is changing and the values of (α, γ) change in the following order: (0.35,1), (0.35, 0), and (0.35,0.5).

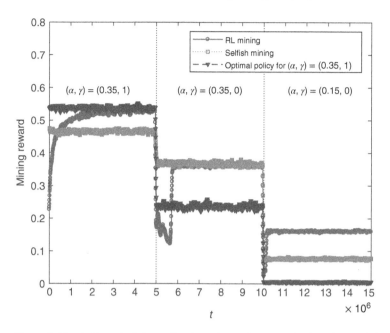

Figure 2.18 Achieved mining gain when the environment is changing and the values of (α, γ) change in the following order: (0.35,1), (0.35, 0), and (0.15,0).

Table 2.2 The optimal policy for the blockchain environment with $(\alpha = 0.35, \gamma = 1)$ when $l^{(a)} \leq 8$ and $l^{(h)} \leq 8$.

$l^{(a)} l^{(h)}$	1	2	3	4	5	6	7	8
1	***	*a*	***	***	***	***	***	***
2	w**	*m*	*w*	*a*	***	***	***	***
3	w**	*oo	w**	*w*	*a*	***	***	***
4	w**	*m*	oo*	w**	*w*	*a*	***	***
5	w**	*mw	*m*	oo*	w**	*w*	*w*	*a*
6	w**	*mw	*mw	*m*	oo*	w**	ww*	*w*
7	w**	*mw	*mw	*mw	*m*	oo*	w**	ww*
8	w**	*mw	*mw	*mw	*mw	*m*	oo*	w**

* represents an unreachable state.

mining is fixed to $T_\varepsilon = 10^3$. The optimal policy mining strategy adopts the optimal policy for the blockchain environment with $(\alpha = 0.35, \gamma = 1)$. We derived the optimal policy for $(\alpha = 0.35, \gamma = 1)$ by iteratively exploiting the MDP solver [40] to search over the policy space, as proposed in [27]. Table 2.2 describes the found optimal policy for $(\alpha = 0.35, \gamma = 1)$ when $l^{(a)} \leq 8$ and $l^{(h)} \leq 8$. The performances of the optimal policy for $(\alpha = 0.35, \gamma = 1)$, and the selfish mining are treated as benchmarks for our RL mining in the changing blockchain environment. blackIn Figures 2.16–2.18, for different values of the parameters (α, γ), the performances of optimal selfish mining are still obtained using the policy of the optimal selfish mining under model parameters $(\alpha, \gamma) = (0.35, 1)$. From the simulation results, we can see that when the environment has changed, the optimal policy mining strategy derived from the MDP model is not optimal anymore; our RL mining can adaptively learn the optimal policy for different environments. This demonstrates the advantage of RL mining over these model-based mining strategies.

2.3 Performance Analysis of the Raft Consensus Algorithm

In this part, we analyze the network splitting performance of the Raft consensus algorithm under a simple model that accounts for protocol details.

2.3.1 Review of Raft Algorithm

This section briefly summarizes the Raft algorithm. A more complete and detailed description of Raft refers to [13]. Raft is a consensus algorithm for managing a replicated ledger at every node. At any given time, each node is in one of the three states: leader, follower, or candidate. Raft algorithm divides time into terms with finite duration. Terms are numbered with consecutive integers. Each term begins with an election, in which one or more candidates attempt to become a leader. If a candidate wins the election, then it serves as a leader for the rest of the term. The state transition is shown in Figure 2.19 [41]. All nodes start from the follower state. If a follower does not hear from the leader for a certain period of time, then it becomes a candidate. The candidate then requests votes from other nodes to become a leader. Other nodes will reply to the vote request. If the candidate gets votes from a majority of the nodes, it will become a leader. This process is called Leader Election. Specifically, if a follower receives a heartbeat within the minimum election timeout of hearing from a current leader, it does not grant its vote to the candidate. This helps maximizing the duration of a leader to keep working and avoiding frequent disruptions from some isolated/removed nodes.[4]

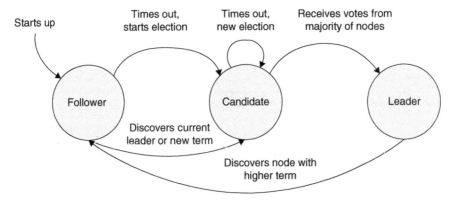

Figure 2.19 State transition model for the Raft algorithm. Source: Howard et al. [41].

---

4 These nodes will not receive heartbeats, so they will time out and start new elections.

In normal operation of Raft, there is exactly one leader and all the other nodes are followers. The leader periodically sends out heartbeats to all followers in order to maintain its authority. All transactions during this term go through the leader. Each transaction is added as an entry in the node's ledger. Specifically, the leader first replicates the received new transaction to the followers. At this time, the entry is still uncommitted and stays in a volatile state. After the leader receives feedbacks from a majority of followers that have written the entry, the leader notifies the followers that this entry is committed. This process is called Ledger Replication.

In Raft algorithm, there are several timeout settings. One of them controls the election process. The election timeout is the amount of time that a follower needs to wait to become a candidate. The election time counter is decreased as long as the follower receives no heartbeat. The follower transfers to candidate state when the election time reaches zero. The election time counter resets to a random value when the follower receives a heartbeat from a leader. The random election timers in Raft help to reduce the probability that several followers transfer to candidates simultaneously.

2.3.2 System Model

In this part, we focus on the analysis for network split probability. Consider a distributed network with N nodes and N is odd.[5] The way of message exchanging among nodes is according to the Raft algorithm. To ensure the system efficiency, the interval between heartbeats is much less than election timeout. The average time between failures of a single node is much larger than election timeout.

A network split happens if more than half of nodes are out of the current leader's control. We try to investigate the relationship between the parameters (such as timeout counter, packet loss rate, and network size) and network split probability. First, the process of one follower node transferring to a candidate is modeled as an absorbing Markov chain. Then, the network spilt probability is derived based on the model. Finally, the expected time for a node transferring from the follower to the candidate and the expected number of received heartbeats for one node are derived. Besides, we also discuss the impact of packet loss on the election performance.

In the following analysis, within one term, the initial state is defined as after a successful received heartbeat, the follower's election counter is reset. Suppose that there is a leader and $N - 1$ followers. We assume that the communication delay is much less than the heartbeat interval. One heartbeat means one step in the Markov chain.

2.3.3 Network Model

Define the packet loss probability as p, and suppose that p is a constant value for a given network. Denote the timeout value for each round of election as E_t, which is initially uniformly chosen from the range $[a, b]$. The interval between two heartbeats is τ. A discrete and integer time scale is adopted. Thus, if a follower fails to receive $K = \lfloor E_t/\tau \rfloor$ heartbeats consecutively, then it assumes that there is no viable leader and transitions to candidate state

---

5 When N is even, the algorithm still works and our claims still hold. Here, for the convenience of analysis, we consider odd N. So that $N - 1$ is even.

to start an election. Note that $K \in \{K_1, K_2, \ldots, K_r\}$, and K is uniformly chosen from the set $\{K_1, K_2, \ldots, K_r\}$, where $K_1 = \lfloor a/\tau \rfloor$ and $K_r = \lfloor b/\tau \rfloor$. In the following analysis, K denotes as the maximum number of heartbeats for an election counter to timeout.

Let $g(n)$ be the stochastic process representing the stage status $\{1, 2, \ldots, r\}$ of a given node at time n. Let $b(n)$ be the stochastic process representing the left steps of election time counter for the node at time n. Once the independence between $g(n)$ and $b(n)$ is assumed, we can model it as a two-dimensional process $\{g(n), b(n)\}$.

We adopt the short notation: $P\{i, k_i - 1 \mid i, k_i\} = P\{g(n+1) = i, b(n+1) = k_i - 1 \mid g(n) = i, b(n) = k_i\}$. In this Markov chain, the only non-null one-step transition probabilities are

$$P\{i, k_i - 1 \mid i, k_i\} = p \tag{2.11}$$

$$P\{i, K_i \mid j, k_j\} = (1 - p)/r \tag{2.12}$$

$$P\{i, 0 \mid i, 0\} = 1, \tag{2.13}$$

where $i, j = 1, 2, \ldots, r$ and $k_i \in \{1, \ldots, K_i\}$. Equation (2.11) relies on the fact that the follower fails to receive a heartbeat from the current leader and its election time counter is decreased by 1. Equation (2.12) shows the fact that the follower receives a heartbeat and reset the election time counter. Equation (2.13) shows that once the election time counter reaches zero, the follower transitions to the candidate state.

Denote $\{i, 0\}$ as the absorbing state. Since $i = 1, 2, \ldots, r$, there are r absorbing states. Denote the other states except state $\{i, 0\}$ in state space of $\{g(n), b(n)\}$ as transient states. There are t transient states, where $t = \sum_{i=1}^{r} K_i$. Let us order the states such that the first t states are transient and the last r states are absorbing. The transition matrix has the following canonical form

$$\mathbf{P} = \begin{pmatrix} \mathbf{Q} & \mathbf{R} \\ \mathbf{0} & \mathbf{I} \end{pmatrix}, \tag{2.14}$$

where $\mathbf{Q}$ is a $t \times t$ matrix, $\mathbf{R}$ is a nonzero $t \times r$ matrix, $\mathbf{0}$ is an $r \times t$ zero matrix, and $\mathbf{I}$ is an $r \times r$ identity matrix. Specifically, the entry q_{ij} of $\mathbf{Q}$ is defined as the transition probability from transient state s_i to transient state s_j, the entry r_{mn} of $\mathbf{R}$ is defined as the transition probability from transient state s_m to absorbing state s_n.

When $K_r - K_1 \ll K_1$ or $b - a < \tau$, the election timeout only has value 1. Thus, $r = 1, t = K$ and then the only non-null one-step transition probabilities in Eqs. (2.11)–(2.13) can be simplified as follows:

$$P\{k - 1 \mid k\} = p, \tag{2.15}$$

$$P\{K \mid k\} = 1 - p, \tag{2.16}$$

$$P\{0 \mid 0\} = 1, \tag{2.17}$$

where $k \in \{1, \ldots, K\}$. Thus, the transition matrix $\mathbf{P}$ becomes a $(K+1) \times (K+1)$ matrix as

$$\mathbf{P} = \begin{pmatrix} 1-p & p & & \mathbf{0} \\ \vdots & & \ddots & \\ 1-p & 0 & & p \\ \mathbf{0} & & & 1 \end{pmatrix}. \tag{2.18}$$

At the nth step, the transition matrix is $\mathbf{P}^n$ and the entry $p_{ij}^{(n)}$ of the matrix $\mathbf{P}^n$ is the probability of being in the state s_j from the state s_i.

For simplicity, suppose that election timeout counter has a fixed value K in the following analysis. It is straightforward to extend the analytical results to the case in which election timeout is a random value, i.e. $r > 1$.

2.3.4 Network Split Probability

Consider a network with one leader and $N-1$ followers. Note that when more than half of N nodes become candidates, the leader will not be qualified and thus the network will split. For network split probability, we have the following proposition.

Proposition 2.1 For a network with N nodes, the transition matrix is given in Eq. (2.18). Then, the probability of network split before the nth step is given by

$$p_n = 1 - \sum_{m=0}^{\lfloor \frac{N}{2} \rfloor} \binom{N-1}{m} \left(p_{1(K+1)}^{(n)} \right)^m \left(1 - p_{1(K+1)}^{(n)} \right)^{N-1-m}, \tag{2.19}$$

where $p_{1(K+1)}^{(n)}$ is the $(1, K+1)$th entry of the matrix $\mathbf{P}^n$.

Proof: The entry $p_{i(K+1)}^{(n)}$ of the matrix $\mathbf{P}^n$ is the probability of being absorbed before the nth step, when the chain is started from state s_i. Thus, $p_{1(K+1)}^{(n)}$ is the probability of the follower starts from the initial state and transits to candidate state before the nth step.

Denote Y_n being the number of nodes that transit to candidate state before the nth step given that all nodes start from the initial state. Thus, $P\{Y_n = m\}$ is the probability that m nodes become candidates before the nth step. Suppose that all nodes are independent. Therefore, Y_n is a binomial distribution random variable with the form:

$$P\{Y_n = m\} = \binom{N-1}{m} \left(p_{1(K+1)}^{(n)} \right)^m \left(1 - p_{1(K+1)}^{(n)} \right)^{N-1-m}. \tag{2.20}$$

Therefore, $P\{Y_n \geq \lfloor \frac{N}{2} \rfloor + 1\}$ is the probability that more than half of the followers become candidates before the nth step. We have

$$p_n = P\left\{ Y_n \geq \left\lfloor \frac{N}{2} \right\rfloor + 1 \right\} = 1 - \sum_{m=0}^{\lfloor \frac{N}{2} \rfloor} P\{Y_n = m\}. \tag{2.21}$$

Thus, Proposition 2.1 is proved. ∎

Based on Proposition 2.1, we derive the following properties of network split probability.

Property 2.1 The transition probability at the nth step $p_{1(K+1)}^{(n)}$ has the following form:

$$
p_{1(K+1)}^{(n)} = \begin{cases} 0, & \text{if } n < K; \\ p^K, & \text{if } n = K; \\ p_{1(K+1)}^{(n-1)} + \left(1 - p_{1(K+1)}^{(n-K-1)}\right)(1-p)p^K, & \text{if } n > K; \end{cases} \tag{2.22}
$$

Proof: According to the transition matrix $\mathbf{P}$ shown in Eq. (2.18), $p_{1(K+1)}^{(n)}$ has the following forms when $n < K$ and $n = K$.

If $n < K$, then a follower cannot transit to the candidate state. Therefore,

$$
p_{1(K+1)}^{(n)} = 0. \tag{2.23}
$$

If $n = K$, by calculating $\mathbf{P}^n$, we obtain

$$
p_{1(K+1)}^{(n)} = p^K. \tag{2.24}
$$

In the following analysis, we derive the form of $p_{1(K+1)}^{(n)}$ when $n > K$. According to $\mathbf{P}$ shown in Eq. (2.18), we have

$$
p_{1(K+1)}^{(n)} = p_{1(K+1)}^{(n-1)} + p \cdot p_{1K}^{(n-1)}, \tag{2.25}
$$

$$
p_{1i}^{(n)} = p \cdot p_{1(i-1)}^{(n-1)}, \quad \forall i \neq 1 \text{ and } i \neq K+1, \tag{2.26}
$$

and

$$
p_{11}^{(n)} = (1-p) \cdot \sum_{i=1}^{K} p_{1i}^{(n-1)}. \tag{2.27}
$$

From Eq. (2.26), we obtain

$$
p_{1K}^{(n)} = p^{K-1} \cdot p_{11}^{(n-K+1)}. \tag{2.28}
$$

Since the row sums of transition matrix $\mathbf{P}$ are equal to 1, we obtain

$$
\sum_{i=1}^{K} p_{1i}^{(n-1)} = 1 - p_{1(K+1)}^{(n-1)}. \tag{2.29}
$$

By combining Eq. (2.29) with Eq. (2.27), we have

$$
p_{11}^{(n)} = (1-p)\left(1 - p_{1(K+1)}^{(n-1)}\right). \tag{2.30}
$$

By combining Eq. (2.28) and Eq. (2.30) to Eq. (2.25), we obtain

$$
p_{1(K+1)}^{(n)} = p_{1(K+1)}^{(n-1)} + \left(1 - p_{1(K+1)}^{(n-K-1)}\right)(1-p)p^K. \tag{2.31}
$$

Thus, Property 2.1 is proved. ∎

Property 2.2 When the network size $N \to \infty$ and $p_{1(K+1)}^{(n)} \to 0$, the number of nodes transferring to candidate state before the nth step Y_n is approximated by Poisson distribution random variable and $Y_n \sim \mathscr{P}\left((N-1)p_{1(K+1)}^{(n)}\right)$. The probability of network split before the nth step is given by

$$p_n = 1 - \sum_{m=0}^{\lfloor \frac{N}{2} \rfloor} e^{(N-1)p_{1(K+1)}^{(n)}} \frac{\left((N-1)p_{1(K+1)}^{(n)}\right)^m}{m!}. \tag{2.32}$$

2.3.5 Average Number of Replies

Since Y_n is a binomial distribution random variable, given the network size N, the expected value of the number of candidates at the nth step is

$$N_C^{(n)} = (N-1)p_{1(K+1)}^{(n)}. \tag{2.33}$$

The expected value of the number of followers at the nth step is

$$N_f^{(n)} = (N-1)\left(1 - p_{1(K+1)}^{(n)}\right). \tag{2.34}$$

Thus, the average number of replies collected by the leader in the nth step is

$$E(N_{reply}) = N_f^{(n)}. \tag{2.35}$$

2.3.6 Expected Number of Received Heartbeats for a Follower

Proposition 2.2 Given that the follower starts from the initial state, then the expected number of received heartbeats before it transfers to candidate state is n_{11}, where n_{11} is the first entry of matrix $\mathbf{N} = (\mathbf{I} - \mathbf{Q})^{-1}$.

Proof: According to the theorem of absorbing Markov chain in [42], the entry of matrix $\mathbf{N} = (\mathbf{I} - \mathbf{Q})^{-1}$ is the expected number of times the chain is in state s_j, given that it starts in state s_i. The detailed proof is given as follows:

Since

$$\mathbf{P} = \begin{pmatrix} \mathbf{Q} & \mathbf{R} \\ \mathbf{0} & \mathbf{I} \end{pmatrix},$$

we have

$$\mathbf{P}^n = \begin{pmatrix} \mathbf{Q}^n & (\mathbf{Q}^{n-1} + \mathbf{Q}^{n-2} + \cdots + \mathbf{Q} + \mathbf{I})\mathbf{R} \\ \mathbf{0} & \mathbf{I} \end{pmatrix}. \tag{2.36}$$

Note that

$$(\mathbf{I} - \mathbf{Q})(\mathbf{I} + \mathbf{Q} + \mathbf{Q}^2 + \cdots + \mathbf{Q}^{n-1}) = \mathbf{I} - \mathbf{Q}^n. \tag{2.37}$$

In Appendix A.2, we prove that the absolute values of the eigenvalues of $\mathbf{Q}$ are all strictly less than 1. Thus, $\mathbf{I} - \mathbf{Q}$ is invertible. Define $\mathbf{N} = (\mathbf{I} - \mathbf{Q})^{-1}$. Multiplying both sides of Eq. (2.37) by $\mathbf{N}$ gives

$$\mathbf{I} + \mathbf{Q} + \mathbf{Q}^2 + \cdots + \mathbf{Q}^{n-1} = \mathbf{N}(\mathbf{I} - \mathbf{Q}^n).$$

Thus, we have

$$\mathbf{P}^n = \begin{pmatrix} \mathbf{Q}^n & (\mathbf{I} - \mathbf{Q})^{-1}(\mathbf{I} - \mathbf{Q}^n)\mathbf{R} \\ \mathbf{0} & \mathbf{I} \end{pmatrix}. \tag{2.38}$$

In the appendix, we also prove that when n goes to infinity, $\mathbf{Q}^n$ goes to $\mathbf{0}$. Therefore, when n goes to infinity,

$$\mathbf{N} = \mathbf{I} + \mathbf{Q} + \mathbf{Q}^2 + \cdots, \tag{2.39}$$

and thus

$$n_{ij} = q_{ij}^{(0)} + q_{ij}^{(1)} + q_{ij}^{(2)} + \cdots, \tag{2.40}$$

where $q_{ij}^{(k)}$ is defined as the (i,j)th entry of $\mathbf{Q}^k$.

Let $X^{(k)}$ be a random variable, and

$$X^{(k)} = \begin{cases} 1 & \text{if the chain is in state } s_j \text{ after } k \text{ steps;} \\ 0 & \text{otherwise.} \end{cases} \tag{2.41}$$

According to $\mathbf{P}^k$, we have

$$P(X^{(k)} = 1) = p_{ij}^{(k)} = q_{ij}^{(k)}, \quad i,j = 1, \ldots, K, \tag{2.42}$$

and

$$P(X^{(k)} = 0) = 1 - q_{ij}^{(k)}. \tag{2.43}$$

These equations hold when $k = 0$ since $\mathbf{Q}^0 = \mathbf{I}$. Therefore, $E(X^{(k)}) = q_{ij}^{(k)}$.

The expected number of times the chain is in state s_j in the first n steps, given that it starts from state s_i,

$$E(X^{(0)} + X^{(1)} + \cdots + X^{(n)}) = q_{ij}^{(0)} + q_{ij}^{(1)} + \cdots + q_{ij}^{(n)}. \tag{2.44}$$

Letting n goes to infinity, we have

$$E(X^{(0)} + X^{(1)} + \cdots) = q_{ij}^{(0)} + q_{ij}^{(1)} + \cdots. \tag{2.45}$$

By comparing Eq. (2.40) with Eq. (2.45), we obtain that the entry of matrix $\mathbf{N}$ is the expected number of times the chain is in state s_j, given that it starts from state s_i. Denoting s_1 as the initial state, n_{11} is the expected number of received heartbeats before a follower transfers to candidate state when it starts from the initial state. ∎

2.3.7 Time to Transition to Candidate

Proposition 2.3 Suppose that a follower starts from the initial state, the expected time for this follower to transition to candidate state is given as

$$t_c = \sum_{j=1}^{l} n_{1j}, \tag{2.46}$$

where n_{1j} is the $(1,j)$th entry of matrix $\mathbf{N}$.

Proof: According to Proposition 2.2, the entry n_{ij} of $\mathbf{N}$ gives the expected number of times that the follower is in the transient state s_j if it is started from the transient state s_i. Therefore, the sum of the entries in the ith row of $\mathbf{N}$ is the expected times in any of the transient states for a given starting state s_i, i.e. the expected time required before the follower transfers to the candidate state. Denoting s_1 as the initial state, we obtain the proposition. ∎

Property 2.3 The average interval of the received heartbeats for a follower in one term is given by

$$
t_{in} = \frac{\sum_{j=1}^{t} n_{1j}}{n_{11}}.
\tag{2.47}
$$

2.3.8 Time to Elect a New Leader

When the leader crashes or network split happens, the node that times out first will get its own vote plus votes from other available nodes. If the candidate gets votes from a majority of the nodes, it will become a leader. If the first candidate's vote request fails to reach a majority of the nodes before they time out due to packet loss or network delays, split votes will happen. When split votes happen, no candidate wins the election. Then, all nodes have to wait for another period of election timeout before starting a new election.

In Section 9 of [43], the analysis of split vote rate with fixed latency was presented. Unlike Ongaro [43], we investigate the impacts of packet loss on the election performance in this part. Specifically, we try to analyze the upper bound of time that a successful leader election will take.

Denote the time to detect the failed leader and re-elect a new leader as $T_{election}$. We have $T_{election} = E_t \times N_{election}$, where E_t is the timeout value uniformly chosen from the range $[a, b]$ and $N_{election}$ is the number of elections needed to detect the failed leader and re-elect a new leader. The upper bound of the time to detect the failed leader or re-start next election is the maximum election timeout value.

In the normal condition, it takes one round of election to re-elect a new leader. When split votes happen, more elections are needed to elect a new leader. Let s be the number of available nodes. Suppose that the average time between failures of a single node is much larger than election timeout so that s is constant during the period of election. Denote N_{vote} as the number of votes obtained by the first candidate in one round of election. The probability of the first candidate's success in one round of election is $P\left(N_{vote} > \left\lfloor \frac{N}{2} \right\rfloor\right)$, is given by

$$
P\left(N_{vote} > \left\lfloor \frac{N}{2} \right\rfloor\right) = \sum_{k=\left\lfloor \frac{N}{2} \right\rfloor}^{s-1} \binom{s-1}{k} (1-p)^{2k}(1-(1-p)^2)^{s-1-k}.
\tag{2.48}
$$

It is highly unlikely that the other nodes that time out later win the election because there are fewer available votes left. Suppose that the node that times out later are unable to collect majority votes. Thus, $P\left(N_{vote} > \left\lfloor \frac{N}{2} \right\rfloor\right)$ is approximately equal to the probability of success in one round of election. It is should be pointed out that the $P\left(N_{vote} > \left\lfloor \frac{N}{2} \right\rfloor\right)$ is smaller than the probability of success in one round of election. Therefore, the expected value of $N_{election}$

is give as

$$E(N_{election}) < \sum_{n=1}^{\infty} n \left(1 - P\left(N_{vote} > \left\lfloor \frac{N}{2} \right\rfloor\right)\right)^{n-1} P\left(N_{vote} > \left\lfloor \frac{N}{2} \right\rfloor\right) = \frac{1}{P(N_{vote} > \left\lfloor \frac{N}{2} \right\rfloor)}.$$

(2.49)

Therefore, we have

$$E(T_{election}) < \frac{E_{t_{max}}}{P\left(N_{vote} > \left\lfloor \frac{N}{2} \right\rfloor\right)},$$

(2.50)

where $E_{t_{max}}$ is the maximum timeout value.

2.3.9 Simulation Results

In this section, we first validate the efficiency of the analysis model, and then, we investigate the impacts of the parameters (such as packet loss rate, election timeout period, and network size) on availability and network split probability in more details.

To validate the model, we have compared its results with those obtained with the discrete event simulator according to [13]. In simulations, each message was assigned a latency chosen randomly from the uniform range of [0.5, 10]ms, and the interval between two heartbeats $\tau = 50$ ms. Furthermore, CPU time should be short relative to network latency, and the speed of writing to disk does not play a significant role anyhow. To derive the statistical results, we run simulated program 10 000 times and then record the network split time of each run. Each run follows the following steps.

Initialization: initialize simulation time $t = 0$; initialize the list of the followers; and initialize the number of candidates to be zero.

While (the number of candidates is less than half of nodes), then do the following:

Step 1: For each follower i, determine whether the follower receives a heartbeat according to the given packet loss rate. If the follower receives a heartbeat, then reset the election time counter. If the follower fails to receive a heartbeat from the current leader, then its election time counter is decreased by 1. If the election time counter reaches zero, then denote follower i as a candidate and delete follower i from the list of followers.

Step 2: Count up the number of candidates. If the number of candidates is larger than half of nodes, then record time t as network split time. Otherwise, update $t = t + \tau$, go back to step 1.

Figures 2.20 and 2.21 plot the cumulative distribution function (CDF) of the number of heartbeats for a network to split. Each CDF summarizes 10 000 simulated trials. The analytical results are calculated based on Eq. (2.19). Given the value of election timeout K, network size N, and packet loss rate p, Figures 2.20 and 2.21 show that analytical results match well with the simulation results. Therefore, one can detect when the network is abnormal by comparing with the reference value given by the analytical model.

From Figures 2.20 to 2.21, we observe that (i) network split probability highly depends on the packet loss rate p and the election timeout value K. As one expected, as p decreases

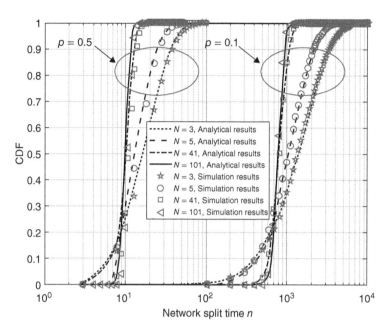

Figure 2.20 CDF of network split time, $K = 3$.

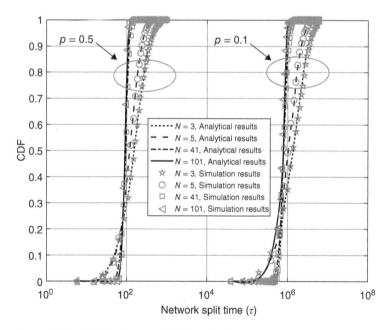

Figure 2.21 CDF of network split time, $K = 6$.

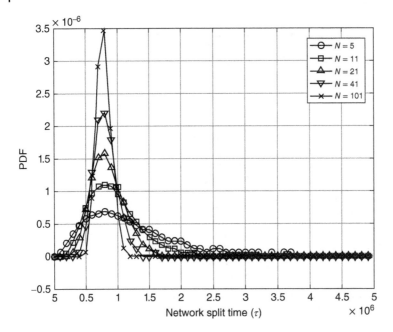

Figure 2.22 PDF of network split time $K = 6$, $p = 0.1$.

or K increases, for the same size of network, the probability of network split before the nth step p_n decreases. (ii) As the network size increases, the CDF curves become steeper. Given p and K, when n is small, p_n decreases with when N increases. This can be observed from p_n in Eq. (2.19). To further understand the second point, we show the PDF of network split time. Figure 2.22 shows the PDF of network split time for different network sizes. We observe that large network has smaller split probability than the one in small network at the beginning of running time. When running time increases over a certain point, the network split probability increases with the size of network. Because the follower's probability of transition to candidate is small at the beginning of running time, the probability that more than half of all nodes become candidates is lower for larger networks. When the follower's probability of transition to candidate gets greater with the running time, the network split probability increases with the network size.

The analytical model given in Section 2.3 is convenient to determine the probability of network split time. Based on the proposed model, we exploit the impacts of the parameters (such as packet loss rate, election timeout period, and network size) on availability and network split performance in more details. Figures 2.23 and 2.24 describe the expectation and variance of network split time for different network sizes, respectively. The results show that the variance of network split time for large network is smaller than that for small network. However, the expectation of network split time for larger network is very close to that for small network. Therefore, a larger network has better stability in terms of network split time.

Figures 2.25 and 2.26 show the expectation and variance of network split time in different packet loss rates, respectively. As Figure 2.25 shows, given the packet loss rate $p = 0.1$ and

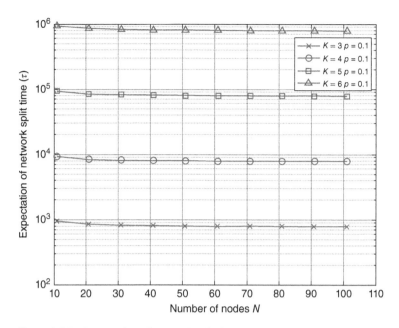

Figure 2.23 Expectation of network split time given different network sizes.

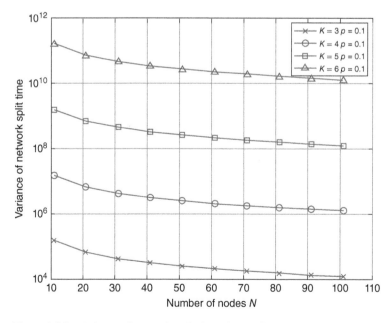

Figure 2.24 Variance of network split time given different network sizes.

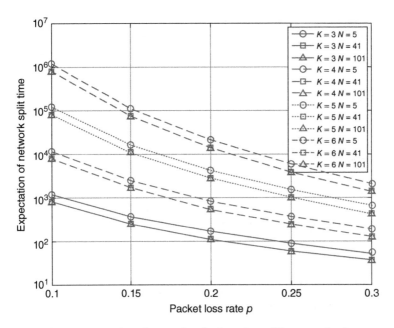

Figure 2.25 Expectation of network split time given different packet loss rates.

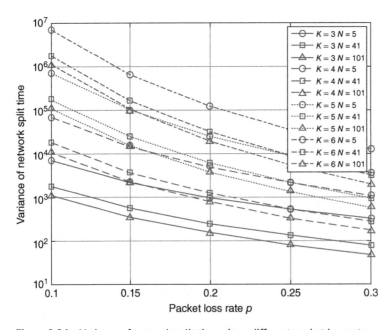

Figure 2.26 Variance of network split time given different packet loss rates.

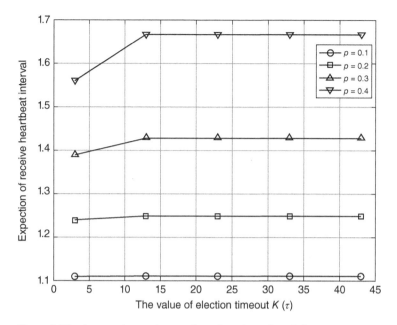

Figure 2.27 Average interval to receive a heartbeat for a follower.

$N = 5$, the expectation of split time is about 1000 and 10 000 when $K = 3$ and $K = 4$, respectively. This result means that the network's stable time is prolonged 10 times by adding one more heartbeat. Given the packet loss rate $p = 0.3$ and $N = 5$, the expectation of split time is about 50 and 110 when $K = 3$ and $K = 4$, respectively. The expectation of network split time is prolonged 1 times by adding one more heartbeat. Increasing election timeout is helpful to lower the network split probability caused by packet loss, especially under smaller packet loss rate.

Figure 2.27 plots the average interval of receiving one heartbeat for one follower per term. The results show that packet loss rate has significant impact on receiver heartbeat interval. On the other hand, election timeout period has insignificant impact on receiver heartbeat interval. Especially, the interval of receiving one heartbeat increases with election timeout period but goes to a constant.

However, prolonging the election timeouts would result in longer time to detect a leader failure and re-elect the new leader in failure cases rises. Consider that there are $N - 1$ nodes available (as the current leader is failed). Figures 2.28 and 2.29 show the probability of the number of elections given different packet loss rates when network size $N = 3$ and $N = 101$, respectively.

The results show that 80% elections complete in the first round of election given $p = 0.1$ and $N = 3$, and nearly, 100% elections complete in the first round of election given $p = 0.1$ and $N = 101$. Given $p = 0.5$, 25% elections complete in the first round of election when $N = 3$ and nearly 0% elections complete in the first round of election when $N = 101$.

Figure 2.30 plots the expectation of the number of elections in different packet loss rates. As Figure 2.30 shows, when $p > 0.3$, the expected value of $N_{election}$ of larger networks increases more rapidly than that of smaller networks. The results mean that (i) when p is

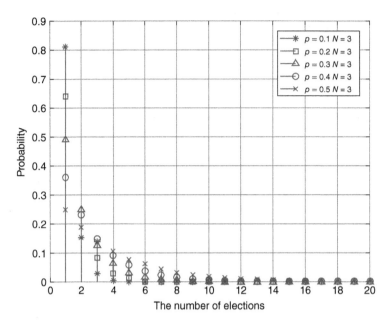

Figure 2.28 Probability of the number of elections, $N = 3$.

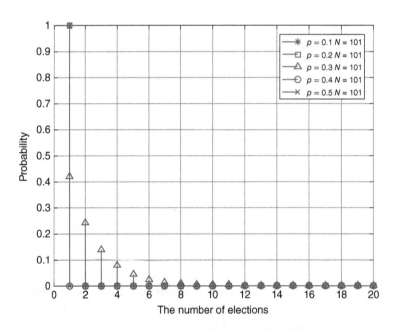

Figure 2.29 Probability of the number of elections, $N = 101$.

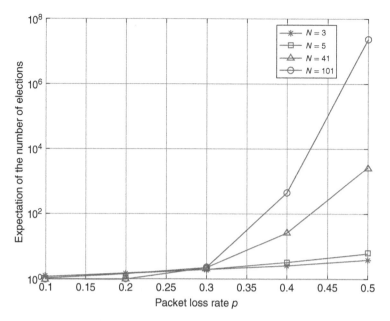

Figure 2.30 Expectation of the number of elections.

small (i.e. $p < 0.2$), the election performance of larger networks does not so highly depends on the packet loss rate p as that of smaller networks. The average $N_{election}$ of larger networks is smaller than that of smaller networks. (ii) As p increases, the election performance is degraded tremendously, especially in the lager network. The availability of larger networks is much poorer than that of smaller network under larger packet loss rate.

2.3.10 Discussion

2.3.10.1 Extended Model

This paper tries to model and analyze the network split process, where one follower node transfers to a candidate that is modeled as an absorbing Markov chain and the packet loss probability p for all the nodes is a constant value, for the sake of simplicity. From a realistic point of view, this paper can be extended in the following ways.

In a network, the packet loss probability of each node may be different, and some nodes may crash. For example, a leader might be more prone to failures or packet loss because of an enhanced workload, congested uplinks, etc. To address those issues, we can extend the proposed model to support biased per-node packet loss probability. For follower s, define the packet loss probability of p2p link between leader and follower s as p_s. Thus, the only non-null one-step transition probabilities of Markov chain are

$$
\begin{cases}
P\{i, k_i - 1 \mid i, k_i\} = p_s & (2.51) \\[2mm]
P\{i, K_i \mid j, k_j\} = (1 - p_s)/r & (2.52) \\[2mm]
P\{i, 0 \mid i, 0\} = 1, & (2.53)
\end{cases}
$$

where $i, j = 1, 2, \ldots, r$ and $k_i \in \{1, \ldots, K_i\}$. Eq. (2.51) relies on the fact that follower s fails to receive a heartbeat from the current leader and its election time counter is decreased by 1. Equation (2.52) shows the fact that follower s receives a heartbeat and reset the election time counter. Equation (2.53) shows that once the election time counter reaches zero, follower s transitions to the candidate state.

For node crashing, take the leader failure as an example. Let X be a random variable, where

$$X = \begin{cases} 1 & \text{if the leader fails;} \\ 0 & \text{otherwise.} \end{cases} \tag{2.54}$$

Then, the packet loss probability of the p2p link between the leader and follower s can be modified as

$$p_s = \begin{cases} p_{s_0} & X = 0; \\ 1 & X = 1. \end{cases} \tag{2.55}$$

where p_{s_0} is the packet loss probability of the p2p link between the leader and the follower s in the normal condition. When the current leader crashes, p_s is equal to 1. Thus, the follower transits to the candidate state within K steps according to the extended model in Eqs. (2.51)–(2.53), where K is the maximum number of heartbeats for an election counter to timeout. The detailed analysis of network split time would be more complex based on the extended model. We would like to work on it in our future work.

2.3.10.2 System Availability and Consensus Efficiency

This paper focuses on the network splitting time, which is only one specific metric of the blockchain network. Several general metrics, such as the system availability and the consensus efficiency, are more interesting.

The system availability can be given as follows:

$$\eta = \frac{T_{normal}}{T_{normal} + T_{election}}. \tag{2.56}$$

where T_{normal} is the period of normal operation (i.e. more than half of nodes are under the control of the current leader) and $T_{election}$ is the period including detection of a failed leader and re-election of a new leader.

To evaluate the system availability, we have to analyze T_{normal} and $T_{election}$. Because the failure of the current leader and communication interruption caused by packet loss are the main reasons of unavailability of the network, T_{normal} not only depends on the time between failures of a single node but also the stable time that the network experiences before network split caused by packet loss. Typical time between failures of a single node is several months or more [13], and the average stable time is about 16 hours ($p = 0.1$, election timeout is 300 ms, the interval between heartbeats is 50 ms). Suppose there is no failure occurs, T_{normal} can be derived by the network split probability that is presented in Section 3.3. However, the analysis of T_{normal} would be more complexly when considering the failure of leader.

It should be pointed out that the speed of the transaction confirmation varies with the number of available nodes[6] within a period of normal operation, i.e. when the number of

---

6 Here, the available node means the node that is under the control of current leader.

available nodes is large, the leader is likely to receive feedbacks from a majority of followers more quickly.

Hence, the consensus efficiency, which can be defined as the average speed of transaction confirmation, is a more important metric of the blockchain network. Based on the proposed model, we can obtain the average number of replies collected by the leader in the given time (see Section 3.6), which is helpful to analyze the speed of transaction confirmation as the confirmation on transactions requires the agreements from a majority of nodes in Raft consensus algorithm.

We would like to work on the analysis about the system availability and the consensus efficiency in our future work.

2.4 Conclusion

Consensus is the basis for blockchains. The PoW consensus is very important for public blockchains because of its security and decentralization abilities. In the first part of this chapter, we employed RL algorithms to solve the mining MDP problem of the PoW-based public blockchains. We showed that, without knowing parameters about the blockchain network model, our RL mining can achieve the mining reward of the optimal policy that requires knowledge of the parameters. Therefore, in a dynamic environment in which the parameter values can change over time, RL mining can be more robust. Raft is a well-adopted consensus algorithm for private blockchains. In the second part of this chapter, an analytical model for Raft consensus algorithm is proposed. The analytical model given in this paper is very convenient to obtain the network split probability and several performance metrics, which provides guidance on how to determinate the parameters. Furthermore, the analytical model is able to monitor the condition of network and detect the abnormal condition by providing reference value of network performance. The simulation results match the analytical results well. Using the proposed model, we have shown the parameters' (such as packet loss rate, election timeout, and network size) impacts on availability. Increasing election timeout is helpful to lower the network split probability caused by packet loss. However, the availability of larger networks is much poorer than that of smaller network under larger packet loss rate as it takes much longer time to elect a leader in larger network. We also observe that larger network has smaller split probability than the one in small network at the beginning of running time and more focused splitting time.

Appendix A.2

(a) The absolute values of eigenvalue of $\mathbf{Q}$ are all strictly less than 1.

Based on Gershgorin circle theorem [10], for $K \times K$ matrix $\mathbf{Q}$, all eigenvalue satisfy

$$|a_k - q_{ii}| \leq \sum_{j \neq i} |q_{ij}|, \quad i, k = 1, 2, \dots, K, \tag{A.2.1}$$

where q_{ij} is the (i, j)th element of $\mathbf{Q}$.

Due to $\sum_{j=1}^{K} q_{ij} \leq 1$ and $q_{ij} \geq 0$,

$$-\sum_{j \neq i} q_{ij} \leq a_k - q_{ii} \leq \sum_{j \neq i} q_{ij}$$

$$-\sum_{j=i}^{K} q_{ij} + 2q_{ii} \leq a_k \leq \sum_{j=i}^{K} q_{ij}, \quad \forall i, k = 1, 2, \ldots, K. \tag{A.2.2}$$

Since **R** in Eq. (2.22) is not zero matrix, there exists one row of **Q** such that $\sum_{j=1}^{K} q_{ij} < 1$.

Thus,

$$-1 < a_k < 1, \quad k = 1, 2, \ldots, K. \tag{A.2.3}$$

That means $| a_k | < 1$.

(b) Proof of $\lim_{n \to \infty} \mathbf{Q}^n = \mathbf{0}$

Note that $\mathbf{Q} = \mathbf{U \Lambda U}^{\mathrm{H}}$, where $\mathbf{U}^{\mathrm{H}} \mathbf{U} = \mathbf{I}$ and

$$\Lambda = \begin{bmatrix} a_1 & & \\ & \ddots & \\ & & a_t \end{bmatrix}. \tag{A.2.4}$$

Since $|a_i| < 1$, we have $\lim_{n \to \infty} \Lambda^n = \mathbf{0}$.

Furthermore, since $\mathbf{Q}^n = \mathbf{U \Lambda}^n \mathbf{U}^{\mathrm{H}}$, we have $\lim_{n \to \infty} \mathbf{Q}^n = \mathbf{0}$.

References

1 S. Nakamoto. Bitcoin: a peer-to-peer electronic cash system, 2008. http://bitcoin.org.

2 Y. Yuan and F. Wang. Blockchain and cryptocurrencies: model, techniques, and applications. *IEEE Transactions on Systems, Man, and Cybernetics: Systems*, 48 (9): 1421–1428, 2018.

3 G. W. Peters, E. Panayi, and A. Chapelle. Trends in crypto-currencies and blockchain technologies: a monetary theory and regulation perspective, 2015. http://dx.doi.org/10.2139/ssrn.2646618.

4 A. Kosba, A. Miller, E. Shi, Z. Wen, and C. Papamanthou. Hawk: The blockchain model of cryptography and privacy-preserving smart contracts. In *Proceedings of IEEE Symposium on Security and Privacy (SP)*, IEEE, 2016, pages 839–858.

5 B. W. Akins, J. L. Chapman, and J. M. Gordon. A whole new world: income tax considerations of the bitcoin economy, 2013. https://ssrn.com/abstract=2394738.

6 Y. Zhang and J. Wen. An IoT electric business model based on the protocol of bitcoin. Paris, France, IEEE, 2015, pages 184–191.

7 M. Sharples and J. Domingue. *The Blockchain and Kudos: A Distributed System for Educational Record, Reputation and Reward*. Lyon, France, Springer Verlag, 2015, pages 490–496.

8 S. King and S. Nadal. Ppcoin: peer-to-peer crypto-currency with proof-of-stake, 2012. https://peercoin.net/assets/paper/peercoin-paper.pdf.

9 Nxtwiki. Whitepaper:Nxt, 2015. http://wiki.nxtcrypto.org/wiki/Whitepaper:Nxt.

10 F. Sonner. Gershgorin circle theorem. https://en.wikipedia.org/wiki/Gershgorin-circle-theorem.

11 M. Castro and B. Liskov. *Practical Byzantine Fault Tolerance*. USENIX Association, 1999, pages 173–186.

12 L. Lamport. The part-time parliament. *ACM Transactions on Computer Systems*, 16 (2): 133–169, 1998.

13 D. Ongaro and J. Ousterhout. *In Search of An Understandable Consensus Algorithm*. Philadelphia, PA, USENIX Association, 2019, pages 305–319.

14 Hyperledger Project. An Introduction to Hyperledger, 2015. https://www.hyperledger.org/.

15 Ethereum. Consortium chain development. https://github.com/ethereum/wiki/wiki/Consortium-Chain-Development.

16 M. Du, X. Ma, Z. Zhang, X. Wang, and Q. Chen. A review on consensus algorithm of blockchain. In *2017 IEEE International Conference on Systems, Man, and Cybernetics (SMC)*, 2017, pages 2567–2572.

17 The Zilliqa Team. The zilliqa technical whitepaper, 2017. https://docs.zilliqa.com/whitepaper.pdf.

18 C. Dwork and M. Naor. Pricing via processing or combatting junk mail. In *Annual International Cryptology Conference*. Springer, 1992, pages 139–147.

19 J. Garay, A. Kiayias, and N. Leonardos. The bitcoin backbone protocol: analysis and applications. In *Annual International Conference on the Theory and Applications of Cryptographic Techniques*. Springer, 2015, pages 281–310.

20 R. Pass, L. Seeman, and A. Shelat. Analysis of the blockchain protocol in asynchronous networks. In *Annual International Conference on the Theory and Applications of Cryptographic Techniques*. Springer, 2017, pages 643–673.

21 C. Lee. Litecoin-open source P2P digital currency, 2011. https://litecoin.com/en/.

22 V. Buterin. Ethereum: a next-generation smart contract and decentralized application platform, 2014. https://github.com/ethereum/wiki/wiki/White-Paper.

23 F. Tschorsch and B. Scheuermann. Bitcoin and beyond: a technical survey on decentralized digital currencies. *IEEE Communication Surveys and Tutorials*, 18 (3): 2084–2123, 2016.

24 I. Eyal and E. G. Sirer. Majority is not enough: bitcoin mining is vulnerable. *Communications of the ACM*, 61 (7): 95–102, 2018.

25 D. Kraft. Difficulty control for blockchain-based consensus systems. *Peer-to-Peer Networking and Applications*, 9 (2): 397–413, 2016.

26 K. Nayak, S. Kumar, A. Miller, and E. Shi. Stubborn mining: generalizing selfish mining and combining with an eclipse attack. In *2016 IEEE European Symposium on Security and Privacy (EuroS&P)*, IEEE, 2016, pages 305–320.

27 A. Sapirshtein, Y. Sompolinsky, and A. Zohar. Optimal selfish mining strategies in bitcoin. In *International Conference on Financial Cryptography and Data Security*. Springer, 2016, pages 515–532.

28 R. S. Sutton and A. G. Barto. *Reinforcement learning: An introduction*. MIT press, 2018.

29 L. P. Kaelbling, M. L. Littman, and A. W. Moore. Reinforcement learning: a survey. *Journal of Artificial Intelligence Research*, 4: 237–285, 1996.

30 C. J. Watkins and P. Dayan. Q-learning. *Machine Learning*, 8 (3–4): 279–292, 1992.

31 V. Mnih, K. Kavukcuoglu, D. Silver, A. A. Rusu, J. Veness, M. G. Bellemare, A. Graves, M. Riedmiller, A. K. Fidjeland, G. Ostrovski, et al. Human-level control through deep reinforcement learning. *Nature*, 518 (7540): 529, 2015.

32 D. Silver, J. Schrittwieser, K. Simonyan, I. Antonoglou, A. Huang, A. Guez, T. Hubert, L. Baker, M. Lai, A. Bolton, et al. Mastering the game of go without human knowledge. *Nature*, 550 (7676): 354, 2017.

33 J. Schulman, S. Levine, P. Abbeel, M. Jordan, and P. Moritz, Trust region policy optimization. In *International Conference on Machine Learning*, 2015, pages 1889–1897.

34 I. Moraru, D. G. Andersen, and M. Kaminsky. There is more consensus in egalitarian parliaments. In *Proceedings of the 24th ACM Symposium on Operating Systems Principles*, Ser. SOSP '13, New York, NY, USA: Association for Computing Machinery, 2013, pages 358–372. https://doi.org/10.1145/2517349.2517350.

35 R. C. Merkle. A digital signature based on a conventional encryption function. In *Conference on the Theory and Application of Cryptographic Techniques*. Springer, 1987, pages 369–378.

36 W. Wang, D. T. Hoang, P. Hu, Z. Xiong, D. Niyato, P. Wang, Y. Wen, and D. I. Kim. A survey on consensus mechanisms and mining strategy management in blockchain networks. *IEEE Access*, 7: 22 328–22 370, 2019.

37 M. Wang, M. Duan, and J. Zhu. Research on the security criteria of hash functions in the blockchain. In *Proceedings of the 2nd ACM Workshop on Blockchains, Cryptocurrencies, and Contracts*, ACM, 2018, pp. 47–55.

38 V. Bagaria, S. Kannan, D. Tse, G. Fanti, and P. Viswanath. Deconstructing the blockchain to approach physical limits. *arXiv preprint arXiv:1810.08092*, 2018.

39 D. Puthal, N. Malik, S. P. Mohanty, E. Kougianos, and G. Das. Everything you wanted to know about the blockchain: its promise, components, processes, and problems. *IEEE Consumer Electronics Magazine*, 7 (4): 6–14, 2018.

40 I. Chadès, G. Chapron, M.-J. Cros, F. Garcia, and R. Sabbadin. MDPtoolbox: a multi-platform toolbox to solve stochastic dynamic programming problems. *Ecography*, 37 (9): 916–920, 2014.

41 H. Howard, M. Schwarzkopf, A. Madhavapeddy, and J. Crowcroft. Raft refloated: do we have consensus? *SIGOPS Operating Systems Review*, 49 (1): 12–21, Jan. 2015. https://doi.org/10.1145/2723872.2723876.

42 Markov Chains. http://www.academia.edu/7549558/Chapter-11-Markov-Chains.

43 D. Ongaro. *Consensus: bridging theory and practice*. PhD Thesis, Stanford University, 2014.

3

A Low Communication Complexity Double-layer PBFT Consensus

Chenglin Feng[1], Wenyu Li[1], Bowen Yang[2], Yao Sun[2], and Lei Zhang[2]

[1]*College of Science and Engineering, University of Glasgow, Glasgow, G12 8QQ, UK*
[2]*James Watt School of Engineering, College of Science and Engineering, University of Glasgow, Glasgow, G12 8QQ, UK*

3.1 Introduction

The blockchain technology has shown its great potential in various sectors [1]. It is widely agreed that this technology will play an important role in financial services, energy trading, supply chain management, Internet of Things (IoT) [2–4], as well as privacy protection in various distributed networks [5, 6]. The consensus mechanism guarantees the orderness, integrity, and consistence of the blocks, leading to a strong relation to the performance of blockchain systems. Therefore, a number of algorithms have been investigated to cater performance demands for different blockchain applications in terms of transaction throughput, latency, node scalability, security level, etc.

Depending on application scenarios and performance requirements, different consensus algorithms can be considered in a blockchain system. Specifically, in the case of a permissionless public blockchain, nodes are allowed to join or leave the network without permission and authentication; therefore, proof-based algorithms such as proof-of-work (PoW) [7], proof-of-stake (PoS) [8], and their variants are commonly used (PoW in Bitcoin, for example). Proof-based algorithms are designed with excellent node scalability performance through node competition, which is essential to deal with the double-spending problem. However, they could be very resource demanding [9]. For instance, recently published estimates of Bitcoin's electricity consumption are wide-ranging, on the order of 20–80 TWh annually, or about $0.1 - -0.3\%$ of the global electricity consumption [10].

Additionally, these consensus mechanisms have other limitations such as long transaction confirmation latency and low throughput. For instance, the transaction per second (TPS) is generally limited to seven in Bitcoin and about 15 in Ethereum, while the transaction confirmation delay is typically as considerable as 10 minutes in Bitcoin and 15 seconds in Ethereum [3]. It is worth pointing out that the computational requirement of proof-based consensus varies from one to another. Although proof-based consensus has been mostly applied to the applications of public blockchain, it has a limited generic distributed system and blockchain coverage, as it is still incrementally resource demanding. Therefore, the search of voting-based consensus for blockchain and new generation distributed system is imminent.

Wireless Blockchain: Principles, Technologies and Applications, First Edition.
Edited by Bin Cao, Lei Zhang, Mugen Peng and Muhammad Ali Imran.
© 2022 John Wiley & Sons Ltd. Published 2022 by John Wiley & Sons Ltd.

Unlike the public blockchains, the private and consortium blockchains prefer to adopt lighter consensus protocols such as PBFT [11], Paxos [12], and Raft [13, 14] to reduce the amount of computational power and improve the transaction throughput [15]. Those consensus protocols have been widely used in general distributed systems for data synchronization. Meanwhile, their properties are also critically important to the application scenarios of the blockchain-enabled IoT ecosystem, which is typically composed of low cost and low power devices.

3.1.1 PBFT Applied to Blockchain

To protect distributed systems from malicious users, Practical Byzantine Fault Tolerance (PBFT) was proposed in [11] as an improved and practical protocol based on original Byzantine Fault Tolerance (BFT) [14, 16]. Compared to the proof-based consensus, such as PoW where the security threshold is 51%, i.e. absolute secure transaction can be achieved if the malicious user(s) occupies no more than half of the overall resource, PBFT requires the number of malicious users under 33% of total participants to ensure the system immune from the malicious attacks [11].

Benefit from the lower complexity and low energy consumption, which is particularly important for wireless IoT applications, PBFT is favored for private and consortium chains [17]. A promising advancement of PBFT can be found in Hyperledger development [18], part of Hyperledger business blockchain frameworks, which has been adopted by tech giants such as IBM or Wall Street Fintech, such as J. P. Morgan [19].

3.1.2 From CFT to BFT

The PBFT is a well-tested algorithm, and variants are developed on top of it. Here, we dive deeply in history and try to draw a clear path of its evolution. Starting from considering the type of failure, fail-stop failure and Byzantine failure are almost two extremes of fault tolerance. For a server that is fail-stop, it works normally until crush. When the server is not crushed yet, it may not be able to response or deliver messages within the designated time window. The tolerance of this kind of fault is called Crush Fault Tolerance (CFT). For a server of byzantine failure, it exhibits arbitrary behaviors including forging original messages and telling "lies" to other servers. The tolerance of this kind of fault is named as BFT. Clearly, CFT is relatively less complex compared with BFT. It is easy to detect a crushed server in synchronize networks, but CFT could be tricky in asynchronies scenarios.

3.1.2.1 State Machine Replication

The whole story began at around 1975 when people required improvement of accessing a database both of reliability and efficiency. Traditionally, data is kept and manipulated in a centralized way, where a host is responsible for storing all data via communicating directly with the clients. However, the server running on that host may be unavailable for the client once there is a failure on it. For instance, the transactions and deposit enquiry server will not be available if a bank's system runs on this architecture and has crashed. More seriously, the data could be loosed forever by malicious attack. One intuitive solution is to replicate servers for fault tolerance (state machine replication or SMR). The replication was proposed by Johnson and Thomas [20] to tackle problems mentioned above. In more detail, we backup data on independent servers also known as replicas (or nodes, backups). The client invokes a request to one of the backups and receives an acknowledgement after the execution is synchronized across the network. When connection delay occurred, one replica may receive duplicated messages since retry. Thus, the timestamp is appended to each message. It is also used to determine whether a message is up to date.

In recent years, with the growing interest in blockchain, many new protocols based on SMR are proposed. In *The Next 700 BFT Protocols*, the authors present a method to simplify the designing of new protocols by introducing Abortable Byzantine faulT toleRant stAte maChine replicaTion (Abstract) as a new way to illustrate BFT [21]. Results show that the proposed protocol using Abstract provides a better performance in terms of both latency and throughput.

3.1.2.2 Primary Copy

The earliest works on replication were based on relatively ideal assumptions, i.e. requests will be eventually delivered to other nodes despite delay; replicas always operate normally and response upon a new request. Soon, this discussion was extended to find a protocol that is resilient to crash on replicas as well as recovery after a crash. Around 1976, the primary copy was introduced by Alsberg and Day [22]. In this technique, one node is designated as the primary to coordinate the protocol. Each consensus reaching process is initiated by the primary and relayed node by node through request and acknowledgement. As some operations on distributed system rely on the results of the previous one, updates across the network must be totally ordered (serial consistency). For that reason, the sequence number is introduced to indicate the order of operations. Besides, the crashed (fail-stop) replica is detected by time-out and followed by reconstruction of the network.

3.1.2.3 Quorum Voting

Except peer-to-peer communication pattern to synchronize data across the network, voting was also proposed in 1976 by Thomas [23], where a replica decides whether to accept a modification of local data by collecting votes from the rest nodes of the network. Later, Gifford in [24] presented the numerical analysis on the number of votes that is required to reach consensus (quorum size) and how it affects the performance. To guarantee that the replica is aware of the previous modification of data upon executing a new request, the voting quorum of two consecutive operations must have an intersection. Assuming that a network consists of $2n$ replicas and the quorum size of the voting is $n + 1$, for two consecutive operations, one replica must be required to participate in both operations. If votes of later operation are valid, i.e. the results collected from the quorum of later voting are identical with intercepted replica and been accepted. The accepted value is guaranteed to be calculated based on the knowledge of previous operations.

The concept of quorum is also borrowed in Ethereum Casper to provide safety for Ethereum 2.0 [25]. It is an additional mechanism on the top of PoS and serves as a finality gadget. The Casper does not generate the block but determines the sequence of blocks on the chain. The designated replicas (validators in Ethereum) vote for a parent–children relationship for two collections of blocks and form a quorum certificate granted by more than $\frac{2}{3}$ of replicas.

3.1.3 Byzantine Generals Problem

The aforementioned works mainly focus on the fail-stop replicas, which is not always the case in reality. Thus, Pease et al. [26] attempted to address the problem in 1980. They proved the optimal 1/3 threshold inductively, i.e. if faulty replicas may tell "lie"' (send arbitrary messages), non-faulty nodes are still able to synchronize data under the case of no more than one-third faulty nodes. The discussion was based on interactive consistency, in which replicas forward the value stored on others, then pick one version that appears most frequently. It is noticeable that the threshold is given under "replicas are attempting to reach unanimity for preloaded values," which is not identical as nowadays algorithms. It is further discussed and formalized as Byzantine Generals Problem in 1982 [16].

3.1.4 Byzantine Consensus Protocols

3.1.4.1 Two-Phase Commit

Probably, one of the most well-known BFT algorithm is the Practical Byzantine Faulty Tolerance [11], yet its phase-wise protocol ancestor can be traced back to CFT era. The two-phase commit protocol [27] is commonly applied to the distributed system where replicas do not send arbitrary faulty messages. In brief, replicas vote on one operation in the first prepared phase to the primary, and the primary multicasts a commit for a valid voting or send abort. The replicas then undo for abort or execute for commit. Interestingly, this protocol [27] was depicted as coordinating generals on a campaign, which may be the origin of using the term "general" for BFT problems.

3.1.4.2 View Stamp

In 1985, a BFT dedicated two-phase protocol (three phases if including initial) was proposed in [28] with different quorum sizes. Although a fault consensus cannot be reached, the protocol is unable to proceed at the present of a faulty primary. That is also the case for all protocols involving primary copy. The view stamp [29] was later introduced to tackle this challenge. A view stamp is an identifier for the current primary. If the replica notices the stamp is altered (a view change happened), it then correspondingly aborts/proceeds current operation and start the new primary election.

3.1.4.3 PBFT Protocol

The PBFT protocol can be regarded as a well-rounded BFT solution with those techniques mentioned above. Given that there are at most f byzantine nodes out of n replicas and $n > 3f$, this protocol exhibits important properties: consensus reaches within a finite period; no faulty consensus be reached; and consensus reaching continues even when view change happens. As an example, we assume that there is one primary node and three state machine replicas. The consensus is triggered by a client sending a request to the nodes' header (Replica 0). Then, consensus will be operated among the nodes, and if an agreement is reached among the replicas, the new record will be committed to the blockchain. The whole consensus process includes three stages, i.e. *pre-prepare*, *prepare*, and *commit*, as shown in Figure 3.1. On receiving the request from the client, the primary node (i.e. Replica 0) broadcasts a *pre-prepare* message to the other replicas. Because

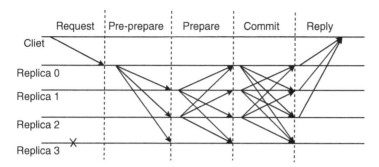

Figure 3.1 Single-layer PBFT consensus processing. Source: Castro et al. [11].

the primary could be byzantine and deliver multiple different requests, replicas then make decisions on what is the correct version of the request by exchanging pre-prepare messages in the prepare phase. Each replica will accept a request if there are no less than $2f + 1$ identical pre-prepare messages received from others. Because at most f messages come from byzantine nodes, this guarantees $f + 1$ messages from non-faulty replicas corresponding to what we have mentioned in Quorum Voting. To keep the execution of consecutive requests in total order (one after another), replicas run another round of messages exchange in *prepare* and *commit* stages. In each stage, a minimum number of consistent messages are required for stepping into the subsequent stage. This additional phase (in comparison with the two-phase commit) fixes the sequence of requests across different views.

It is defined that the consensus is reached when the commit phase is successful among the majority of non-faulty nodes. Specifically, the client must receive at least $f + 1$ replies from the replicas to validate the final consensus with a total number of $3f + 1$ replicas. This ensures that at least one non-faulty replica replies to this operation. In the case of the client fails to collect $f + 1$ replies, the client may resend the request to the primary for retry. Upon receiving the same request again, if the consensus was already reached on the commit phase, replicas just resend the final stage messages. If the consensus has not been reached, the network goes over the protocol again.

Another bottleneck of PBFT is the scalability. As mentioned in Section 3.1, the communication complexity limits the performance of protocols. To solve this problem, the HotStuff leverages threshold signature to reduce communication complexity [30]. To be specific, in each phase, the primary broadcasts messages and each replica responses a valid message with a partial signature. The primary collects the responses with a partial signature more than $\frac{2}{3}$ replicas and combine into a digital signature. This signature is broadcast by replica again, which serves as a quorum certificate. Unlike HotStuff, the multi-layer PBFT proposed improved node scalability by the reorganizing network structure, where the threshold signature can also be applied to further reduce the communication complexity. Moreover, it is noticeable that as the network scales up, the primary in HotStuff has to collect and broadcasts an increasing number of messages and combines more partial signatures. This workload can be barked down by implementing our tree-like structure. Also, the pace of HotStuff is affected by the primary as it waits for the aggregation of partial signature to advance.

3.1.5 Motivations

Although PBFT has shown good performance in terms of latency, resource requirement, and nodes complexity, the node scalability, which is a metric to measure how well the network reflects the capacity of the system to handle the increasing number of nodes, is a bottleneck of PBFT as it requires heavy inter-node communications. Thus, from the communication complexity perspective, the PBFT-based blockchain hardly scales up to 100 nodes [31].

Variant PBFT-based consensuses have been proposed to solve the problem of its poor scalability. For example, PBFT with short-lived signature [32] minimizes the consensus time for signature verifying. Another significant evolution path is sharding, where shards have their own consensus group; hence, the transactions can be confirmed within a short time because of the smaller size network [33, 34]. However, the trade-off between the increased communication costs and lowered security levels should be carefully considered. For instance, every shard keeps its own data, which is not shared with other shards. In the case of users requesting contracts on several shards, inter-node communications go up quickly. Meanwhile, putting segments of data into different repositories without proper redundancy and recoverability is risky. Losing control of any individual shard will interrupt the blockchain completely, either causing untraceable and irreversible records to future records or forking the chain from the breaking point [35].

The communication complexity issues in the PBFT system is due to the exhaustive peer-to-peer communication among the nodes. To reduce the cost of communication, intuitively, one can construct a hierarchical multi-layer PBFT by refraining the communication within their layers or sub-groups. The sub-consensus can be performed per group, and the overall consensus can be defined as exceeds the number of groups achieved the sub-consensus. This multi-layer PBFT system model is initially proposed in [36], providing a brief analysis of communication complexity based on a particular case. However, there are still many challenges to be addressed before this idea can be applied to a real system. First, the complexity analysis derived in [36] cannot be applied to general situations as it is developed under the premise that the number of nodes in each sub-group is equal. To better represent the practical situation, detailed analysis should be provided considering various node allocation scenarios. Also, the security analysis in [36] is derived under the hypothesis that faulty nodes only exist in the bottom layer, which does not apply to real situations where faulty nodes are randomly distributed into all layers. Therefore, new security analysis should also be provided to verify the reliability of a multi-layer PBFT system. Finally, a new complete protocol is urgently required to ensure liveness and safety of the network. Yet in this chapter, we focus on the double-layer scenario.

3.1.6 Chapter Organizations

In this chapter, we propose a double-layer PBFT system, where we assume m replicas in the first layer and n sub-layer replicas in the second layer. It gives a total number of $1 + m + mn$ nodes in the network. This double-layer model can be further extended to a more general X-layer case ($X > 2, X \in \mathbb{I}$), which is able to accommodate at most $\lfloor \frac{n}{3} \rfloor \times \lfloor \frac{m}{2} \rfloor$ faulty nodes to ensure absolute safety of the system under malicious attacks. Meanwhile, only $m^2 + mn^2$

inter-node messages, instead of $O(Z^2)$ messages in a traditional single-layer PBFT network consensus, are required. However, it is inevitable that the scalability is improved at the cost of a longer delay as the consensus-reaching process goes through more than one layer.

The main contributions of this chapter are summarized below:

- A novel double-layer PBFT model is constructed, which is scalable because it reduces the inter-node communications to $C \approx 1.9Z^{\frac{4}{3}}$, compared to the traditional PBFT system of $O(Z^2)$.
- The analytical security performance of the proposed system is derived and analyzed mathematically. It proves that under certain conditions, the maximum number of faulty nodes can increase from $\lfloor \frac{m}{3} \rfloor$ to $\lfloor \frac{n}{3} \rfloor \times \lfloor \frac{m}{2} \rfloor$.
- In addition, a new double-layer PBFT protocol is based on which consensus can be reached among nodes in different layers.

The rest of this chapter is organized as follows. In Section 3.4, the double-layer PBFT model is proposed. Then, the communication complexity is analyzed and compared to the original PBFT. In Section 3.5, the security threshold is derived based on the double-layer system. Section 3.2 proposes the protocols for double-layer PBFT system and Section 3.6 concludes the chapter.

3.2 Double-Layer PBFT-Based Protocol

In this section, we propose a practical protocol dedicated to the double-layer PBFT, where replicas in the first layer are denoted as r_i^{L1} (superscript $L1$ for layer number, subscript i for replica index). r_i^{L1} act as leaders for the corresponding second-layer replicas (r_i^{L2}). One leading r_i^{L1} and its corresponding r_i^{L2}s, all together, form a consensus group, resulting in a tree-like topology structure.

When each group reaches consensus, group members reply to their leader instead of the client. Then, the leader collects replies and sends them to the client on behalf of that consensus group. The client accepts the results only agreed by more than half consensus groups. The protocol overview is illustrated in Figure 3.2. Meanwhile, there is a group configuration GP that describes the allocation group members and their leader. It should be updated when the network structure is changed. In brief, the new protocol inserts successive *pre-prepare*, *prepare*, and *commit* phases before starting the *commit* phase in the upper layer.

3.2.1 Consensus Flow

3.2.1.1 The Client

From the client perspective, the consensus is achieved under primary's proxy. A client c sends a request message $[o, t, c]_{request}$ to primary. This request invokes an operation o with timestamp t. Timestamps are ordered by time, so the stamp of later operation contains higher values. The request is sent to the replica. The identity of the replica is extracted from the view number contained in replies from previous operations. On receiving the request, primary multicasts messages using the protocol stated below.

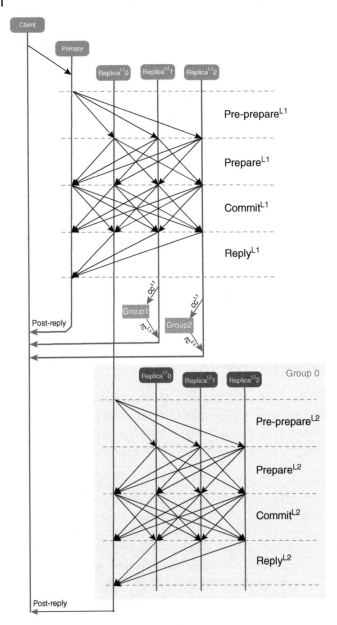

Figure 3.2 Topology of the proposed double-layer PBFT system. (Note that we give double-layer system as an example here, the idea proposed in this chapter can be extended to arbitrary-layer PBFT.)

All group leaders reply results to the client directly. The *poset – reply* has the form $[o, t, c, i, r, rc^{L1}, v, GP]_{poset-reply}$, where v is the current view number, i is the replica number, r is the result of the execution, rc is the reply certificate, and GP describes the replicas allocation.

Assuming that there are m replicas in the first layer and n in the second, and the number of faulty replicas in a consensus group is f^g (to distinguish from f). If leaders have received $f^g + 1$ matching valid replies from the same consensus group, this group is said to have reached consensus. The network reaches consensus when more than half of the groups have the same replies. The client only accepts results replied from group leaders when at least half of them are consistent.

3.2.1.2 First-Layer Protocol

In the first layer, the primary and m replicas form a consensus group. When the primary p receives a request $M = [o, t, c]_{request}$ from the client, it authenticates the request and client's identity. Then, primary assigns a sequence number α to M. After that, the primary steps into *pre-prepare* phase by multicasting $[M, d, \alpha, v]_{pre-prepare^{L1}}$, where d is the digest of M (superscript is to distinguish *pre-prepare*^{L1} in the first layer from *pre-prepare*^{L2} in the second layer). The primary multicasts messages only among r^{L1}. Thus, only r^{L1} reacts on *pre-prepare*^{L1}. The propagation is restricted because the protocol runs layer by layer.

For *pre-prepare*, *prepare*, and *commit* messages, a replica r_i^{L1} accepts the one with the same view v; the authenticity is then verified; α is between watermark h and H. The watermark is introduced to ensure a weak synchronization and defined in Section 3.2.3.

With the conditions above, a replica i in the first layer accepts a $pre - prepare^{L1}$ message from primary only when there is none different request with the same view v and sequence number α is accepted. Then, it multicasts $[d, \alpha, i, v]_{prepare^{L1}}$ messages to all r_i^{L1} in the first layer. It records both $pre - prepare^{L1}$ and $prepare^{L1}$ messages to its log. During the *prepare* phase, each r_i^{L1} replica collects $2f$ messages with matching sequence number α, view v, and request M. With the received $pre - prepare^{L1}$ messages, they form $prepared - certificate^{L1}$, which indicates that a particular r_i^{L1} replica has prepared the request.

For prepared r_i^{L1}, it multicasts $[d, \alpha, i, v]_{commit^{L1}}$ and waits for more than $2f + 1$ matching $commit^{L1}$ messages with the same view, sequence, and digest from different r^{L1} replicas. Received messages form $commit-certificate^{L1}$ (cc^1), and this request is said to be committed on replica r_i^{L1}. Then, the replica pauses the execution and initiates another round of protocol in the second layer as described in Section 3.2.1.3.

The committed replicas send $[o, t, i, r, v]_{reply^{L1}}$ to their group leader, i.e. the primary in the first layer. The primary confirms that this group has reached consensus by checking more than half of group members reply consistent $reply^{L1}$ messages, including itself. The group leader collects $reply^{L1}$ and forms a reply certificate rc^{L1}. After that, this primary replies to the client with $[o, t, c, i, r, rc^{L1}, v, GP]_{poset-reply}$. Notice that it is not necessary for primary to reply to the client on behalf of consensus, but we require primary to do so to keep the algorithm the same on all replicas in case of a massive deployment.

3.2.1.3 Second-Layer Protocol

A committed r_i^{L1} multicasts new *pre-prepare* message to r^{L2} within the same consensus group, where another round of PBFT protocol is implemented. All group members reply to the leader in a similar manner in Section 3.2.1.2 when the request is committed again. For a replica r_p^{L1} which acts as primary, it multicasts a similar $[M, d, \alpha, v, cc^{L1}]_{pre-prepare^{L2}}$ to r_i^{L2} replicas in the same consensus group, where cc^{L1} is the $commit - certificate^{L1}$. The v,

α, and M are inherited from the previous process. A replica r_i^{L2} in the consensus group will accept the request if the condition mentioned in the first-layer protocol is satisfied, in addition to the presence of cc^{L1}.

On receiving valid $pre - prepare^{L2}$ message, the pre-prepared r_i^{L2} multicasts $[d, \alpha, i, v]_{prepare^{L2}}$ messages to all r_i^{L2} in the same consensus group. It adds both $pre - prepare^{L2}$ and $prepare^{L2}$ messages to its log. In the $prepare^{L2}$ phase, each r^{L2} replica collects $2f$ messages with a matching sequence number α, view v, and request M. With received $pre - prepare^{L2}$ message, it forms a quorum-prepared $certificate^{L2}$, which indicates that this r_i^{L2} replica has prepared the request.

Then, the prepared replicas r_i^{L2} and their r_i^{L1} leader multicast $[d, \alpha, i, v]_{commit^{L2}}$ and collect $2f + 1$ matching $commit^{L2}$ messages with the same view, sequence, and digest form different r_i^{L2} replicas. These $commit^{L2}$ form $commit - certificate^{L2}$, and this request is said to be committed. Replica then executes the message that has been committed. After the execution, all group members reply to the result to their group leader, and the leader replies to the client in a similar manner in Section 3.2.1.2.

3.2.2 Faulty Primary Elimination

3.2.2.1 Faulty Primary Detection

The most commonly used condition for initiating a view change is by detecting whether the primary is responding, i.e. the replicas keep a timer which will be reset each time a new request is received. However, a faulty primary that assigns different *pre-prepare* to different replicas cannot trigger time-out. Thus, we present a possible mechanism without a timer to detect faulty primary nodes that multicast random messages during *prepare* phase. Because one replica may skip several operations when the connection is lost, we require the detection to be independent without the prerequisite of total order and continuous sequence number α, i.e. replicas accept discrete α as long as it is consistently increasing/decreasing. To facilitate the understanding of the mechanism, we first assume that the primary is not faulty. As there are at most f faulty replicas, the collected $2f + 1$ messages contain at most f faulty messages in the worst situation. Therefore, the rest $f + 1$ messages must be identical as they are obtained from non-faulty replicas. Those faulty messages (or the digest of the messages) could be unmatched with each other. Or the faulty messages are identical, but those faulty messages are different from those matching messages among non-faulty nodes. For simplicity, we discuss that in the latter case, there are two versions of messages in $2f + 1$ collections, which are kept among non-faulty replicas and faulty replicas, respectively.

When primary is non-faulty and each replica collects $2f + 1$ messages, the number of different versions n_v falls in the interval 2 and $f + 1$. That is because, in the worst case mentioned above, each faulty replica multicasts different versions of messages, resulting in f different versions among faulty replicas. Besides, there is an additional version kept by $f + 1$ non-faulty replicas. At the presence of faulty primary that multicasts random messages, the messages in *pre-prepare* received by non-faulty replicas should differ. In other words, n_v exceeds the upper bound $f + 1$, and the primary is detected to be faulty. This condition, along with time-out, triggers the *view-change* phase.

3.2.2.2 View Change

In the conventional PBFT [11], replicas invoke view change in the prepare phase. Changes are made to adapt to our multi-layer model. As shown in Figure 3.3, a replica is the primary of its sub-layer replicas. Thus, in this protocol, replicas in a specific layer detect their faulty primary in the upper layer and invoke cross-layer view change. Each replica, which suspects the primary to be faulty, multicasts a *view-change* messages with the stable checkpoint to new primary (the new primary may be determined by election mechanism based on the current view number). The new primary decides whether to lunch a *new-view*.

Suppose a replica is in layer L as a member of group K, and notice the group leader is the group member of group in layer $L - 1$. If it suspects the group leader is faulty for not responding or delivering *pre-prepare* messages with invalid sequence number, it stops accepting requests and starts view change that moves the view of this group from v into $v + 1$ by multicasting $[v + 1, \alpha, C, P, i]_{view-change}$, where α is the sequence number for the latest checkpoint for this replica and C is the $2f$ matching certificate for this checkpoint. P is the collection of *prepare − certificate* ($2f + 1$ matching prepare requests) for each *pre-prepare* request that is higher than α. i is the identification of the sender.

If the new primary p in new view $v + 1$ has received $2f − 1$ valid *view-change* messages from other group members. It multicasts $[v + 1, GP, \gamma, O]_{new-view}$, where γ is a set of $2f$ matching *viewChange − certificates* and O is the set of *pre-prepare* messages needed to be multicasted. This is because they are failed to reach a consensus in the last view v. The sequence of *pre-prepare* in O is ranging from the latest checkpoint known to the new leader and the latest α in P. GP is the update of itself describing how p allocate its group member in layer $L - 1$ to the rest members in group K. Member replicas accept and execute a new valid view.

However, taking over the leader of group K in layer L implies that this replica becomes the member of a group in layer $L - 1$ (for instance, group J). Thus, the primary p multicasts $[join, v, \gamma, i]_{join}$ to group J after the *view-change*. The v is the view number for group J because it should remain unchanged if there is no view change in J. p extracts view in J from the *commit − certificate* that passed by the original group leader from group K. Matching v

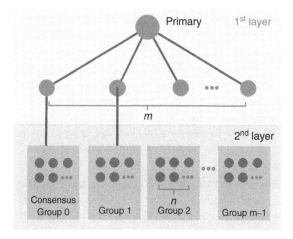

Figure 3.3 Implementation flow chart for double-layer PBFT model.

and γ prove the validity. Then, this replica directly commits the current operation because it must have been committed in layer $L - 1$; otherwise, the consensus protocol will not be executed in layer L. This *join* message informs the member in group J that there is a change of replica and group J updates local GP.

Also, a member in layer L is the leader in layer $L + 1$. To reallocate its former members (since p is no longer their leader), it multicasts $[v + 1, G, \gamma, i, r]_{redistribute}$ to replica r, where G is the new designated group in layer $L - 1$ for replica r. The replica r then redirects itself to group G and update local GP. The replicas governed by group K in layer deeper than $L - 1$ are reaching consensus in seam sequence as they are (indirectly) lead by group K.

3.2.3 Garbage Cleaning

One may notice that the operation recorded on replicas increase its size as the protocol run. To discard unnecessary operations that have already reached a consensus, we implement Castro's [11] garbage collection.

Replicas multicast *checkpoint* messages with the sequence number of latest committed operations. Sequence number with $f + 1$ *checkpoint* (including its own) is seated as low watermark h. To prevent a replica who encounters with the transmission delay from going too far, a logic size L is set such that the replica only executes operation between h and $H = L + h$.

3.3 Communication Reduction

Undeniably, the communications are increased in the proposed protocol (Section 3.2) because more phases are implemented. To reduce the extra complexity and latency, the *redistribute* messages are embedded into GP in *new-view* message. Then, it is extracted by the current members of group K and passes to their former group in layer $L + 1$ by piggybacking in *pre-prepare*. Those replicas in layer $L + 1$ redirect themselves before responding to new *pre-prepare*. Thus, only the join message attributes to extra communication expenditure which, in the two layer case, the size m of layer 1.

To facilitate the assessment of view change complexity of our double-layer protocol, suppose that the number of replicas in the first layer is m, and for each group in the second layer is n (as depicted in Figure 3.3). Note that the view change can be triggered from either the first layer or the second layer. First, we consider the case that the view change is triggered in the first layer. As described in the protocol, the first part of our view change process runs within this layer (which is similar to original PBFT). For this part, the complexity is $O(m^3)$ with view change. As one of the members in the first layer is selected as the new primary, the extra complexity is introduced by multicasting the *redistribute* messages to its group in the second layer. Because the certificate is generated in the first layer, the complexity of this part is $O(m^2 n)$. The total complexity is $O(m^3 + m^2 n)$.

In the next step, we consider that a view change is triggered in one of the groups in the second layer. Similarly, the complexity is $O(n^3)$ for the first part. The extra complexity is introduced by multicasting the *join* messages to the first layer. However, the certificate is now generated in one of the groups in the second layer. For this reason, the additional complexity is $O(n^2 m)$ and the total complexity is $O(n^3 + n^2 m)$.

For comparison, the complexities are $O((m + mn)^3)$ for the original PBFT and $O(m + mn)$ for HotStuff when considering the same total number of replicas. It can be seen that in both first layer or second layer triggered cases, the complexity of the proposed double-layer PBFT is not as good as HotStuff; however, it reduces the complexity in large scale when compared with original PBFT.

3.3.1 Operation Synchronization

The precondition for entering the next phase does not require synchronization across all replicas. Because of the loss of connection or temporary crash, one replica may skip some operations. Although the consensus is still reached, the sequence number α is not necessarily continuous.

However, for clients who also access data from the network, operations are preferably synced on each replica. The asynchronous replica can be detected by the discrete sequence number and the reached watermark. A replica may extract the missing operations by inquiring them from other replicas in real time or at a constant interval.

Once the replica decides to extract a missing operation from the rest of the network, it multicasts $[n\alpha, i, v]_{extract-request}$. If there are no fewer than $2f + 1$ valid *extract-reply*, the replica accepts the reply as its missing operation. For replica who received *extract-request*, it replies $[O, \alpha, i, v]_{extract-reply}$ if the request is valid.

3.3.2 Safety and Liveness

The group-wide operation is inconsistence with the original PBFT. The network is weakly synchronized such that the time t for a messaged been received after sending does not go infinity. The Byzantine replica is assumed unable to subvert cryptography. The safety and liveness are retained within/across groups with the modified view change protocol. Suppose there are no more than $\frac{1}{3}$ of the nodes in a group are faulty, and the network is weakly synchronized such that the time t for a messaged been received after sending does not go infinity.

Because the protocol requires more than $\frac{2}{3}$ replicas to communities before advanced into next operation (to ensure the number of responses from non-faulty replicas is always greater than that from faulty), there is at least one non-faulty replica overlap for two consecutive operations. All non-faulty replicas agree on each other, and then they agree on total order of operations, providing safety. Also, the bound of faulty replicas indicates that the protocol always collects sufficient responses to proceed for liveness. The consensus in the first layer is the precondition to invoke protocol in sub-groups. Thus, safety is guaranteed across groups. The modified view change protocol replaces the faulty group leader. Thus, liveness is guaranteed across groups. This consistency with PBFT in safety and liveness within and across groups leads to the consistency for the whole network.

3.4 Communication Complexity of Double-Layer PBFT

The double-layer PBFT model is proposed in Figure 3.3, where the first-layer leader controls m replicas, each of which serves as a primary node of the n sub-layer replicas in the second layer. Therefore, there are $1 + m + mn$ nodes in the system. For pear-to-pear communication, the communication complexity is the square of node number. Thus, the required complexity C to reach consensus is $C = (m + 1)^2 + m(n + 1)^2$. Note that here we assume each sub-group in the second layer has the same number of nodes.

Then, we investigate the replicas arrangement with the minimum communication complexity for a double-layer system. Suppose the overall number of nodes is Z and denote $Y = Z - 1$, thus $Y = m + mn$. The total complexity is then represented as:

$$C = \left(\frac{Y}{1+n} + 1\right)^2 + Y(1 + n). \tag{3.1}$$

Because the second-order derivative of this equation is always positive, the minimum value of C is achieved at the zero point of the first-order derivative. In another word, for a double-layer system containing Z nodes in total, the minimum communication complexity can be achieved when n equals to the nearest integer to the real positive root of the following equation:

$$n^3 + 3n^2 + n = 2Z - 1, \quad \left(3 \leq n \leq \frac{Z-4}{3}\right). \tag{3.2}$$

Note that it is based on the assumption that the system is full (the number of sub-layer replicas in each sub-group is equal) and the number of sub-layer replicas in each sub-group is equal. If the double-layer system is not full, the communication complexity reaches a smaller value when the vacancies are equally distributed into the sub-groups. The minimum value can be reached by distributing vacancies to the minimum number of sub-groups. However, when m and n are fixed by optimal allocation and the system is full, the relationship between communication complexity C and total node number Z can be written as

$$C \approx 1.9Z^{\frac{4}{3}}. \tag{3.3}$$

Through double-layer PBFT, complexity can be significantly shrunk compared to the original single-layer PBFT at the cost of longer delay. In addition, with such a topology, the security performance should be analytically investigated to guide the actual system deployment.

3.5 Security Threshold Analysis

In the double-layer PBFT, nodes in both layers participate in the consensus reaching process. The first layer is a classic PBFT model that tolerates no more than $\lfloor \frac{m}{3} \rfloor$ faulty nodes based on the conclusion $Z \geq 3f + 1$ [11]. In the second layer, as there are m PBFT consensus groups, we need to analyze the threshold of consensus-reached sub-groups required to ensure the security and liveness of the whole system. During the consensus-reaching process, as the leader of each sub-group directly send *post-reply* to the client, each consensus sub-group is regarded as a whole. While any individual node in PBFT systems can be divided into three categories, including consensus reached, not reached, and faulty node. A consensus network may only be in two situations: consensus reached and not reached. In other words, a consensus sub-group would also be in two situations, either consensus reached or not. In this case, a system tolerates at most $\lfloor \frac{m}{2} \rfloor$ failed groups to reach consensus, i.e. the security threshold of consensus-reached sub-groups is $\lfloor \frac{m}{2} \rfloor$.

Table 3.1 Frequently used notations.

Notation	Definition
P_P	Consensus success rate in FPD model
P_f	Faulty probability of nodes
$P(A)$	Probability of Condition A
$P(B|A)$	Probability of Condition B under the condition of A happening
P_N	Consensus success rate in FND model
K	Total number of faulty nodes

To facilitate the deployment of distributed system in different scenarios, we analyze the success rate in two models under malicious attacks and the frequently used notations are summarized in table 3.1. The faulty probability determined (FPD) model is used when the probability of every single faulty node is fixed, and the faulty number determined (FND) model is used when the number of faulty nodes in the system is fixed. In these two models, different initial conditions are given to analyze the security performance of the system. More specifically, we assume the faulty nodes in the FPD model are independent with each other, and they have the same faulty probability. Conversely, in the FND model, as the total number of faulty nodes is fixed, whether one node is faulty depends on the situation of other nodes. In addition, these two models have different application scenarios. The FND model, which is more similar to the traditional PBFT, is suitable for small systems where the number of faulty nodes can be easily estimated. However, it is more appropriate to use FPD model to evaluate the performance of large systems where node failure is estimated by probability. For example, in manufacture, it is normally given the reject rate of one product instead of the specific number.

3.5.1 Faulty Probability Determined

Let us assume that P_f is the faulty probability of each node. To find out the relationship between the success rate P_P and P_f, we shall first define two important conditions, i.e. Conditions A and B as listed below, under which consensus can be reached.

- *Condition A*: no more than $\lfloor \frac{m}{3} \rfloor$ faulty nodes in the first layer,

- *Condition B*: no more than $\lfloor \frac{m}{2} \rfloor$ groups fail in the second layer.

In addition, Condition A and Condition B are not independent. If one replica in the first layer is faulty, it will be impossible for the corresponding sub-group to reach consensus. Therefore, we have $P_P = P(A) \times P(B|A)$.

We can formulate the function of the system consensus success rate P_P against P_f as follows

$$P_P = \sum_{i=0}^{\lfloor \frac{m}{3} \rfloor} \left(C_m^i (1 - P_f)^{(m-i)} P_f^i \right.$$

$$\left. \sum_{j=0}^{\lfloor \frac{m}{2} \rfloor - i} P_g^j (1 - P_g)^{(m-i-j)} \right). \tag{3.4}$$

P_g represents the probability of a group, with a non-faulty leader, failing to reach consensus. We assume that there are g faulty nodes in one single group. To make this group fail, $\lfloor \frac{n}{3} \rfloor + 1 \leq g \leq n$ since PBFT group tolerates up to $\lfloor \frac{n}{3} \rfloor$ faulty nodes. Therefore, we have

$$P_g = \sum_{g=\lfloor \frac{n}{3} \rfloor+1}^{n} C_n^g P_f^g \left(1 - P_f\right)^{n-g}. \tag{3.5}$$

To verify the closed-form expression derived, a simulation is performed based on random sampling by using MATLAB. In the simulation process, we take the faulty probability P_f and node number m, n as the input and use a random array consisting $m + mn$ numbers to represent the status of nodes. Each number in this random array is either 1 or 0, representing faulty and non-faulty node, respectively. On top of that, we set that each array element has a probability of P_f to be 1 (faulty); otherwise, it is 0 (non-faulty). In this case, by counting the faulty nodes in each layer and sub-group and comparing the results with the thresholds, we determine whether the consensus can be reached or not. Then, the above-mentioned process is repeated over 10 000 times and the simulation success rate for one value of P_f can be obtained by taking the ration of success times to the total repeating times. Finally, by increasing P_f from 0 to 1, we get the simulation curve and compare it to the analytical result of the closed-form expression. The simulation design of the FND model is similar while the only difference is that FND takes the faulty node number instead of faulty probability as the input.

Note that the purpose of the simulation is to examine the accuracy of the theoretical derivations. Therefore, the complex peer-to-peer communication process is temporarily ignored in the simulation performed. However, we are also working on a system simulation, which takes every communication and view change into consideration, to further test the multi-layer PBFT system.

Figure 3.4 shows that the analytical curve of success rate matches perfectly with the simulation points as the probability of faulty nodes increases. This verifies the effectiveness of the analytical result we derived.

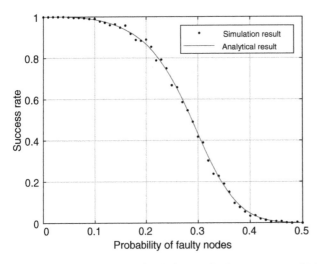

Figure 3.4 Analytical and simulation results for success rate in the FPD model. ($m = n = 10$).

3.5.2 Faulty Number Determined

In the FND model, we assume that there are K faulty nodes in the whole system. We aim to find the relationship between K and the success rate P_N. Meanwhile, we use the same assumption of Conditions A and B to calculate P_N, $P_N = P(A) \times P(B|A)$.

Unlike the FPD model, $P(A)$ and $P(B|A)$ are calculated using the hypergeometric model [37] because the FND model is based on the prerequisite of a fixed number of faulty nodes.

Also, there should be at most $\lfloor \frac{m}{2} \rfloor - i$ failed sub-groups with non-faulty leaders, which means at most $\lfloor \frac{m}{2} \rfloor - i$ sub-groups have more than $\lfloor \frac{n}{3} \rfloor$ faulty nodes. However, in the FND model, the number of faulty nodes in each group affects situations in the other groups so that it will be extremely complicated to consider m groups together. Therefore, a simplified binomial distribution model on the group level is adopted, assuming every group has the same faulty probability of P_{g2}.

P_{g2} represents the probability of a group with a non-faulty leader failing in the second layer. It can be calculated as follows:

$$P_{g2} = \sum_{g=\lfloor \frac{n}{3} \rfloor + 1}^{n} \frac{C_n^g C_{mn-n-1}^{K-g}}{C_{m+mn-1}^{K}}. \tag{3.6}$$

Then, we derive the probability P_N against K as

$$P_N = \frac{1}{C_{m+mn}^{K}} \sum_{i=0}^{\lfloor \frac{m}{3} \rfloor} \left(C_m^i C_{mn}^{K-i} \sum_{j=0}^{\left(\lfloor \frac{m}{2} \rfloor - i \right)} P_{g2}^j \left(1 - P_{g2} \right)^{(m-i-j)} \right). \tag{3.7}$$

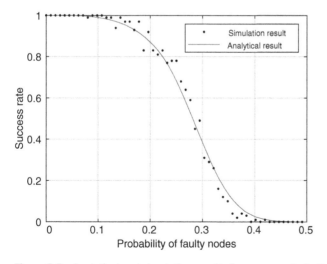

Figure 3.5 Analytical and simulation results for success rate in the FND model ($m = n = 10$).

In Eq. (3.7), $\sum_{i=0}^{\lfloor\frac{m}{3}\rfloor} C_m^i C_{mn}^{K-i}$ requires $K - i > 0$ because the combinatorics must be positive. When $K - i > 0$, i.e. $K < \lfloor\frac{m}{3}\rfloor$, the success rate can be simply calculated as

$$P_N = \frac{1}{C_{m+mn}^K} \sum_{i=0}^{K} C_m^i C_{mn}^{K-i}. \tag{3.8}$$

Figure 3.5 compares the analytical curve of success rate with the simulation points under different proportions of faulty nodes. It shows that, in the high success region where $P_N > 0.85$, the analytical curve matches the simulated curve. When the success rate of the system is lower, there is a slight difference between our calculation and simulation results. The rationale behind is that a part of the model (group level in the second layer) is simplified from hypergeometric distribution to the binomial distribution by using P_{g2} as a failure rate for every group in Eq. (3.6). However, the difference is negligible and will approach zero as the total number of the nodes in the system increases because the hypergeometric distribution approaches the binomial distribution in large systems [37].

3.6 Conclusion

In this chapter, a scalable double-layer PBFT mechanism is introduced to reduce the communication complexity of the original single-layered PBFT. The results reveal that the communication complexity of the proposed double-layer PBFT system is significantly reduced to a minimum of $C \approx 1.9Z^{\frac{4}{3}}$ at system's maximum optimized capacity. To reach the minimum communication complexity, the optimal values of m and n are discussed. In addition, the analysis of the security threshold shows that the success rate sinks significantly when the proportion of faulty nodes exceeds $\frac{1}{3}$ of the total and the threshold that keeps success rate at 100% rises from $\lfloor\frac{m}{3}\rfloor$ to $\lfloor\frac{n}{3}\rfloor \times \lfloor\frac{m}{2}\rfloor$ in advanced model. These results show that the security performance of the double-layer system is largely determined by the first layer and is improved over operation time.

On the other hand, some improvement can be further implemented into the proposed double-layer PBFT mechanism. For example, a system model to differentiate nodes in the first and second layer could be proposed as a trade-off between FND/FPD and the advanced model. Another potential topic is the deployment of multi-layer PBFT system. It is worth to mention that some application scenarios such as financial service are sensitive to both latency and scalability; thus more, advanced research could be conducted to solve the issue.

References

1 M. H. Miraz and M. Ali. Applications of blockchain technology beyond cryptocurrency. *Annals of Emerging Technologies in Computing*, 2 (1): 1–2, 2018.

2 Y. Sun, L. Zhang, G. Feng, B. Yang, B. Cao, and M. A. Imran. Blockchain-enabled wireless Internet of Things: performance analysis and optimal communication node deployment. *IEEE Internet of Things Journal*, 6 (3): 5791–5802, 2019.

3 B. Cao, Y. Li, L. Zhang, L. Zhang, S. Mumtaz, Z. Zhou, and M. Peng. When Internet of Things meets blockchain: challenges in distributed consensus. *IEEE Network*, 33 (6): 133–139, Nov. 2019. ISSN 1558-156X. https://doi.org/10.1109/MNET.2019.1900002.

4 D. Yu, W. Li, H. Xu, and L. Zhang. Low reliable and low latency communications for mission critical distributed industrial Internet of Things. *IEEE Communications Letters*, 6 (1): 313–317, 2020.

5 H. Xu, L. Zhang, O. Onireti, Y. Fang, W. Buchanan, and M. Imran. BeepTrace: blockchain-enabled privacy-preserving contact tracing for COVID-19 pandemic and beyond. *IEEE Internet of Things Journal*, 09, 2020. https://doi.org/10.1109/JIOT.2020 .3025953.

6 P. V. Klaine, L. Zhang, B. Zhou, Y. Sun, H. Xu, and M. Imran. Privacy-preserving contact tracing and public risk assessment using blockchain for COVID-19 pandemic. *IEEE Internet of Things Magazine*, 3 (3): 58–63, 2020.

7 S. Nakamoto. White paper: bitcoin: a peer-to-peer electronic cash system. 2008. http://www.bitcoin.org/bitcoin.pdf.

8 P. Vasin. White paper: BlackCoin's proof-of-stake protocol v2. 2014. https://blackcoin.co/blackcoin-pos-protocol-v2-whitepaper.pdf.

9 L. Zhang, H. Xu, O. Onireti, M. A. Imran, and B. Cao. How much communication resource is needed to run a wireless blockchain network? *arXiv preprint arXiv:2101.10852*, 2021.

10 C. Bendiksen, S. Gibbons, and E. Lim. The bitcoin mining network-trends, marginal creation cost, electricity consumption and sources. *CoinShares Research*, 3–19, 2018.

11 M. Castro, B. Liskov, et al. Practical byzantine fault tolerance. *OSDI 1999: Proceedings of the 3rd Symposium on Operating Systems Design and Implementation*, 99: 173–186, 1999.

12 L. Lamport. The part-time parliament. *ACM Transactions on Computer Systems*, 16 (2): 133–169, 1998.

13 D. Ongaro and J. Ousterhout. In search of an understandable consensus algorithm. *USENIX*, 305–320, 01 2014.

14 H. Xu, L. Zhang, Y. Liu, and B. Cao. RAFT based wireless blockchain networks in the presence of malicious jamming. *IEEE Wireless Communications Letters*, 2020. ISSN 2162-2345. https://doi.org/10.1109/LWC.2020.2971469.

15 M. Vukolić. The quest for scalable blockchain fabric: proof-of-work vs. BFT replication. In *International Workshop on Open Problems in Network Security*, pages 112–125. Springer, 2015.

16 L. Lamport, R. Shostak, and M. Pease. The byzantine generals problem. *ACM Transactions on Programming Languages and Systems*, 4 (3): 382–401, 1982.

17 O. Onireti, L. Zhang, and M. Imran. On the viable area of wireless practical byzantine fault tolerance blockchain networks. *IEEE GLOBECOM*, 1–6, 2019.

18 E. Androulaki, A. Barger, V. Bortnikov, C. Cachin, K. Christidis, A. De Caro, D. Enyeart, C. Ferris, G. Laventman, Y. Manevich, S. Muralidharan, C. Murthy, B. Nguyen, M. Sethi, G. Singh, K. Smith, A. Sorniotti, C. Stathakopoulou, M. Vukolic, S. W. Cocco, and J. Yellick. Hyperledger fabric: A distributed operating system for permissioned blockchains. *Proceedings of the 13th EuroSys Conference*, pages 30:1–30:15, 2018. https://doi.org/10.1145/3190508.3190538.

19 J. P. Morgan. White paper: blockchain and the decentralization revolution. https://www.jpmorgan.com/jpmpdf/1320745566550.pdf.

20 P. R. Johnson and R. Thomas. *RFC0677: Maintenance of duplicate databases.* RFC Editor, 1975.

21 R. Guerraoui, N. Knežević, V. Quéma, and M. Vukolić. The next 700 BFT protocols. In *Proceedings of the 5th European Conference on Computer systems*, pages 363–376, 2010.

22 P. A. Alsberg and J. D. Day. A principle for resilient sharing of distributed resources. In *Proceedings of the 2nd International Conference on Software Engineering*, pages 562–570, 1976.

23 R. H. Thomas. A solution to the update problem for multiple copy data bases which uses distributed control. Technical report, BOLT BERANEK AND NEWMAN INC CAMBRIDGE MA, 1976.

24 D. K. Gifford. Weighted voting for replicated data. In *Proceedings of the 7th ACM Symposium on Operating Systems Principles*, pages 150–162, 1979.

25 V. Buterin and V. Griffith. Casper the friendly finality gadget. *arXiv preprint arXiv:1710.09437*, 2017.

26 M. Pease, R. Shostak, and L. Lamport. Reaching agreement in the presence of faults. *Journal of the ACM (JACM)*, 27 (2): 228–234, 1980.

27 J. N. Gray. Notes on data base operating systems. In *Operating Systems*, pages 393–481. Springer, 1978.

28 G. Bracha and S. Toueg. Asynchronous consensus and broadcast protocols. *Journal of the ACM (JACM)*, 32 (4): 824–840, 1985.

29 B. M. Oki and B. H. Liskov. Viewstamped replication: a new primary copy method to support highly-available distributed systems. In *Proceedings of the 7th Annual ACM Symposium on Principles of Distributed Computing*, pages 8–17, 1988.

30 M. Yin, D. Malkhi, M. K. Reiter, G. G. Gueta, and I. Abraham. HotStuff: BFT consensus with linearity and responsiveness. In *Proceedings of the 2019 ACM Symposium on Principles of Distributed Computing*, pages 347–356, 2019.

31 H. Sukhwani, J. M. Martinez, X. Chang, K. S. Trivedi, and A. Rindos. Performance modeling of PBFT consensus process for permissioned blockchain network. *2017 IEEE 36th Symposium on Reliable Distributed Systems*, pages 253–255, 2017.

32 X. Fan. Scalable practical byzantine fault tolerance with short-lived signature schemes. *Proceedings of the 28th Annual International Conference on Computer Science and Software Engineering*, pages 245–256, 2018.

33 M. Zamani, M. Movahedi, and M. Raykova. White paper: RapidChain: scaling blockchain via full sharding dfinity Palo Alto, CA. 2018. https://eprint.iacr.org/2018/460.pdf.

34 L. Luu, V. Narayanan, C. Zheng, K. Baweja, S. Gilbert, and P. Saxena. A secure sharding protocol for open blockchains. *Proceedings of the 2016 ACM SIGSAC Conference on Computer and Communications Security*, pages 17–30, 2016. https://doi.org/10.1145/2976749.2978389.

35 P. J. Sadalage and M. Fowler. *NoSQL Distilled: A Brief Guide to the Emerging World of Polyglot Persistence.* Pearson Education, 2013.

36 W. Lv, X. Zhou, and Z. Yuan. Design of tree topology based byzantine fault tolerance system. *Journal of Communications*, 38 (Z2): 143–150, 2017.

37 J. A. Rice. *Mathematical Statistics and Data Analysis.* Cengage Learning, 2006.

4

Blockchain-Driven Internet of Things

Bin Cao, Weikang Liu, and Mugen Peng

State Key Laboratory of Networking and Switching Technology, Beijing University of Posts and Telecommunications, Beijing, China

4.1 Introduction

Blockchain has shown great potential in Internet of Things (IoT) ecosystems for establishing trust and consensus mechanisms without the involvement of any third party. Understanding the relationship between communication and blockchain as well as the performance constraints posing on the counterparts can facilitate designing a dedicated blockchain-enabled IoT system [1, 2]. This section starts by introducing the challenges and issues of the IoT system and illustrating the advantages of blockchain for the IoT system.

4.1.1 Challenges and Issues in IoT

The Internet of Things (IoT) has been identified as one of the most disruptive technologies of this century. It has attracted much attention from the society, industry, and academia as a promising technology that can enhance day-to-day activities; the creation of new business models, products, and services; and as a broad source of research topics and ideas. Although the first idea of IoT emerged no more than two decades ago, and many IoT ecosystems have been generated since then; some unsolved and important issues still remain as follows:

Trust: IoT cloud servers are closed systems. For one thing, the service providers have the ability to illegally control the IoT devices. For another, it is hard to build the cooperation and trust relationship among different IoT business agencies.

Security: The IoT data center is vulnerable because it is easy to be attacked by hackers using a Distributed Denial of Service attack (DDoS), and when it happens, all IoT services may be affected because of the centralized topology.

Overhead: The current centralized model has a high maintenance cost; that is, it is expensive to timely update the software of millions of IoT devices.

Scalability: The poor scalability of the centralized topology cannot meet the needs of massive IoT device connections; that is, a large delay might be caused by a surge of service requests.

Wireless Blockchain: Principles, Technologies and Applications, First Edition.
Edited by Bin Cao, Lei Zhang, Mugen Peng and Muhammad Ali Imran.
© 2022 John Wiley & Sons Ltd. Published 2022 by John Wiley & Sons Ltd.

4.1.2 Advantages of Blockchain for IoT

As a brand of new distributed ledger technology (DLT), blockchain was originally designed for the digital currency Bitcoin in 2009 [3]. With decades of operation in a decentralized network, Bitcoin did not encounter serious security incidents. This can be largely attributed to the advantage of its consensus mechanism, which uses the computing power of the whole network to ensure the immutability of the data. As a secure and decentralized solution, blockchain is expected to transform IoT ecosystems by making them smart and more efficient. According to IDC (International Data Corporation) report, by 2019, 20% of IoT deployments will have basic levels of blockchain-enabled services [2].

Blockchain is a peer-to-peer (P2P) DLT for establishing trust and consensus in decentralized networks. On the one hand, to address the challenges in a trustless distributed environment, consensus mechanism is adopted in blockchain in a decentralized way to reach agreement for transactions among individual users. On the other hand, using digital signature and hash algorithm-based encryption, security can be assured in the decentralization blockchain system [4].

Blockchain ledger has three basic concepts: transaction, block, and chain. The "transaction" in blockchain is not restricted to trading. In fact, all the valuable information can act as a transaction to be broadcast in the blockchain network. The blocks are storage units to record transactions, which are created and broadcast by those users authorized by consensus mechanism. Each block is identified uniquely by its hash value, which is referenced by the block that came after it. This establishes a link between the blocks, thus creating a chain of blocks, called the ledger. With the blocks accumulating sequentially in a consensus process, the cost of attack and malicious modification would be increased exponentially [2].

First, using blockchain-based decentralization, the burden of hot spot and the probability of single point of failure (SPF) can be reduced significantly. Second, the consensus mechanism and encryption algorithm in blockchain can be leveraged to strengthen IoT security. In addition, by using smart contact [5], IoT devices can carry out trading and execute actions autonomously. As a public-distributed ledger where stored information can be audited by all the users, blockchain provides a trust platform for IoT business cooperation.

4.1.3 Integration of IoT and Blockchain

Currently, the implementation of IoT and blockchain is on the agenda in industry, and there are already promising solutions and initiatives in several areas. In the supply chain industry, the work in [5] provides a blockchain-enabled supply chain model. In this model, the information stored in the blockchain can serve as a log of delivery for container shipments. All the movements of the container from source to destination can be tracked by any supply chain entities, so that the shipment delay can be minimized and the missing asset can be tracked accurately. In the healthcare domain, the work in [6] provides a user-centric model for processing personal health data using a blockchain network, ensuring the data ownership of individuals, as well as data integrity. By enforcing access control policies, the system makes sure that users can handle their personal data without worrying about the privacy issues. Blockchain is also available in the other IoT applications, such as remote software updates and insurance for vehicles [7].

In particular, blockchain plays an important role in energy trading for IoT applications in the energy Internet. Nowadays, there exist some blockchain technologies that have investigated how to promote energy sharing among IoT devices to increase the efficiency of energy utilization. Taking the Internet of Vehicles (IoVs) as an example, electric vehicles have the ability to absorb excessive energy during the non-peak period and provide energy as distributed generators during the peak period. To enable secure energy trading, the authors in [8] propose a localized P2P electricity trading framework, in which consortium blockchain is exploited to improve the security of transactions without relying on a third party. To improve the trading efficiency, the authors in [9] propose a credit-based payment scheme, which supports fast and frequent trading among energy nodes by establishing virtual credit banks. Some digital currency has been presented for renewable trading based on blockchain, such as "Specoin" [10].

As shown in Figure 4.1, to operate a blockchain-enabled IoT system, the main steps are illustrated as follows:

- All IoT devices operate on the same blockchain network.
- An IoT device generates a transaction for payment (or recording significant information) and broadcasts it to the network.
- The IoT devices receive the information and transactions in the network and validate them.
- All IoT devices perform a hash algorithm to elect a winner whose candidate block will be broadcast and validated as a new block.
- All IoT devices insert the identical copy of the new block into their local ledgers.
- The transaction stored in blockchain ledger triggers the smart contract in the IoT device.
- The IoT device carries out a specific task, that is, the movement of containers in a supply chain scenario, and supplying power in a smart energy scenario.

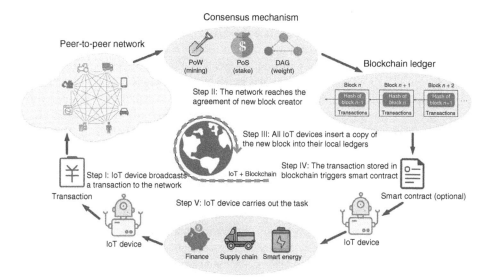

Figure 4.1 An example of implementing blockchain in the IoT system.

According to Figure 4.1, we can see that a consensus mechanism is the cornerstone in blockchain-enabled IoT systems, which builds a bridge between the raw data from the infrastructure and the confirmed information for performing various applications. In Section 4.2, we illustrate the main idea of different types of consensus mechanisms and list their advantages and disadvantages in the IoT ecosystem.

4.2 Consensus Mechanism in Blockchain

In this section, we discuss different types of consensus mechanisms in blockchain and consider whether the design criteria of corresponding consensus mechanisms can meet the needs of IoT.

A consensus mechanism plays an indispensable role in blockchain to resolve trust concerns by answering the question "who will be the one who has the right to insert the next block into the blockchain." [11]. With a consensus mechanism, the information can be announced orderly to all users without the involvement of a third party. Nowadays, various consensus mechanisms have been proposed; Proof of Work (PoW) and Proof of Stake (PoS) are the most widely used ones. However, the two consensus mechanism-based traditional blockchains face significant challenges when applied to IoT systems. We introduce Practical Byzantine Fault Tolerance (PBFT) and Direct Acyclic Graph (DAG) based consensus mechanisms as an effective solution.

4.2.1 PoW

PoW was proposed in the original blockchain application (e.g. Bitcoin). Bitcoin use transactions to move coins from one user wallet to another. In particular, the coins are represented in the form of transactions, more specifically, a chain of transactions [12].

The core idea of PoW is the competition for computing power [2]. The node performing the consensus mechanism (called the miner) uses its computing resource for the hashing operation to compete for the right to generate the new block with bonuses. The winner is the first one who obtains a hash value lower than the announced target.

In central bank, all the transactions are verified, processed, and recorded in a centralized private ledger. While in Bitcoin, every user acts as a bank and it keeps a copy of the ledger. In Bitcoin, the role of the distributed ledger is given to the so-called blockchain. Because of the storage of multiple copies of blockchain at multiple nodes in the network, new vulnerabilities arise such as keeping a consistent global view of the blockchain in the whole system. For instance, a user (say Alice) could simultaneously generate two different transactions, using the same set of coins, to two different receivers (say, Bob and Carol). This type of malicious behavior by a user is known as double spending. Now, if both the receiver processes the transaction independently based on their local view of the blockchain, and the transaction verification is successful, then it leaves the blockchain into an inconsistent state. The primary requirements to avoid the above problem are two-folded: (i) distribute the transaction verification process to ensure the correctness of the transaction and (ii) everyone in the network should quickly know about a successfully processed transaction to ensure the consistent state of the blockchain. To fulfill the requirements mentioned above, Bitcoin uses the concept of PoW and a probabilistic distributed consensus protocol.

The distributed transaction verification process ensures that a majority of miners will verify the legitimacy of a transaction before it is added in the blockchain. Whenever the blockchain goes into an inconsistent state, all the nodes update their local copy of blockchain with the state on which a majority of miners agree; in this way, the correct state of the blockchain is obtained by election. However, this scheme is vulnerable to Sybil attacks [13]. With Sybil attack, a miner creates multiple virtual nodes in the network, and these nodes could disrupt the election process by injecting false information in the network such as voting positive for a faulty transaction. Bitcoin counters the Sybil attacks by making use of a PoW-based consensus model, to verify a transaction, in which the miners have to perform some computational task to prove that beyond their virtual representatives, there are real entities. The PoW consists of a complex cryptographic math puzzle, similar to Adam Back's Hashcash [14]. In particular, PoW involves scanning for a value (called nonce) that when hashed such as with SHA-256, the resulting hash begins with the required number of zero bits. The number of zero bits required is set by the target. The resulting hash has to be a value less than the current target and so it must consist of a certain number of leading zero bits to be less than that. In this way, PoW imposes a high level of computational cost on the transaction verification process, and the verification will depend on the computing power of a miner instead of the number of (possibly virtual) identities. The main idea is that it is much harder to fake the computing resources than it is to perform a Sybil attack in the network. On the other hand, the high computing difficulty would cause deteriorated and meaningless energy consumption. Note that the available resources of IoT devices are very limited. Therefore, PoW is not a good option for IoT systems.

4.2.2 PoS

Operation of the Bitcoin protocol is such that security of the network is supported by physically scarce resources: specialized hardware needed to run computations and electricity spent to power the hardware. This makes Bitcoin inefficient from a resource standpoint. To increase their share of rewards, Bitcoin miners are compelled to participate in an arms race to continuously deploy more resources in mining. While this makes the cost of an attack on Bitcoin prohibitively high, the ecological unfriendliness of the Bitcoin protocol has resulted in proposals to build similar systems that are much less resource intensive [15].

One possible decentralized ledger implementation with security not based on expensive computations relies on proof of stake (PoS) algorithms. The idea behind proof of stake is simple: instead of mining power, the probability to create a block and receive the associated reward is proportional to a user's ownership stake in the system. An individual stakeholder who has p fraction of the total number of coins in circulation creates a new block with p probability. The rationale behind proof of stake is the following: users with the highest stakes in the system have the most interest to maintain a secure network, as they will suffer the most if the reputation and price of the cryptocurrency would diminish because of the attacks. To mount a successful attack, an outside attacker would need to acquire most of the currency, which would be prohibitively expensive for a popular system.

Because winning probability is directly determined by coinage, PoS is beneficial for the wealthy miner and might cause oligopolies or near-monopolies, and then results in the generation of a powerful third party. From this sense, the PoS consensus mechanism may not fit well to establish a smart distributed IoT system.

4.2.3 Limitations of PoW and PoS for IoT

PoW and PoS are two typical traditional consensus mechanisms that work on a "single chain" (forking is illegal) architecture. To avoid forking and maintain a single version of the blockchain ledger among all users, the consensus mechanism must slow down the access rate of new blocks. This might cause some significant bottlenecks in applying to IoT systems.

4.2.3.1 Resource Consumption

To slow down the access rate of new blocks and protect the blockchain network from attack, the traditional consensus process consumes too many resources (i.e. computing power in PoW, coinage in PoS), which is too costly for the resource-limited IoT devices.

4.2.3.2 Transaction Fee

A transaction fee is needed in traditional consensus mechanisms to feed the miners, which might cause a heavy burden on the IoT system where most tradings are micropayments.

4.2.3.3 Throughput Limitation

Because the capacity of a new block is limited, transaction per second (TPS) is limited to dozens usually (e.g. seven TPS in Bitcoin and 20–30 TPS in Ethereum), which is unable to respond to the exponential growth of IoT devices.

4.2.3.4 Confirmation Delay

Because of the low access rate of new blocks, the confirmation delay is too long for IoT applications (e.g. 60 minutes in Bitcoin and three minutes in Ethereum).

4.2.4 PBFT

Both PoW and PoS are mainly used in public chains. Unlike public chains, private blockchains prefer to adopt lighter algorithms [16] such as Paxos [17] and Raft [18] to reduce the amount of computational power and improve the transaction throughput. This property is critically important to the application scenarios of blockchain-enabled IoT ecosystem, where it is typically composed of low-cost and low-power devices. Although the consensus used by private chain does not protect the integrity of transactions from malicious attacks, they enable the Crash Fault Tolerance (CFT) for the applying system [18]. To protect the system from malicious users, PBFT was proposed in [19] as an improved and practical protocol based on Byzantine Fault Tolerance (BFT). Compared to the Proof-based consensus, where the security-bound threshold is 51%, i.e. absolute secure transaction can be achieved if the malicious users occupy no more than half of the overall resource, PBFT requires the number of malicious users under 3% of total participants to ensure the system immune from the faulty and Byzantine nodes [19]. PBFT is favored for private and consortium chains, thanks to the lower complexity and low

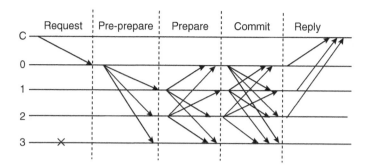

Figure 4.2 Three stages of PBFT consensus processing. Source: Castro et al. [19].

energy consumption, a well-known example of PBFT implementation is the Hyperledger Fabric, part of Hyperledger business blockchain frameworks, which has been adopted by tech giants such as IBM or Wall Street Fintech, such as J. P. Morgan [20].

Figure 4.2 shows the protocol diagram of the original single-layer PBFT, where as an example, we assume that there is one primary node and three state machine replicas. The consensus is triggered by the client when it has a request to be recorded into the blockchain. It starts from the client sending the request to the blockchain network nodes' header (or broadcast to every node). Then, consensus will be operated among the nodes, and if the consensus is reached, the new record will be committed to the chain, vice versa. The whole consensus process includes three stages: pre-prepare, prepare, and commit (as shown in Figure 4.2). On receiving the request from the client, the primary node broadcasts a pre-prepared message to the other nodes. In prepare and commit stages, all replicas send peer-to-peer (P2P) messages to the check integrity of received messages. In each stage, a minimum number of consistent messages are required for stepping on the subsequent stage. The consensus is technically reached when commit phase is successful. Specifically, the client must receive at least $f + 1$ replies (f denotes the number of Byzantine faulty nodes in the group) from the nodes to validate the final consensus. In the case of client failed to collect $f + 1$ replies, the client may resend the request to primary for retrying. Upon receiving the same request again, if the consensus is already reached on commit phase, replicas just resend the final stage messages. If the consensus is not reached, the network goes over the protocol again.

From Figure 4.2, we can see that PBFT is a very communication-demanding protocol. The pre-prepare stage requires $a - 1$ times of communications. The prepare stage and commit stage require $a^2/2$ times of communications, respectively. Therefore, PBFT uses about a^2 times of inter-node communication for a system with a nodes for a completed consensus. Obviously, the system is not scalable because the complexity burden is non-affordable when a is large (e.g. thousands). Although PBFT has shown good performance in terms of transaction throughput and latency, however, the scalability, which is a metric to measure how well the network reflects the capacity of the system to handle the increasing amount of work, is a bottleneck of PBFT because it relies on heavy inter-node communications. As the size of the system grows, the communication cost increases rapidly, resulting in low efficiency and poor scalability. More specifically, PBFT takes a^2 times of inter-node communications for a system with a nodes to reach a consensus. Thus, from the communication complexity perspective, the PBFT-based blockchain hardly scales up to 100 nodes [21]. Li et al. [16] proposed a scalable multi-layer PBFT-based blockchain to solve the problem of poor scalability.

4.2.5 DAG

The DAG architecture and its consensus mechanism is proposed to overcome the short-comings of traditional consensus for IoT [22]. Some typical DAG consensus processes are shown in Figures 4.3 and 4.4. A DAG-based consensus mechanism allows users to insert their blocks into the blockchain at any time, as long as they process the earlier transactions. In this way, many branches would be generated simultaneously, which is called forking. This phenomenon is usually regarded as an issue in many traditional consensus processes because it would cause "double-spending" [2]. However, the DAG-based consensus mechanism designs innovative protocols and algorithms to address the double-spending problem and allows any new arrival transactions access to the blockchain network in a forking topology. As a result, the confirmation rate and TPS will no longer be limited. Moreover, because the data stored in DAG is protected by massive forking blocks, resource consumption can

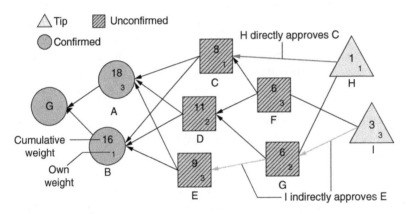

Figure 4.3 An example of Tangle.

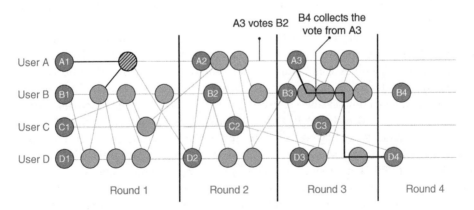

Figure 4.4 An example of Hashgraph.

be very low for a single user to create a new block. Accordingly, the professional miner disappears and low or no transaction fees are possible, which is critically important to the IoT ecosystem.

In this section, we introduce the consensus mechanism in Tangle and Hashgraph, respectively, which are the two typical DAG-based approaches.

4.2.5.1 Tangle

Tangle is the mathematical foundation of IOTA [23], a cryptocurrency for the IoT industry. As shown in Figure 4.3, Tangle is a DAG-based distributed ledger for recording transactions. It allows different branches to eventually merge into the chain, resulting in a much faster overall throughput. In Tangle, to access the ledger as a new vertex for storing a transaction, it has to approve a number of tips (typically two [23]). Thanks to this, the higher the arrival rate of new transactions, the faster earlier transactions can be confirmed. On the other hand, because tips are the childless vertexes in Tangle, the new vertex selects tips and covers them, which could limit the branch to a reasonable scale. Moreover, because the workload to create a new vertex is light, all users can issue their transactions at any time without transaction fee, which is critical to IoT application scenarios.

The consensus in Tangle relates to cumulative weight. As shown in Figure 4.3, the cumulative weight of a specific transaction is the sum of a vertex's own weight (proportional to the PoW that the issuing node invested into it [23]) and the overall weights of the vertices directly and indirectly approve it. Because the transactions stored in Tangle are secured by computing power, the cumulative weight of a transaction means its validity in the network and acts as a decisive criterion to address the double-spending problem.

In order to issue a new transaction and let the other users in the whole system accept it (i.e. win enough cumulative weight to reach an agreement for the consensus), the main procedures are listed as follows. (i) A user creates a unit as a candidate vertex in the DAG graph to store its transaction and uses its private key to sign the transaction. (ii) The user selects two tips with no-conflict according to a Markov Chain Monte Carlo (MCMC) algorithm [23] and adds the hash of the selected tips into its storage unit. (iii) The user finds a nonce to solve a cryptographic puzzle to meet the difficulty target. It is similar to PoW but with a very low difficulty-of-work, which can avoid spamming. After that, the user broadcasts the storage unit to others. (iv) When the other users receive it, they should check whether it is legal or not based on the digital signature and PoW-based nonce. The successfully checked new storage unit would be added as a new tip in Tangle and wait for confirmation through direct approval and indirect approval until its cumulative weight reaches the predefined threshold.

In a public ledger, building forking (or branch) and redoing the work is the only way to tamper with data and conduct double-spending. To address this problem, the single-chain-based consensus mechanism (e.g. PoW) uses the longest chain as the criterion. To this end, to guarantee and maximize its own profit, a rational user should choose the longest chain to work when forking occurs because the longest chain has the lowest probability to be orphaned. Similarly, Tangle uses the MCMC tip selection algorithm to select the branch with the largest overall cumulative weight. With the assistance of distributed and parallel approval in Tangle, the overall computing capability of honest users in large-scale IoT systems could be powerful to prevent double-spending, where the branch generated by an attacker is hard to outweigh the honest one. Meanwhile, any single user does not need to consume much power on computing for security.

4.2.5.2 Hashgraph

Hashgraph [24] is proposed for replicated state machines with a guarantee of BFT, it is asynchrony, decentralization, no PoW, eventual consensus with probability 1, and high speed in the consensus process. The gossip protocol and virtual voting are two key elements in Hashgraph. Using the gossip protocol, every transaction will be known by all users. After that, the agreement of the order of transactions will be reached through a virtual voting algorithm. In order to get a better understanding of Hashgraph, we will briefly introduce how the gossip protocol and virtual voting work.

According to the gossip protocol, in a fixed interval, each user in Hashgraph should randomly choose another one to announce all the transactions it knows. For example, the shadow unit in Figure 4.4 representing user B sends some information to user A that A does not know, so A creates the event that links A and B to store the unknown information. In this way, every event will be known by all participants eventually. Note that the gossip protocol is a low-cost method; the overhead to exchange a storage unit is very small, which includes positional information (3–6 bytes), signature (64 bytes), and transactions within the unit (about 100 bytes).

To achieve the consensus, the system needs to select the "famous witnesses" through virtual voting (all users perform the voting algorithm based on the graph connectivity). The famous witnesses are elected from witnesses that are the first events in each round (the dark gray units in Figure 4.4). An electing process includes voting and checking. As shown in Figure 4.4, the witnesses in round 3 vote for the witnesses in round 2. Then, the witnesses in round 4 will collect the votes in round 3. If the voting in round 3 and checking in round 4 succeed, the witnesses in round 2 would become famous. The events in round 1 voted by the famous witnesses in round 2 will be confirmed. The creation time of the confirmed events will be accepted by all users, which acts as a proof to prevent double-spending.

4.3 Applications of Blockchain in IoT

Originally proposed as the backbone technology of Bitcoin [3], Ethereum, and many other digital currencies, blockchain has become a revolutionary decentralized data management framework that establishes consensuses and agreements in a trustless and distributed environment [25]. In addition to the soaring in the finance sector, blockchain has been used in the field of Internet of Things, which attracted much attention from many other major industrial sectors ranging from supply chain [26], smart city [27], to E-healthcare [28, 29], etc. In addition, because the typical application in E-healthcare, BeepTrace [28], has been covered in this book, it will not be repeated here.

4.3.1 Supply Chain

4.3.1.1 Introduction

IoT has huge potential to impact global food supply chain (FSC) by increasing productivity in terms of supply chain performance. Among many challenges, agri-food safety and its impact on the environment due to food wastage are of major concerns. The US Center for Diseases Control (CDC) estimates that 48 million people get sick from foodborne illness, 128 000 are hospitalized, and 3000 die each year in the United States alone. Apart from illness, economically and criminally motivated food adulteration is also a growing concern due to globalization and wide growing supply chain networks [30]. Real-time monitoring of the food quality and visibility of that quality index would prevent outbreak of foodborne illnesses, economically motivated adulteration, contamination, food wastage due to misconception of the labeled expiry dates, and losses due to spoilage, which have broad impacts on the food security [31]. In order to improve safety and prevent wastage, modern IoT-based technologies are required to monitor the food quality and increase the visibility level of the monitored data.

Blockchain has emerged as a decentralized public consensus system that maintains and records transactions of events that are immutable and cannot be falsified [32]. Blockchain technology has attracted attention beyond cryptocurrency because of its ability to provide transparent, secure, and trustworthy data in both private and public domains [33]. The technology is based on a distributed ledger, which is not owned or controlled by a single entity. Data in the public ledger is visible publicly and any authorized entities can submit a transaction, which is added to the blockchain upon validation. The advantage of blockchain technology can be applied in FSC to improve the digital data integrity, which is obtained as the product passes through different entities of the FSC. The complete food product visibility across different entities of the supply chain can become a reality with the integration of sensor-based Radio Frequency Identification (RFID) technology and blockchain-based data management systems [34]. The key benefits of applying blockchain technology in FSC are (i) real-time tracking and sensing of food products throughout the FSC and allowing identification of key bottlenecks; (ii) discouraging adulteration of food products and identifying weak links on occurrence; (iii) determining the shelf life of food products leading to reduced waste; (iv) providing end-to-end information to the consumer; and 5) allowing specific and targeted recalls.

4.3.1.2 Modified Blockchain

Implementation of the blockchain would provide an advantage to meet the transparency criteria of an FSC. However, implementing the conventional blockchain for the FSC would be disadvantageous from cost perspective. Mining is a computationally extensive procedure and a miner would be rewarded after a successful mining. If mining is introduced in the supply chain, the miner would be rewarded by the selling price from the product and it would increase the price of the product, which is not desired for a sustainable FSC.

Furthermore, the cyber data network should be immune to malicious attacks restricting unwanted transaction additions to the blockchain. There exists a fundamental difference between cryptocurrency and supply chain. In supply chain, there is the presence of a physical object that does not exist in cryptocurrency. Hence, if someone with a physical object can cryptographically prove it, then it will negate the purpose of miners. A proof-of-object (PoO)-based new consensus algorithm is proposed in this modified blockchain architecture.

1. *Proof-of-object:* In cryptocurrency, PoW means enough computation has been performed to bolster the validity of a block in the presence of other competing blocks. Similar to

PoW, PoO means any node that claims the possession of the physical object has to cryptographically prove it. Once a node claims for PoO, other participating nodes verify the authenticity of the claim. Likewise, in blockchain, the new block is added once consensus is reached about the authenticity of the block.

2. *Dual addressing:* In this architecture, dual addressing is proposed.
   - *Cyber address:* A public address in the blockchain memory, which can be viewed by any nodes in the network.
   - *RFID address:* A physical address, which is specific to a food package and is not shared publicly. Additionally, the physical and cyber addresses should be linked to each other in such a way that the cyber address can be derived from the physical address, but the inverse operation is computationally expensive.

4.3.1.3 Integrated Architecture

First, a few nomenclatures are provided, which will be followed throughout this paper. The sensor along with the RFID is termed together as a "sensorID". The sensorID can be a passive or an active type, a single sensor, or a multiple sensor type. The data collecting and processing node that scans a sensorID is termed as a "terminal". The common network shared by all the terminals is termed as "shared network". The scan of a sensorID by a terminal and enlisting the data is termed as a "transaction". Once a transaction is validated based on the consensus of participating terminals, the transaction is converted into a "block" and included in the blockchain. Apart from terminals, there exists another type of node, a "manager," that is responsible for policy making and processing requests based on consensus with other nodes. Finally, there exists a third type of node, called "agent," that requests information about a sensorID from the blockchain by providing a proper cyber address. "Address collision" is referred to the existence of a minimum of two identical cyber or physical addresses.

A typical food-based supply chain is shown in Figure 4.5. Each packaged food product with an embedded sensorID travels through multiple stages of transactions at different terminals starting from packaging through transportation, storage, and finally to a consumer for purchase. A data block is created containing the information about the package at each valid transaction. Once the transaction is verified, the transaction of the sensorID

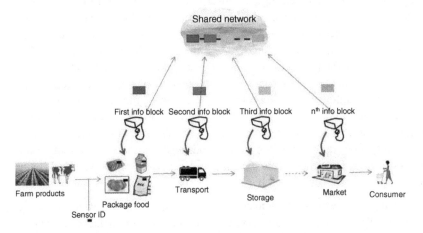

Figure 4.5 Architecture of blockchain implementation for FSC.

is converted into a block of information and appended to its pre-existing data blocks, thus forming a chain of information blocks and thus a blockchain.

An agent can obtain earlier transaction information from blockchain by using the cyber address. A typical example of an agent is a consumer, who wants to know more details about a food product before purchase. In that case, the agent can scan the sensorID and obtain the product details such as physical conditions of the package at different locations and time from the blockchain. A detailed report can be generated based on the single or multiple sensor values obtained at different times and corresponding locations. The sensor values can be further used to forecast on the lifetime of packaged products. Additionally, if any node in the supply chain fails to maintain the operating conditions, it can be readily identified. When a food package is scanned at terminal 1, the transaction is broadcast to every other terminal in the network. Upon discovery, terminal 2 sends back its acknowledgment, which is authenticated randomly by terminal 3. Once authentication is done, the transaction is accepted with a certain confidence level. Once the transaction is validated into a block, it is added to the blockchain and broadcast to all nodes in the network. The complete flowchart of the proposed algorithm is shown in Figure 4.6.

4.3.1.4 Security Analysis

Security is an important feature that needs to be addressed for the proposed architecture. The primary types of vulnerabilities are (i) tampering, (ii) spamming, (iii) privacy stealing, (iv) physical layer attack, and (v) preferential treatment. Tampering means unauthorized addition of information blocks in the blockchain. The blockchain data structure is integrated cryptographically and a 51% attack is required for tampering of existing blockchain data. Additionally, terminal to terminal communication data is cryptographically protected and hence the transaction verification stage is tamper-proof. However, tampering of acknowledged transactions is possible, which can be prevented using secure communication channel implementation among the terminals. All traffic to and from the terminals can be encrypted over the Secure Sockets Layer/Transport Layer Security protocol, as proposed in [35] for confidential IoT-based application services.

Privacy stealing is another problem in recent IoT architecture, where a malicious terminal steals information and impersonates a real terminal. In [36], different level of privacy information is categorized and respective security goals are provided. If a malicious node can infiltrate the local storage of a terminal, then it will have access to the previously stored physical addresses. As a prevention, the local data storage should be encrypted and isolated from the Web-based IoT application layer, such as Constrained Application Protocol, Hyper Text Transfer Protocol, Message Queuing Telemetry Transport [35]. Randomized spamming can be another problem in the decentralized architecture, where unwanted broadcast information is stored digitally. For example, if a fake manufacturing terminal registers a fake sensorID in the network, it will not alter the information of other legitimate sensorID information but will occupy storage memory. This problem can be solved by a certificate-based authentication where a trusted third party will issue certificates to the legit manufacturer for registering a sensorID. Physical layer information, such as sensorID's RFID address, and scanning node IP address are important information, which should be protected. Hence, extra security features at the hardware level should be implemented to prevent attacks.

Preferential treatment to some transactions over others can happen when a biased group of terminals are present in the network. However, as majority of the transaction information

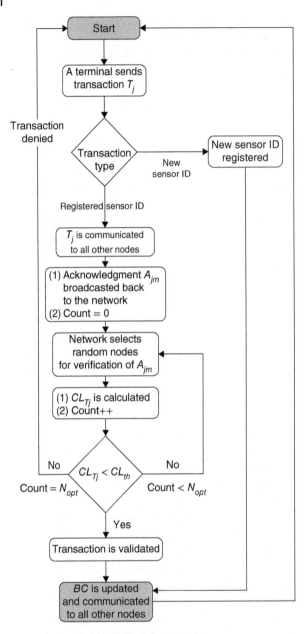

Figure 4.6 Flowchart of the proposed architecture.

are communicated cryptographically with other nodes' consensus, preferential treatment is least likely to happen in this scenario.

4.3.2 Smart City

4.3.2.1 Introduction
Smart city, which brings intelligence to various aspects of our life such as healthcare, logistic, and transportation, has been recognized as a representative IoT application.

To actualize the vision of smart cities, a huge amount of resources (e.g. sensors, actuators, and data) need to be deployed in cities to facilitate decision making and task execution. These resources help make our life increasingly intelligent and convenient, while their vulnerability to unauthorized access puts our property and safety in danger [37].

Access control has been regarded as an effective solution to prevent unauthorized resource access and has found various applications in smart cities. Each sub-system in smart cities, such as health, education, and transportation [38], has its own access control system. As a result, it is extremely difficult to implement conventional centralized and unified access control in smart cities, as the centralized access control server can easily become a bottleneck [39]. Also, the server itself turns out to be a SPF and would sabotage the whole access control system once it is destroyed by man-made or natural disasters. More importantly, centralization renders the access control system vulnerable to attacks because attackers may easily compromise the whole system by attacking the server only. As mentioned above, access control in smart cities needs to be decentralized and trustworthy (i.e. robust to attacks) to cope with the large-scale and distributed nature.

Apart from being decentralized and trustworthy, access control in smart cities also needs to be dynamic. This means that access control systems should be able to automatically adapt to the dynamically changing access context information, such as time, location, and situations. For example, normally permitted access to a smart building should be prohibited when the building is in a fire accident. Access to an electronic health record should be restricted to requests from only a certain IP address scope. If the IP address is beyond the scope, the access request must be denied. Another critical requirement of access control in smart cities is that it must be sufficiently fine-grained to achieve high flexibility and accuracy. Smart cities usually contain various objects (i.e. resources to be accessed) with different access control requirements and subjects (i.e. entities sending access request) with different attributes. Thus, fine-grained is one of the must satisfy requirements for the access control in smart cities. Thus, it is necessary to design a decentralized, trustworthy, dynamic, and fine-grained access control scheme to prevent unauthorized access to IoT resources in smart cities.

4.3.2.2 Smart Contract System

Figure 4.7 illustrates the Attribute-Based Access Control (ABAC) framework, which is composed of four smart contracts, i.e. Subject Attribute Management Contract (SAMC), Object Attribute Management Contract (OAMC), Policy Management Contract (PMC), and Access Control Contract (ACC). The SAMC, OAMC, and PMC are used to store and manage (e.g. update, add, and delete) the attributes of subjects, the attributes of objects, and the ABAC policies, respectively. The ACC performs the access control by interacting with the other three smart contracts. We introduce the details of the four smart contracts as follows.

(1) Subject attribute management contract: The SAMC is responsible for the storage and management of the subject attributes in the smart campus. The SAMC is usually deployed by the administrators of subjects on the blockchain and only the administrators have permissions to execute it. In the case of a smart campus, the administrators can be the executive office if the subjects are students and staffs, while if the subjects are IoT devices, the administrators can be their owners. To distinguish between different subjects in the system, each of them is assigned a unique identifier (i.e. ID). In addition, each ID is linked

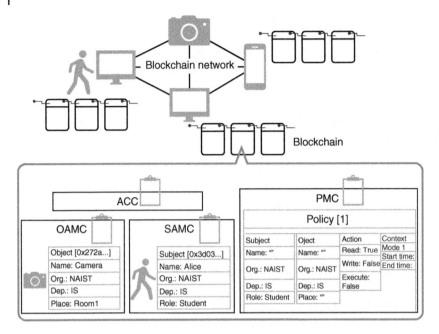

Figure 4.7 The ABAC framework.

with multiple attributes to indicate that the subject with that ID possesses those attributes. We use Ethereum account addresses as such ID information throughout this paper. In addition to the storage of the subject attributes, the Application Binary Interfaces (ABIs)[1] of `subjectAdd()` and `subjectDelete()` are also provided to add/update and delete the subject attributes, respectively.

(2) Object attribute management contract: Similar to the SAMC, the role of the OAMC is storing and managing the attributes of the objects, which can be executed only by the object administrators. Each object also has multiple attributes linked with its ID, i.e. an Ethereum account address in this paper. In addition to the storage of the object attributes, the ABIs of `objectAdd()` and `objectDelete()` are also provided to add/update and delete the object attributes, respectively.

(3) Policy management contract: The PMC is responsible for the storage and management of the ABAC policies defined in this paper. Similar to the SAMC and OAMC, only the administrators (e.g. the object owners) of the policies have permissions to execute the PMC. A policy is a combination of a set SA of subject attributes, a set OA of object attributes, a set A of actions, and a set C of context information. This combination states that the subjects with attributes in SA can perform the actions in A on the objects with attributes in OA under the context in C. For simplicity, we adopt time as the simple context information for dynamic access control in this paper. We use three parameters to represent time, which are Mode, *startTime* and *endTime*. The Mode parameter indicates whether dynamic access control is used. If the Mode is set to 0, dynamic access control is not applied. If the Mode is

---

1 A smart contract consists of variables as its states and functions called Application Binary Interfaces to view and change the states.

set to 1, dynamic access control is applied and the parameters *startTime*and *endTime* need to be further specified to indicate the start time and end time of the allowed access session. That is, access is allowed only if it is during the period between the *startTime*and *endTime*.

(4) Access control contract: The core of the access control system is the ACC, which is responsible for controlling the access requests from the subjects to the objects. To execute the ACC, a transaction that contains the required request information (e.g. subject ID, object ID, and actions) must be sent to the `accessControl()` ABI. When the ACC receives the transaction, it will retrieve the corresponding subject attributes, object attributes, and the policies from the SAMC, OAMC, and PMC, respectively. Based on the attributes and policies, the ACC decides if the subject is allowed to perform the requested actions on the object. Finally, such decision results will be returned to both the subject and object for reference.

4.3.2.3 Main Functions of the Framework

The ABAC framework provides the following main functions.

(1) Adding, updating, and deleting subject/object attributes: The basic functions provided by the proposed ABAC framework are adding, updating, and deleting the attributes of the subjects/objects. For example, to add/update the attributes of a subject, the subject administrator can send a transaction, which contains the subject's ID and the attributes to add/update, to the `subjectAdd()` ABI of the SAMC. If the ABI finds a matched entry for the presented subject ID in the subject list, it will update the attributes of the subject. Otherwise, the ABI will create a new entry for the subject ID in the subject list. When deleting some attributes of a subject, the subject administrator can send another transaction, which contains the subject's ID and attributes to delete, to the `subjectDelete()` ABI. Examples of adding, updating, and deleting the attributes of an object are quite similar to the above ones and are thus omitted here.

*(2) Searching policies:*Because the PMC stores the policies as a list (i.e. array), we need policy search to delete, update, and retrieve a certain policy. The proposed ABAC framework provides the ABIs of `findExactMatchPolicy()` and `findMatchPolicy()` to implement two patterns of policy search, respectively, i.e. search by complete match and search by partial match.

*(3) Adding, updating, and deleting policies:*A policy is defined as a logical combination of subject attributes, object attributes, actions, and contexts. In general, policy administrators can define more fine-grained policies by including more attributes, thus achieving more flexible and dynamic access control. Similar to the attribute management, this framework also provides functions for policy management, including adding, updating, and deleting policies. When adding a new policy, the policy administrator first needs to execute the policy search by partial match (i.e. `findMatchPolicy()`) to find the similar policies. When any similar policies are returned, the administrator then needs to ensure that the new policy to add does not conflict with any of the returned similar policies. Any possible conflicts must be resolved by the administrator. After the conflict resolution, the policy is finally added to the policy list by the administrator through the `policyAdd()` ABI of the PMC.

Similarly, to update a policy, the administrator also needs to apply the search by partial match to find the target policy. Note that other similar policies will also be returned by the search as well. Conflicts (if any) between the new policy used for update and the similar policies must be resolved, after which the target policy is then replaced by the new one

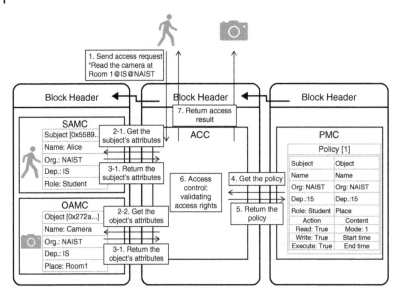

Figure 4.8 Access control process.

through the `policyUpdate()` ABI. Note that when existing policies cover the new ones to add/update, adding/updating policies is not required, which can reduce the monetary cost and the storage overhead of the framework. Different from adding/updating policies, the policy administrator needs to apply the policy search by complete match to find the target policy, when deleting a policy. If the target policy is found, the administrator then deletes it from the policy list by executing the `policyDelete()` ABI.

(4) Access control: Access control is the core function of the proposed ABAC framework. To illustrate the typical access control flow, we show in Figure 4.8 a subject Alice with attributes Role: Student, Dep: IS, and Org: NAIST (i.e. a student belonging to the IS department of NAIST) wishes to access an object camera with attributes Place: Room1, Dep: IS, and Org: NAIST (i.e. an object located in Room 1 of the IS department of NAIST).

4.3.2.4 Discussion

The throughput of the ABAC framework depends heavily on the throughput (i.e. number of transactions included in the blockchain per second) of the underlying blockchain systems. We applied Ethereum 1.0 as the underlying blockchain system in our implementation, the throughput of which is about 15 transactions per second [40]. In addition, the access request processing unit ACC in our framework needs to communicate with other contracts through messages, which further introduces latency to the access control process and thus reduces the throughput of the framework. One of the main reasons behind the low throughput is the consensus algorithm. Our implementation is based on the popular proof-of-work (PoW) algorithm, which requires a huge amount of calculations to add one block of transactions into the blockchain. Note that the proposed framework is actually independent of the underlying blockchain systems and consensus algorithms, as long as the blockchain systems support smart contract-like functionality. This means that the throughput of the proposed framework can be improved by being implemented upon faster blockchains. One promising solution is Ethereum 2.0, which changes the consensus algorithm from PoW to proof-of-stake (PoS) and adopts the technique of sharding to greatly improve the throughput performance. It is expected that Ethereum 2.0 will enable 64 to several hundred times more throughput than Ethereum 1.0.

4.4 Issues and Challenges of Blockchain in IoT

Although the convergence of blockchain and IoT brings a number of opportunities in upgrading the industry, there are many challenges to be addressed before the potentials of blockchain in IoT can be fully unleashed. In this section, we identify several major challenges in incorporating blockchain into IoT and discuss the potential solutions.

4.4.1 Resource Constraints

Most of the IoT devices are resource-constrained. For example, sensors, RFID tags, and smart meters have inferior computing capability, limited storage space, low battery power, and poor network connection capability. However, the decentralized consensus algorithms of blockchains often require extensive computing power and energy consumption. For example, PoW in BTC is shown to have high energy consumption [41]. Therefore, the consensus mechanisms with huge energy consumption may not be feasible to low-power IoT devices.

On the other hand, the bulky size of blockchain data also results in infeasibility of fully deploying blockchains across IoT. For example, the BTC blockchain size almost reaches 185 GB by the end of September 2018. It is impossible to fully store the whole blockchain at each IoT device. Meanwhile, the massive IoT data generated in nearly real-time manner makes this status quo even worse. Moreover, blockchains are mainly designed for a scenario with the stable network connection, which may not be feasible for IoT that often suffers from the poor network connection of IoT devices and the unstable network because of the failure of nodes (e.g. battery depletion).

4.4.2 Security Vulnerability

Although incorporating blockchain technologies into IoT can improve the security of IoT via the encryption and digital signature brought by blockchains, the security is still a major concern for blockchain in IoT because of the vulnerabilities of IoT systems and blockchain systems.

On the one hand, there is a growing trend in deploying wireless networks into industrial environment because of the feasibility and scalability of wireless communication systems. However, the open wireless medium also makes IoT suffer from the security breaches, such as passive eavesdropping [42], jamming, and replaying attacks [43]. Moreover, because of the resource constraints of IoT devices, conventional heavy-weighted encryption algorithms may not be feasible to IoT [44]. In addition, it is also challenging to manage the keys (which are crucial to encryption algorithms) in distributed environment.

4.4.3 Privacy Leakage

Blockchain technologies have some mechanisms to preserve a certain data privacy of transaction records saved in blockchains. For example, transactions are made in Bitcoin (BTC) via IP addresses instead of users' real identities, thereby ensuring a certain anonymity. Moreover, one-time accounts are generated in BTC to achieve the anonymity of users. However, these protection schemes are not robust enough. For example, it is shown in [45] that user pseudonyms can be cracked via learning and inferring the multiple transactions associated with one common user. In addition, the full storage of transaction data on blockchain can also lead to the potential privacy leakage as indicated in [46].

4.4.4 Incentive Mechanism

An appropriate incentive mechanism is a benign stimulus to blockchain systems. For example, a number of BTCs will be rewarded to a miner who first solves the computationally difficult task. Meanwhile, a transaction in Ethereum will be charged with a given fee (i.e. gas) to pay the miners for the execution of contracts. Therefore, there are two issues in designing incentive mechanisms in blockchains: (i) the reward for proving (or mining) a block and (ii) the compensation for processing a transaction (or a contract).

However, it is challenging to design a proper incentive mechanism for BCoT to fulfill the requirements of different applications. Take digital currency platforms as an example, where miners are keen on the price of digital currency. For instance, the BTC reward for a generated block will be halved every 210 000 blocks [47]. The reward decrement will discourage miners to contribute to the solution of the puzzle consequently migrating to other blockchain platforms. How to design a proper rewarding and publishing mechanism of digital currency is necessary to ensure the stability of blockchain systems.

4.5 Conclusion

This section introduces the challenges and issues that exist in the IoT, and for this purpose, the advantages of blockchain to deal with these challenges are produced. Next, in order to better integrate the blockchain into the IoT environment, this section introduces the advantages and limitations of several mainstream consensus mechanisms of blockchain in IoT applications. Based on this, this section also introduces several classic scenarios of blockchain applications in the IoT, such as supply chains, smart cities, and e-health. Finally, some remaining problems are discussed as the open issues.

References

1 L. Zhang, H. Xu, O. Onireti, M. A. Imran, and B. Cao. How much communication resource is needed to run a wireless blockchain network? https://arxiv.org/abs/2101.10852, 2021.

2 B. Cao, Y. Li, L. Zhang, S. Mumtaz, Z. Zhou, and M. Peng. When internet of things meets blockchain: challenges in distributed consensus. *IEEE Network*, 33 (6): 133–139, 2019. https://doi.org/10.1109/MNET.2019.1900002.

3 S. Nakamoto. Bitcoin: a peer-to-peer electronic cash system. *Cryptography Mailing list at* https://bitcoin.org/bitcoin.pdf, 03 2009.

4 I-SCOOP. Blockchain and the internet of things: the IOT blockchain opportunity and challenge. https://www.iscoop.eu/blockchain-distributed-ledger-technology/blockchain-iot, 2018.

5 K. Christidis and M. Devetsikiotis. Blockchains and smart contracts for the internet of things. *IEEE Access*, 4: 2292–2303, 2016. https://doi.org/10.1109/ACCESS.2016.2566339.

6 X. Liang, J. Zhao, S. Shetty, J. Liu, and D. Li. Integrating blockchain for data sharing and collaboration in mobile healthcare applications. pages 1–5, 2017. https://doi.org/10 .1109/PIMRC.2017.8292361.

7 A. Dorri, M. Steger, S. S. Kanhere, and R. Jurdak. Blockchain: a distributed solution to automotive security and privacy. *IEEE Communications Magazine*, 55 (12): 119–125, 2017. https://doi.org/10.1109/MCOM.2017.1700879.

8 J. Kang, R. Yu, X. Huang, S. Maharjan, Y. Zhang, and E. Hossain. Enabling localized peer-to-peer electricity trading among plug-in hybrid electric vehicles using consortium blockchains. *IEEE Transactions on Industrial Informatics*, 13 (6): 3154–3164, 2017. https://doi.org/10.1109/TII.2017.2709784.

9 Z. Li, J. Kang, R. Yu, D. Ye, Q. Deng, and Y. Zhang. Consortium blockchain for secure energy trading in industrial internet of things. *IEEE Transactions on Industrial Informatics*, 14 (8): 3690–3700, 2018. https://doi.org/10.1109/TII.2017.2786307.

10 K. Kotobi and S. G. Bilen. Secure blockchains for dynamic spectrum access: a decentralized database in moving cognitive radio networks enhances security and user access. *IEEE Vehicular Technology Magazine*, 13 (1): 32–39, 2018. https://doi.org/10.1109/MVT .2017.2740458.

11 B. Cao, Z. Zhang, D. Feng, S. Zhang, L. Zhang, M. Peng, and Y. Li. Performance analysis and comparison of PoW, PoS and DAG based blockchains. *Digital Communications and Networks*, 6 (4): 480–485, 2020.

12 M. Conti, E. S. Kumar, C. Lal, and S. Ruj. A survey on security and privacy issues of bitcoin. *IEEE Communication Surveys and Tutorials*, 20 (4): 3416–3452, 2018. https://doi.org/10.1109/COMST.2018.2842460.

13 J. (JD) Douceur. The sybil attack. In *Proceedings of 1st International Workshop on Peer-to-Peer Systems (IPTPS)*, January 2002. https://www.microsoft.com/en-us/research/publication/the-sybil-attack/.

14 A. Back. Hashcash - a denial of service counter-measure. 09 2002.

15 G. BitFury. Proof of stake versus proof of work. *White paper, Sep.*, 2015.

16 W. Li, C. Feng, L. Zhang, H. Xu, B. Cao, and M. A. Imran. A scalable multi-layer PBFT consensus for blockchain. *IEEE Transactions on Parallel and Distributed Systems*, 32 (5): 1146–1160, 2021. https://doi.org/10.1109/TPDS.2020.3042392.

17 L. Lamport. The part-time parliament. *ACM Transactions on Computer Systems*, 16: 133–169, 1998. https://doi.org/10.1145/279227.279229.

18 D. Ongaro and J. Ousterhout. In search of an understandable consensus algorithm. *USENIX*, pages 305–320, 2014.

19 M. Castro and B. Liskov. Practical byzantine fault tolerance. *OSDI*, page 173–186, 1999.

20 J. P. Morgan and O. Wyman. Unlocking economic advantage with blockchain. *A guide for Asset Managers*, 481: 482, 2016.

21 H. Sukhwani, J. M. Martínez, X. Chang, K. S. Trivedi, and A. Rindos. Performance modeling of PBFT consensus process for permissioned blockchain network (Hyperledger Fabric). In *2017 IEEE 36th Symposium on Reliable Distributed Systems (SRDS)*, pages 253–255, 2017. https://doi.org/10.1109/SRDS.2017.36.

22 Y. Li, B. Cao, M. Peng, L. Zhang, L. Zhang, D. Feng, and J. Yu. Direct acyclic graph-based ledger for internet of things: performance and security analysis. *IEEE/ACM Transactions on Networking*, 28 (4): 1643–1656, 2020.

23 S. Popov. The tangle. *cit. on*, page 131, 2016.

24 L. Baird. The Swirlds hashgraph consensus algorithm: fair, fast, byzantine fault tolerance. *Swirlds, Inc. Technical Report SWIRLDS-TR-2016*, 1, 2016.

25 G. Zyskind, O. Nathan, et al. Decentralizing privacy: using blockchain to protect personal data. In *2015 IEEE Security and Privacy Workshops*, pages 180–184. IEEE, 2015.

26 S. Mondal, K. P. Wijewardena, S. Karuppuswami, N. Kriti, D. Kumar, and P. Chahal. Blockchain inspired RFID-based information architecture for food supply chain. *IEEE Internet of Things Journal*, 6 (3): 5803–5813, 2019.

27 Y. Zhang, M. Yutaka, M. Sasabe, and S. Kasahara. Attribute-based access control for smart cities: a smart contract-driven framework. *IEEE Internet of Things Journal*, 8 (8): 6372–6384, 2020.

28 H. Xu, L. Zhang, O. Onireti, Y. Fang, W. B. Buchanan, and M. A. Imran. BeepTrace: blockchain-enabled privacy-preserving contact tracing for COVID-19 pandemic and beyond. *arXiv preprint arXiv:2005.10103*, 2020.

29 W. Liu, B. Cao, L. Zhang, M. Peng, and M. Daneshmand. A distributed game theoretic approach for blockchain-based offloading strategy. In *ICC 2020-2020 IEEE International Conference on Communications (ICC)*, pages 1–6. IEEE, 2020.

30 D. I. Ellis, V. L. Brewster, W. B. Dunn, J. W. Allwood, A. P. Golovanov, and R. Goodacre. Fingerprinting food: current technologies for the detection of food adulteration and contamination. *Chemical Society Reviews*, 41 (17): 5706–5727, 2012.

31 S. Herschdoerfer. *Quality Control in the Food Industry*, volume 2. Elsevier, 2012.

32 S. Apte and N. Petrovsky. Will blockchain technology revolutionize excipient supply chain management? *Journal of Excipients and Food Chemicals*, 7 (3): 910, 2016.

33 S. Underwood. Blockchain beyond bitcoin, 2016.

34 Z. Pang, Q. Chen, W. Han, and L. Zheng. Value-centric design of the internet-of-things solution for food supply chain: value creation, sensor portfolio and information fusion. *Information Systems Frontiers*, 17 (2): 289–319, 2015.

35 M. Ammar, G. Russello, and B. Crispo. Internet of things: a survey on the security of IoT frameworks. *Journal of Information Security and Applications*, 38: 8–27, 2018.

36 X. Lu, Q. Li, Z. Qu, and P. Hui. Privacy information security classification study in internet of things. In *2014 International Conference on Identification, Information and Knowledge in the Internet of Things*, pages 162–165. IEEE, 2014.

37 F. A. Alaba, M. Othman, I. A. T. Hashem, and F. Alotaibi. Internet of things security: a survey. *Journal of Network and Computer Applications*, 88: 10–28, 2017.

38 Y. Zhang, Y. Li, R. Wang, M. S. Hossain, and H. Lu. Multi-aspect aware session-based recommendation for intelligent transportation services. *IEEE Transactions on Intelligent Transportation Systems*, 1–10, 2020.

39 M. Li, X. Sun, H. Wang, Y. Zhang, and J. Zhang. Privacy-aware access control with trust management in web service. *World Wide Web*, 14 (4): 407–430, 2011.

40 V. Buterin. On sharding blockchains. *Sharding FAQ*, 2017.

41 A. Reyna, C. Martín, J. Chen, E. Soler, and M. Díaz. On blockchain and its integration with IoT. Challenges and opportunities. *Future Generation Computer Systems*, 88: 173–190, 2018.

42 X. Li, H. Wang, H.-N. Dai, Y. Wang, and Q. Zhao. An analytical study on eavesdropping attacks in wireless nets of things. *Mobile Information Systems*, 2016: 1–10, 2016.

43 J. Lin, W. Yu, N. Zhang, X. Yang, H. Zhang, and W. Zhao. A survey on internet of things: architecture, enabling technologies, security and privacy, and applications. *IEEE Internet of Things Journal*, 4 (5): 1125–1142, 2017.

44 Y. Yang, L. Wu, G. Yin, L. Li, and H. Zhao. A survey on security and privacy issues in internet-of-things. *IEEE Internet of Things Journal*, 4 (5): 1250–1258, 2017.

45 M. Conti, E. S. Kumar, C. Lal, and S. Ruj. A survey on security and privacy issues of bitcoin. *IEEE Communication Surveys and Tutorials*, 20 (4): 3416–3452, 2018.

46 A. Dorri, S. S. Kanhere, and R. Jurdak. MOF-BC: a memory optimized and flexible blockchain for large scale networks. *Future Generation Computer Systems*, 92: 357–373, 2019.

47 K. Saito and M. Iwamura. How to make a digital currency on a blockchain stable. *Future Generation Computer Systems*, 100: 58–69, 2019.

5

Hyperledger Blockchain-Based Distributed Marketplaces for 5G Networks

Nima Afraz[1,3], Marco Ruffini[1], and Hamed Ahmadi[2]

[1]*CONNECT Center, Trinity College Dublin, 2, Ireland*
[2]*Department of Electronic Engineering, University of York, York YO10 5DD, UK*
[3]*School of Computer Science, University College Dublin, D04V1W8 Dublin, Ireland*

5.1 Introduction

Widespread deployment of the fifth generation (5G) networks will require the network operators to assure new revenue streams to compensate for the capital expenditure incurred by provisioning the new infrastructure. New services will require novel business models as they are unlikely to fit within current network ownership models, where an over-the-top (OTT) or a vertical industry has to undergo manual negotiations to acquire network resources to deliver services to its customers. Therefore, automated business processes become vital as they can facilitate the utilization of the network infrastructure for new services as they appear.

A wide range of industries have already adopted blockchain technology to automate complex business processes and workflows [1, 2]. Blockchain technology helps these enterprises to move away from the Business Process Management (BPM) models where a third-party organization stores the business information in a central repository and controls the transactions in cross-industry environments. This way, they both avoid the single point of failure and allow enterprises to gain control of their data.

As seen earlier in this book, the main innovation of blockchain technology in the context of cryptocurrencies is preventing double-spending while offering a distributed alternative to the costly real-world central bank system, which provides the required mechanisms to avoid double-spending. In short, blockchain technology provides a robust solution for trustable book-keeping. Blockchain technology is being considered as the primary trust solution when a trustworthy central bookkeeper is absent. Example applications are government and private management, electronic voting, authorship, ownership, etc. [3], while more are currently under study, such as applications in pharmaceutical supply chain [4], consumer electronics [5], smart cities [6], privacy protection in health care [7], and network security [8].

It was later realized that the cryptocurrency applications do not fully utilize the potential of blockchain technology. In other words, the same decentralized consensus mechanism could be used to maintain the same level of trust for the logic-enabled transaction rather than simple book-keeping. Therefore, the idea of blockchain was further complemented by the concept of smart contracts, which introduced blockchain technology not only as a robust book-keeping tool but also as an effective automation platform that can address many trust-related concerns in multi-party business ecosystems. A smart contract is an immutable piece of logic (computer program) that enhances the distributed ledger technology with self-enforcing pre-negotiated agreements. Therefore, a smart contract can automate the enforcement of business processes without relying on a central authority.

Telecommunication ecosystems are not an exception in that the operators and other enterprises involved in the ecosystem use many complex business processes including contracts that are often manually negotiated and enforced. One example is the multi-tenant multi-service access network where a number of Infrastructure Providerss (InPs), Virtual Network Operator (VNO), and service providers have to steer through complex business processes to serve the end-users. In this chapter, we are particularly interested in showing how to automate the process of dedicating a set of network resources to a (VNO) on-demand for a given period of time. Furthermore, it gives the ability to the Virtual Network Operators (VNOs) to trade their excess resources with others. More specifically, we are interested in studying the implementation of bilateral resource trading markets using blockchain as the data storage structure solution and smart contracts as the tool for the execution of the market mechanisms and inter-carrier financial settlement. The examples of such bilateral trade markets in telecommunications industry include resource allocation in Network Function Virtualization (NFV) markets [9], promoting femtocell access [10], mobile crowdsensing with budget constraint [11], spectrum allocation [12], and multi-tenant Passive optical networkss (PONs) [13]. The common challenge among these resource markets is the fact that they rely on a third party to both store the private financial information of the participants and execute the auction mechanisms that directly determine the allocation of the resources. However, in real business ecosystems, often no central entity has the full trust of all the partners because of competition. This motivated us to explore an alternative scenario where no central entity can be trusted to provide fair and impartial infrastructure management. We argue that the blockchain technology can be exploited to hold the market in a distributed fashion as an alternative to centralized control. To analyze the feasibility of such a distributed market, we developed smart contracts that implement an auction algorithm capable of allocating the resources to the participants without a central trusted market mediator. We use the open-source framework Hyperledger Fabric to develop the blockchain application. The nodes of the blockchain network are then distributed across multiple cloud-hosted virtual machine instances to allow more realistic and precise experimentation. We use common metrics such as transaction latency and throughput to evaluate the performance of the designed marketplace application. Furthermore, we study the computing resources required to run the blockchain application.

5.2 Marketplaces in Telecommunications

A market emerges when a scarce resource (commodity or service) has to be allocated under competitive conditions. Throughout the history of telecommunications as the demand for access to the network heightened, the resources constructing the network infrastructure have become more valuable either because they are naturally scarce (e.g. electromagnetic spectrum) or costly to produce/deploy (e.g. laying optical fiber and deploying wireless base stations). This leads to the marketization of multiple resource allocation processes that we will briefly review in the remaining of this section.

5.2.1 Wireless Spectrum Allocation

As a natural resource, the electromagnetic spectrum is scarce. As the demand for this resource increased, various methods were used to allocate chunks of the spectrum. Initially, simpler solutions such as *lotteries* [14] or *beauty contests* [15] that relied solely on chance or the merits of the candidates were used. Soon, it was realized that such simplistic approaches make way for manipulative and inefficient use of the spectrum. Therefore, economists formulated market mechanisms such as auctions and other mechanisms that root into the game theory to efficiently allocate radio spectrum while preventing malicious players from manipulating the market. The economists P. Milgrom and R. Wilson had won the 2020 Nobel prize for:

> " … their insights into design new auction formats for goods and services that are difficult to sell in a traditional way, such as radio frequencies." [16]

Spectrum auctions (conducted by government) allocate the electromagnetic spectrum that is divided into radio frequency bands to the candidates based on their submitted bids. Recently, in some countries, the regulator allowed the owners (successful bidders) of the spectrum bands to dynamically reallocate the spectrum when it is not used. This reallocation is normally performed using auction mechanisms too.

5.2.2 Network Slicing

What differentiates 5G from its predecessor generations of cellular communications is that it goes beyond merely multiplying the network capacity and speed and promises an ambitious vision where various services with vastly different functional requirements are seamlessly hosted over the same physical infrastructure without affecting each others' performance [17]. This vision, in addition to many technical and standardization challenges that need to be addressed, demands a new approach to business and ownership models of network infrastructure [18], where automated resource orchestration and provisioning mechanisms handle on-demand resource requirements of the service providers. The most prominent model of resource allocation for 5G networks is slicing, which allows building customized logical networks on top of a shared physical infrastructure [19]. Thanks to the virtualization technologies, it is now possible to allocate virtual instances of the physical infrastructure while assuring seamless functionality using slice isolation techniques. As envisioned by third-Generation Partnership Project (3GPP) [20] in a highly heterogeneous network sharing ecosystem, network slices could be created to accommodate different functional and performance requirements of the network operators. This vision could see the emergence of a new market where a wide range of network operators, infrastructure providers, or public authorities carry out highly frequent transactions involving the exchange of resources, financial commitments, and post-deal operations. Network slicing provides a solution to the diverse infrastructure/resource requirements of modern telecommunication networks. This is done through generating on-demand virtual instances of an end-to-end network on a physical infrastructure [21]. This enables service providers to serve their end-users with the utmost flexibility. In [22], the authors have reviewed the business requirements and standards in the context of multi-tenant mobile networks. They have introduced in detail the architecture of the 5G network slice broker. The idea of the on-demand capacity broker is to enable allocating a portion of the network capacity for a specific time slot to a secondary resource user (Mobile Virtual Network Operator (MVNO) or (OTT) provider). A distributed slice brokering based on the model introduced in this chapter was proposed in [23].

5.2.3 Passive optical networks (PON) Sharing

Conventional telecommunication infrastructure ownership models are being challenged as new market players are rising through the 5G evolution. This evolution involves the need for higher capacity; therefore, higher network infrastructure investment and, in particular, the access network that provides the last-mile connectivity to the end-users [18]. In the fixed access network domain, PONs are at the core of this ownership evolution, as Passive optical networks (PON) sharing (across services and tenants) is a main enabler of high-density, high-capacity data transport in 5G networks [24]. PONs are fiber-optical telecommunication access network solutions that enable sharing of costly fiber and network end points across several users [25]. This is enhanced by the passive nature of their Optical Distribution Network (ODN) network, which does not require any active component in the (ODN) and their wide coverage (typically 20 km but over 100 km in long-reach PON [26]). PONs are one of the most widely deployed access solutions that traditionally provide broadband access using Fiber-to-the-Home (FTTH) and fiber-to-the-curb architectures.

The ideal situation for network sharing is an open-access model, where multiple competing VNOs share a network owned by an independent third party (left-hand side of Figure 5.1).

In a highly dynamic resource-sharing scenario, VNOs and InPs need to exchange network capacity using automatic auctioning mechanisms. The Infrastructure Providers (InP)

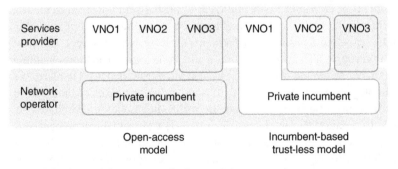

Figure 5.1 Access infrastructure sharing models.

can act as an auctioneer, while the VNOs can buy/sell capacity on-demand to maintain the capacity and latency performance required for some of their services. These conventional ownership models would rely on a central trusted authority (the InP) to invest in deployment, oversee, and regulate the operations and provide revenue assurance. Today, however, often the InP is a private entity (typically the incumbent operator) that is also an operator using the same shared infrastructure to serve its own customers (shown in the right-hand side of Figure 5.1). In this more typical incumbent-based model, because the InP is both auctioneer and (VNO) (thus, it is not an independent third party), the other VNOs cannot trust it to operate the market (i.e. the resource redistribution mechanism).

5.2.4 Enterprise Blockchain: Hyperledger Fabric

Blockchain technology is most popular as the distributed ledger technology behind Bitcoin and other cryptocurrencies. However, blockchain has been progressively re-invented as more and more industries have shown interest in its potential for enabling novel business models. The initial blockchains, including Bitcoin's, came as public blockchains, indicating that the reading access to the ledger is not limited to any particular group as the privacy of the users is protected by the pseudo-anonymous nature of the network. Although public blockchains are suitable for applications such as cryptocurrencies, in enterprise ecosystems, this could become an issue because of the confidential essence of the information. Therefore, private blockchains were introduced to preserve the privacy of the participants. Quite similarly, the difference between permissionless and permissioned blockchains relies on the distinct way they control the contribution of the participants (writing access) to the ledger. In private and permissioned blockchains, a form of Membership Service Provider MSP is used to control and authorize access to the blockchain. Hyperledger Fabric [27] is the most prominent permissioned open-source blockchain platform designed for enterprise ecosystems, and it is maintained by the Linux Foundation. The major components of a blockchain application/network are introduced in the rest of this section with a focus on the Hyperledger Fabric framework, which we use for the experiments throughout this chapter.

A number of purpose-specific frameworks and tools are being developed under Hyperledger's umbrella for the use cases ranging from finance and banking to the Internet of Things, supply chain management, and manufacturing. Hyperledger Fabric is one of the frameworks that provides a permissioned distributed ledger technology for cross-industry applications, i.e. where only specific entities are allowed to participate. The main features of the Hyperledger Fabric platform are as follows:

1. It has a modular architecture that allows plug-and-play implementations of different functions/components such as consensus, membership service, etc.
2. It uses open-source container technologies (i.e. Docker) to host different components of the blockchain network.
3. It makes use of permissioned membership management and access to the ledger.
4. It allows for chaincode (smart contracts) to be written in various programming languages (e.g. Go, JavaScript, or Java), while other blockchain platforms only allow specific languages (e.g. Solidity in Ethereum).

A blockchain application is composed of multiple nodes shown in Figure 5.2 with various roles, communicating with each other to reach an agreement on a final state of the ledger.

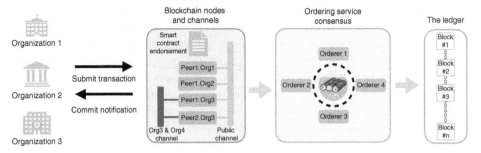

Figure 5.2 Hyperledger fabric architecture.

The major concepts associated with a Hyperledger Fabric network are introduced in the following paragraphs.

5.2.4.1 Shared Ledger

The biggest innovation of blockchain technology is the distributed record-keeping capability that it provides. In the core of every blockchain, there is an append-only shared ledger. This ledger is distributed on multiple nodes over the network and is kept in synchronization using the consensus protocol. The novelty of blockchain's record-keeping process is in the method data is stored as a series of blocks chained to each other using cryptographic hashes. A fabric network might contain multiple ledgers.

5.2.4.2 Organizations

The blockchain network consists of one or multiple organizations who are contributing resources to the network while being able to process their transactions with other participants. Organizations host peers and other components of the network and each maintain a copy of the ledger(s).

5.2.4.3 Consensus Protocol

The heart of the ledger appending process is the consensus protocol, where the blockchain nodes come to an agreement on whether or not to add a block proposed by one of the parties to the ledger. Early blockchains used proof of work (PoW)-based consensus that is proven to be extremely resource consuming. Therefore, numerous new consensus protocols have been proposed. While their description falls outside the scope of this chapter, we outline the Raft protocol [28] used in the experiments of this chapter. Raft solves the problem of achieving an agreement between multiple participants on a shared state even in the face of failures if the majority of the nodes are still up. In Raft [28], the participating nodes choose a leader with a certain term during which the other nodes are the followers of the leader. During its term, the leader continuously sends a heartbeat to its followers to maintain authority. An election is triggered when one of the followers faces timeout waiting for a heartbeat from the leader. Figure 5.3 depicts the different states of the nodes during the elections in the Raft protocol.

5.2.4.4 Network Peers

The network peers (nodes) have different responsibilities depending on their role and the blockchain framework. In (PoW)-based blockchains, these nodes are typically divided into

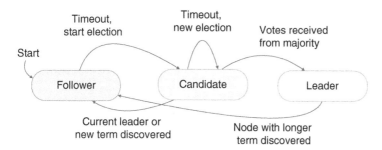

Figure 5.3 Leader election in raft consensus.

miner nodes and validation nodes, with the former taking part in solving the (PoW) hash problem on top of its validation role. In Hyperledger Fabric, the main types of peers are endorser peers (which receive the transaction proposal, run the smart contract, and endorse the transaction) and the orderers that are tasked with ordering the transaction and reaching the consensus on the final block.

5.2.4.5 Smart Contracts (chaincodes)
The logic of the blockchain operation is implemented as a smart contract (or chaincode in Hyperledger Fabric). This is a piece of code that either defines the terms of the transaction or an enforceable function depending on the outcome of a transaction (e.g. a penalty for not meeting a certain clause in the contract).

5.2.4.6 Channels
Channels in Hyperledger Fabric allow the participant organizations in the network to have a virtual blockchain network within the broader blockchain network without needing to replicate the nodes (e.g. hardware resources). This enables further privacy measures for more complex transactions and logic in the blockchain. In addition, Hyperledger Fabric allows multiple ledgers per channel.

5.3 Distributed Resource Sharing Market

The traditional roles of InP and VNOs are being challenged with new market players such as over-the-tops (OTTs) and vertical market services (e.g. automotive, e-health, etc.) that are considered to be the major revenue generation sources for the future 5G networks. These are typical scenarios where network investment and 5G deployment would be very costly or unattractive for legacy operators; instead, verticals expect significant advantage (i.e. there is a large private value for specific applications that require 5G type of connectivity).

Moving from conventional static sharing toward on-demand/on-the-fly dynamic multi-tenancy [22] requires a network sharing management architecture that enables capacity brokering. To understand the importance of automating bilateral market business processes, it is necessary to know how the current process works. A bilateral trade market is a business environment where multiple traders in both seller and buyer roles can exchange commodities (e.g. network resources). A typical bilateral market trade can involve the following:

- The manual negotiation of the terms of trade between VNOs. This includes price setting, which, if happens manually, will not allow dynamic high-frequency trading of the resources.
- Different interpretations of the negotiated terms. In such a case, a third-party authority (e.g. regulator, InP, etc.) could be summoned to solve the dispute; however, this implies additional delays and costs for VNOs.
- Lack of trust among VNOs and the absence of a trusted central authority holding the market by enforcing the terms.

These issues might lead to VNOs having no incentive to participate in a dynamic resource trading market.

Many resource/infrastructure sharing problems in the communications sector are modeled as bilateral trade markets as these markets are capable of supporting multiple participants on both sides of the market. The majority relies on solutions based on game theory, in order to match supply and demand [9–13]. One of the most prominent solutions is the double auction, which focuses on allocating commodities (i.e. resources) to the participants with the highest demand. The end goal is typically to achieve the highest social welfare (i.e. maximizing the aggregate of all participants' utilities) in the market.

In [13], we proposed a double-auction mechanism originally designed to incentivize inter-operator resource sharing in multi-tenant PONs. The auction mechanism is capable of providing an allocation scheme for the resources while assuring trust among the participants (i.e. providing positive incentives to avoid manipulative market behavior). However, we made the assumption that this market model depends solely on a central third-party authority (the InP), which is trusted by all of the market participants. This central authority is thus in charge of both record-keeping of the market data and conducting the auction on behalf of all participants.

Considering, however, that it is quite typical for an InP that shares its network to also offer services to customers, in competition with other VNOs, it is unrealistic to assume that VNOs will trust the InP. Being in competition with other VNOs, the InP could benefit from manipulating the market data or the process of the auction mechanism.

In this section, we show how a distributed model for bilateral trade markets can eliminate the reliance on an impartial central authority. This is achieved making use of the two following features of blockchain technology.

- Distributed record-keeping of all the transactions and participants' data using the distributed ledger technology.
- Conducting the auction in a distributed fashion rather than centralized, enabled by the smart contract technology.

A high-level view of the proposed distributed market model is illustrated in Figure 5.4. The application sends the transaction proposal to the orderer to be broadcast to the peers in the channel. These peers are distributed across the VNOs' servers and are all part of the blockchain network. A **transaction** in the context of this market model is the process of receiving the bids/asks from the traders and conducting the double auction, matching

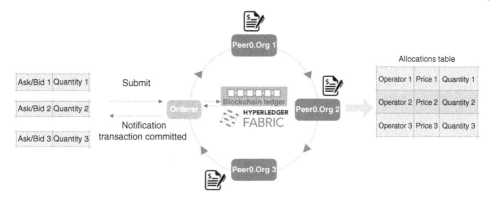

Figure 5.4 Distributed market model. Source: The Linux Foundation.

eligible sellers and buyers, and issuing the results of the resource allocation. The peers proceed with the endorsement of the transactions (i.e. the auction) based on the predefined endorsement policy. The endorsed transaction is then returned to the application and sent to the orderer to finalize the ordering of the transactions into a block (using one of the consensus protocols available, e.g. Solo, Kafka, and Raft).

Distributed markets have been previously studied in contexts such as advertisement marketplaces [29, 30]. In [31], the authors propose a decentralized uniform-price double auction for the real-time energy market. Their solution is implemented using an Ethereum blockchain. They evaluate their model using metrics such as efficiency and the overall blockchain overhead cost.

In Section 5.3.1, the details of the double-auction mechanism are introduced.

5.3.1 Market Mechanism (Auction)

The proposed market mechanism is a sealed-bid homogeneous item double auction that determined the winners and sets the trading prices. In a sealed-bid auction, all traders simultaneously submit sealed asks/bids to the auctioneer, and no trader is aware of the ask/bid of any other participant.

5.3.2 Preliminaries

The market consists of a set of M sellers $\mathbb{S} = \{s_1, s_2, \ldots, s_i\}$, $i \in M$ and a set of N buyers $\mathbb{B} = \{b_1, b_2, \ldots, b_j\}$, $j \in N$ and one auctioneer. This constructs a two-sided market in which a number of traders are competing to trade identical items in a way that it will maximize their payoff. There is also an auction maker (broker) present that is responsible for operating the auction. In our market, the InP plays the role of the auctioneer. In the distributed implementation of this market mechanism, the smart contract replaces the central role of the broker. Once the auction is initiated, each seller announces the quantity of the items offered q_i^S along with the per-item ask value v_i^S to the auctioneer. Simultaneously, the buyers will send their pair of the number of items required q_j^B and the bid value v_j^B to the auctioneer. The ask and bid values (v_i^S, v_j^B) are $\in [0, 1]$. The auction mechanism is common knowledge.

The mechanism provides a matching service to multiple buyers and sellers in a bilateral trade environment. We assume rational traders whose effort is focused on maximizing their payoff by trading the identical items. A market maker, or the auctioneer, is responsible for operating this market and conducting the auctions while having no or incomplete information about the true valuations of the traders. There is a finite number of alternative resource allocation combinations. Each combination would bring a different individual and social payoff for each trader and the entire market. We chose a sealed-bid auction because of the latency constraints in telecommunications networks we cannot afford multiple rounds of communication among the sellers, auctioneer, and the buyers. Therefore, we have to minimize these communications. Using the sealed-bid version of auction helps us to eliminate the need for any additional round of communication among the agents as the traders only send the ask/bid values once along with their Bandwidth Map (BMap) (the suggested allocation schedule). The algorithmic representation of the proposed double-auction mechanism is given in Algorithm 5.1 [13].

Algorithm 5.1: Multi-item double auction. Source: Afraz and Ruffini [13] / with permission of IEEE.

---

1 Sort sellers ascending based on their bid values so $v_1^B > v_2^B > \cdots > v_m^B$

2 Sort buyers descending based on their ask values so $v_1^S < v_2^S < \cdots < v_n^S$

3 Find $max(S_L, B_K) \; \forall \; v_L < v_K$ **and** $\sum_1^K q_j^B \leq \sum_1^L q_i^S$

4 $\gamma = \frac{1}{2} \times (v_{L+1} + v_{K+1})$

5 **if** $\gamma \in [v_L, v_K]$ **then**

6 $\qquad \Theta_{Pr} = min(\sum_1^{i=L} q_i, \sum_1^{j=K} q_j)$

7 $\qquad p^B = p^S = \gamma$

8 **else if** $\gamma \notin [v_L, v_K]$ **then**

9 $\qquad \Theta_{Pr} = min(\sum_1^{i=L-1} q_i, \sum_1^{j=K-1} q_j)$

10 $\qquad p^B = v_K$

11 $\qquad p^S = v_L$

We assume that there are no restrictions on the sets of buyers and sellers that may trade with one another nor any preferences over trade between any of the traders.

5.4 Experimental Design and Results

In this section, we provide the information regarding the design of our experimental environment to study the performance of the distributed marketplace application. We introduce our evaluation methodology along with the tools used for evaluating the performance. Finally, we report the results of our experiments.

5.4.1 Experimental Blockchain Deployment

An enterprise blockchain application enables several business partners to make collaborative decisions using smart contracts and keep secure logs of the transaction on a distributed ledger. This is done through the partner organizations contributing infrastructure to form the blockchain network. Because the main aim of blockchain is to decentralize the process of decision making and record-keeping, the most likely network deployment scenario is a distributed cloud environment. In other words, the participating organizations will dedicate a particular amount of resources to host the blockchain application components, including the peers, orderers, and Certification Authoritys (CAs). However, in the academic literature, the distributed nature of blockchain networks is often overlooked. The majority of studies related to the applications of blockchain technology in communications carry out experiments with limited resources, where the entire blockchain network (and the components) is hosted in a single machine or often on multiple virtual machines (VMs) in a server. This implementation, however, is far from a realistic production scenario as it will not feature parameters such as the network propagation, which could become a bottleneck for highly frequent transactions carried out on the blockchain.

Here, we describe a step beyond merely developing proof-of-concept blockchain solutions and instead focus on pragmatic experiment design and deployment that reflects the realistic capabilities of the proposed blockchain-based solutions. Therefore, we have exploited a range of enterprise-grade software solutions and cloud infrastructure to design the experiments that are briefly introduced in the remainder of this section.

5.4.1.1 Cloud Infrastructure

The leading cloud providers are competing to gain a bigger share of the future cloud market for blockchain applications, with a handful of them already offering Blockchain as a Service (BaaS) and being actively involved in the open-source blockchain communities [32]. We deploy our blockchain solution using multiple VMs on the Google Cloud Computing Engine. These VMs are collocated at the same geographical region/zone (us-central1-a). The experiments are conducted by isolating one virtual machine (VM) for the monitoring and benchmarking purposes and the other VMs hosting the participating organizations' blockchain components.

5.4.1.2 Container Orchestration: Docker Swarm

Hyperledger Fabric utilizes Docker containers to host the blockchain components. Therefore, a blockchain application implemented using Hyperledger Fabric often comprised several containers per organization. This makes manual initiation and the management of the containers very difficult. To automate this process, we use docker-compose scripts along with the Docker Swarm technology to orchestrate the containers during the experiments. This allows us to seamlessly create numerous containers using pre-defined images and deploy them across the Swarm overlay network.

5.4.2 Blockchain Performance Evaluation

A blockchain application operates over an underlying network of different components. The blockchain application handles the transactions submitted by the participating clients and proceeds to the verification and ordering process, throughout which, a block of transactions is generated and the transaction outcome is written on the distributed ledger. In the context of Hyperledger Fabric blockchain framework, the performance of the blockchain application is closely tied to the performance of each component (e.g. peers, orderers, CAs, etc.) and the network that interconnects them. The performance of a blockchain application/network can be measured using the following metrics:

- *Transaction Throughput*, measured in transactions per second (TPS): The number of transactions that are processed by the blockchain and written on the ledger in a given second.

$$Transaction\ Throughput = \frac{Total\ Transactions}{Total\ time\ in\ seconds} \tag{5.1}$$

- *Transaction Latency*: The amount of time taken from the moment when a transaction is submitted until the moment when it is confirmed and is available on the blockchain. This includes the propagation time and the processing time because of the consensus/ordering mechanism.

$$Transaction\ Latency = t_{Confirmation} - t_{Submission} \tag{5.2}$$

- *Computing Intensity*: The amount of computing resources consumed by the blockchain operation throughout the operating time, including the processing power, memory, storage, I/O, and network. This metric is of great importance as it could determine the cost efficiency of a blockchain application. Furthermore, besides the capital expenditure for providing the computing capacity, blockchain networks could require huge amounts of energy to operate. Therefore, the computing intensity would also affect the operating costs of the blockchain.

The performance of three major blockchain frameworks is compared in Table 5.1. A more in-depth study of the performance metrics and evaluation methods is presented by the Hyperledger Performance and Scale Working Group [34, 35].

5.4.3 Benchmark Apparatus

In this section, we briefly introduce the tools used for benchmarking the proposed blockchain application. The benchmark tool stack is depicted in Figure 5.5 where each benchmarking experiment produces two sets of results associated with the resource usage and blockchain performance. First, the report generated by Hyperledger Caliper (shown in the top right corner of Figure 5.5) regarding the blockchain network performance indicators introduced in Section 5.4.2. Second, a container-specific visual resource monitoring report generated by Grafana (bottom right corner of Figure 5.5) reflecting the network and system-level resource consumption. We also provide an insight into the cloud-based deployment environment and implementation details.

Table 5.1 Performance of blockchain frameworks.

Platform/Metric	Bitcoin	Ethereum	Fabric
Average latency	$\approx$ 10 Min	$\approx$ 12.5 Sec	$\approx$ MilliSec
Throughput (TPS)	7	10–30	20 000 [33]

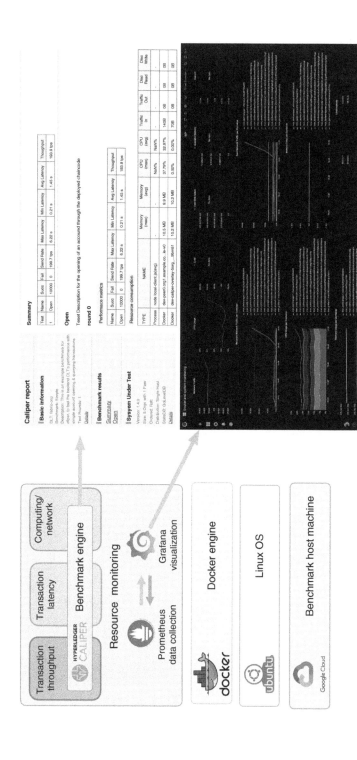

Figure 5.5 Blockchain benchmark tool stack. Source: The Linux Foundation, Grafana Labs, Docker Inc., Canonical Ltd., Google LLC.

5.4.3.1 Hyperledger Caliper

Hyperledger Caliper [36] is one of the sub-projects under the Linux Foundation's blockchain initiative, which is designed to provide a benchmarking and performance evaluation tool for various blockchain frameworks (e.g. Sawtooth, Fabric, Etherium, and more). In this work, we use Hyperledger Caliper to evaluate the performance of our proposed distributed sharing market.

5.4.3.2 Data Collection: Prometheus Monitor

Prometheus [37] is an open-source monitoring and alerting toolkit that collects real-time metrics from the running jobs on the nodes spread over the network (i.e. the Docker overlay

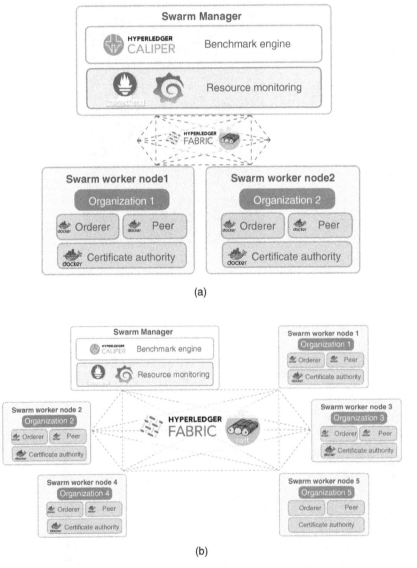

(a)

(b)

Figure 5.6 The experimental Hyperledger blockchain network. (a) Network topology with two operators and (b) network topology with five operators. The Linux Foundation, Grafana Labs.

network) and reports the results back to the central host. The collected metrics are kept as a time series database, which could be accessed through the flexible query language provided by Prometheus.

5.4.4 Experimental Results

In this section, we present the results of the experiments conducted to evaluate the performance of the proposed distributed marketplace application. Figure 5.6 shows the experimental blockchain network deployed to run the experiments related to this section.

5.4.4.1 Maximum Transaction Throughput

A major element influencing the performance of a blockchain application is the amount of computing resources available to it. This includes processing Central Processing Unit (CPU), memory, storage technology (SSD, HDD, etc.), and network connectivity among the nodes. As opposed to public blockchains, private blockchains do not necessarily need to be connected to the internet and could operate in a sub-network that connects the participant organizations. Figure 5.7 depicts the transaction throughput and latency resulting from running experiments using various SUTs. As depicted in Figure 5.7, the number of Virtual Central Processing Unitss (vCPUs) directly affects the performance of the distributed marketplace application. In both the two and five organization (operator) scenarios allocating more vCPUs leads to lower latency under various transaction send rates. In the case of five organizations, the latency converges to infinity when the send rate goes beyond 200 and 300 (TPS) for 8 and 32 vCPUs, respectively.

5.4.4.2 Block Size

As previously described, one of the steps affecting the performance of the blockchain is the consensus. In Hyperledger Fabric, the consensus protocol includes the ordering of the transactions and forming consecutive blocks of transactions. The maximum number of transactions allowed in a block is a design parameter referred to as the block size. As shown in Figure 5.8, increasing the block size will lead to a corresponding increase in the latency because Fabric will wait until the number of transactions equal to the block size is gathered (or the timeout is passed) before the block is formed and is written on the ledger.

5.4.4.3 Network Size

A distributed marketplace blockchain network is made of a number of operators who operate peers (nodes) that execute/verify smart contracts, transactions, and blocks. Figure 5.9

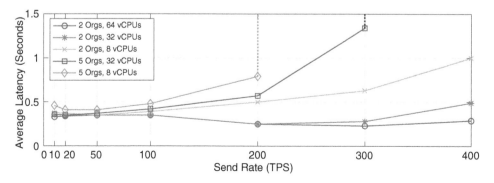

Figure 5.7 Transaction throughput vs. latency with different System under testss (SUTs).

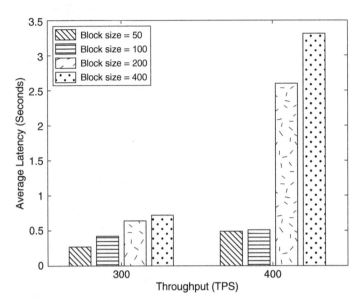

Figure 5.8 Block size vs. performance.

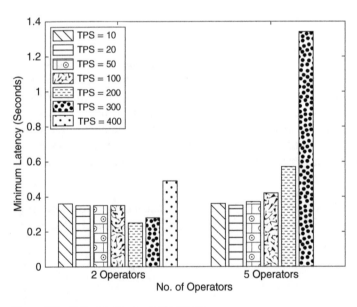

Figure 5.9 Network size vs. performance.

depicts the influence of increasing the number of organizations on the performance of the blockchain-based marketplace. Increasing the number of organizations from 2 to 5 (while using the same resources allocated to the SUTs) increases the latency. This is due to the increased network latency, as more nodes/orderers have to communicate to reach consensus, and higher smart contract execution time, as more peers should execute and sign the contract outcome.

5.5 Conclusions

In this chapter, we presented an alternative approach to the centralized conduct of the resource sharing market in telecommunications where no central entity is trusted by all the parties to manage the market. We used the blockchain technology to develop a distributed marketplace that relies on a collaborative consensus rather than centralized control. The auction mechanism is implemented as a smart contract that requires all (or a subset) of the market players to endorse and verify the transactions occurring due to the auction. Once agreed upon, this smart contract is immutable the same way that all the transaction information recorded on the distributed ledger are immutable. Furthermore, we acknowledged that replacing the centralized market management with a distributed approach will impose certain overheads to the system. We designed practical experiments to evaluate the performance of the distributed market mechanism under varying transaction send rates. The results of this study indicated that our proposed distributed marketplace could process up to 400 (with 2 operators) (TPS) and 300 (TPS) (with five operators) while maintaining a latency below one second. These promising results indicate that a wide range of telecom use cases could seamlessly operate on currently available permissioned blockchain technology. However, use cases that have more stringent latency budgets will need further investigation.

A blockchain application's performance depends on a range of tunable parameters. These parameters include block size, state database choice (e.g. GoLevelDB vs. CouchDB), the blockchain peers' geographical distribution, the endorsement policy, and the consensus protocol. Each one of these parameters could significantly impact the performance indicators of the blockchain network. For instance, a bigger block size could increase transaction throughput while increasing the latency. Therefore, considering the vast number of design choices and implementation options, the performance of the distributed marketplace will depend on its particular features such as trading frequency and latency constraints. It is noteworthy that the performance reported in this chapter is based on the current implementation of the enterprise blockchain platforms while studies suggest that by optimizing certain processes, this performance could considerably improve up to 20 000 (TPS) [33].

References

1 G. Fridgen, S. Radszuwill, N. Urbach, and L. Utz. Cross-Organizational Workflow Management using Blockchain Technology-Towards Applicability, Auditability, and Automation. 2018.

2 F. Milani, L. García-Ba nuelos, and M. Dumas. Blockchain and business process improvement. *BPTrends Newsletter (October 2016)*, 2016.

3 J. Golosova and A. Romanovs. Overview of the blockchain technology cases. In *2018 59th International Scientific Conference on Information Technology and Management Science of Riga Technical University (ITMS)*, pages 1–6, Oct. 2018. https://doi.org/10.1109/ITMS.2018.8552978.

4 T. Bocek, B. B. Rodrigues, T. Strasser, and B. Stiller. Blockchains everywhere - a use-case of blockchains in the pharma supply-chain. In *2017 IFIP/IEEE Symposium on Integrated Network and Service Management (IM)*, pages 772–777, May 2017. https://doi.org/10.23919/INM.2017.7987376.

5 J. Lee. Blockchain technologies: blockchain use cases for consumer electronics. *IEEE Consumer Electronics Magazine*, 7 (4): 53–54, July 2018. ISSN 2162-2248. https://doi.org/10.1109/MCE.2018.2816278.

6 P. K. Sharma, S. Rathore, and J. H. Park. DistArch-SCNet: blockchain-based distributed architecture with Li-Fi communication for a scalable smart city network. *IEEE Consumer Electronics Magazine*, 7 (4): 55–64, July 2018. ISSN 2162-2248. https://doi.org/10.1109/MCE.2018.2816745.

7 H. Wu and C. Tsai. Toward blockchains for health-care systems: applying the bilinear pairing technology to ensure privacy protection and accuracy in data sharing. *IEEE Consumer Electronics Magazine*, 7 (4): 65–71, July 2018. ISSN 2162-2248. https://doi.org/10.1109/MCE.2018.2816306.

8 M. Hajizadeh, N. Afraz, M. Ruffini, and T. Bauschert. Collaborative cyber attack defense in SDN networks using blockchain technology. In *2020 6th IEEE Conference on Network Softwarization (NetSoft)*, pages 487–492, 2020. https://doi.org/10.1109/NetSoft48620.2020.9165396.

9 W. Borjigin, K. Ota, and M. Dong. In broker we trust: a double-auction approach for resource allocation in NFV markets. *IEEE Transactions on Network and Service Management*, 15 (4), Dec. 2018. ISSN 1932-4537. https://doi.org/10.1109/TNSM.2018.2882535.

10 L. Jiang, Q. Wang, R. Song, B. Ye, and J. Dai. A truthful double auction framework for promoting femtocell access. *IEEE Access*, 7, 2019. ISSN 2169-3536. https://doi.org/10.1109/ACCESS.2019.2904548.

11 Y. Liu, X. Xu, J. Pan, J. Zhang, and G. Zhao. A truthful auction mechanism for mobile crowd sensing with budget constraint. *IEEE Access*, 7: 43933–43947, 2019. ISSN 2169-3536. https://doi.org/10.1109/ACCESS.2019.2902882.

12 Q. Wang, J. Huang, Y. Chen, C. Wang, F. Xiao, and X. Luo. *PROST*: privacy-preserving and truthful online double auction for spectrum allocation. *IEEE Transactions on Information Forensics and Security*, 14 (2): 374–386, Feb. 2019. ISSN 1556-6013. https://doi.org/10.1109/TIFS.2018.2850330.

13 N. Afraz and M. Ruffini. A sharing platform for multi-tenant PONs. *Journal of Lightwave Technology*, 36 (23): 5413–5423, Dec. 2018. ISSN 0733-8724. https://doi.org/10.1109/JLT.2018.2875188.

14 A. M. Youssef, E. Kalman, and L. Benzoni. Technico-economic methods for radio spectrum assignment. *IEEE Communications Magazine*, 33 (6): 88–94, 1995. https://doi.org/10.1109/35.387555.

15 A. Rymanov. LTE spectrum allocation: a beauty contest scenario. In *Ifost*, volume 2, pages 359–362, 2013. https://doi.org/10.1109/IFOST.2013.6616913.

16 The Sveriges Riksbank Prize in Economic Sciences in Memory of Alfred Nobel 2020. *NobelPrize.org.* https://www.nobelprize.org/prizes/economic-sciences/2020/press-release/.

17 A. Sgambelluri, F. Tusa, M. Gharbaoui, E. Maini, et al. Orchestration of network services across multiple operators: the 5G exchange prototype. In *EUCNC*, June 2017. https://doi.org/10.1109/EuCNC.2017.7980666.

18 N. Afraz, F. Slyne, H. Gill, and M. Ruffini. Evolution of access network sharing and its role in 5G networks. *Applied Sciences*, 9 (21): 4566, Oct. 2019. ISSN 2076-3417. https://doi.org/10.3390/app9214566.

19 K. Sparks, M. Sirbu, J. Nasielski et al. 5G Network Slicing. Whitepaper, FCC Technological Advisory Council (TAC), 2018. https://www.fcc.gov/general/tac-reports-and-papers.

20 3GPP. TR 28.801 Study on management and orchestration of network slicing for next generation network. Technical report 28.801, 2018. https://portal.3gpp.org.

21 A. Sgambelluri, F. Tusa, M. Gharbaoui, et al. Orchestration next-generation services through end-to-end networks slicing. 2018.

22 K. Samdanis, X. Costa-Perez, and V. Sciancalepore. From network sharing to multi-tenancy: the 5G network slice broker. *IEEE Communications Magazine*, 54 (7): 32–39, July 2016. ISSN 0163-6804. https://doi.org/10.1109/MCOM.2016.7514161.

23 N. Afraz and M. Ruffini. 5g network slice brokering: a distributed blockchain-based market. In *2020 European Conference on Networks and Communications (EuCNC)*, pages 23–27, 2020. https://doi.org/10.1109/EuCNC48522.2020.9200915.

24 J. S. Wey and J. Zhang. Passive optical networks for 5G transport: technology and standards. *Journal of Lightwave Technology*, 37 (12): 2830–2837, June 2019. ISSN 1558-2213. https://doi.org/10.1109/JLT.2018.2856828.

25 M. Ruffini, A. Ahmad, S. Zeb, N. Afraz, and F. Slyne. Virtual DBA: virtualizing passive optical networks to enable multi-service operation in true multi-tenant environments. *IEEE/OSA Journal of Optical Communications and Networking*, 12 (4): B63–B73, 2020. https://doi.org/10.1364/JOCN.379894.

26 M. Ruffini. Multidimensional convergence in future 5G networks. *Journal of Lightwave Technology*, 35 (3): 535–549, Feb. 2017. ISSN 0733-8724. https://doi.org/10.1109/JLT.2016.2617896.

27 E. Androulaki, A. Barger, V. Bortnikov, C. Cachin, et al. Hyperledger Fabric: a distributed operating system for permissioned blockchains. In *EuroSys*, 2018.

28 D. Ongaro and J. Ousterhout. In search of an understandable consensus algorithm. In *USENIX Conference*, pages 305–320, 2014. ISBN 978-1-931971-10-2.

29 V. P. Ranganthan, R. Dantu, A. Paul, P. Mears, and K. Morozov. A decentralized marketplace application on the ethereum blockchain. *2018 IEEE 4th International Conference on Collaboration and Internet Computing (CIC)*, pages 90–97, 2018.

30 N. Afraz and M. Ruffini. A distributed bilateral resource market mechanism for future telecommunications networks. In *2019 IEEE Globecom Workshops (GC Wkshps)*, pages 1–6, 2019. https://doi.org/10.1109/GCWkshps45667.2019.9024612.

31 M. Foti and M. Vavalis. Blockchain based uniform price double auctions for energy markets. *Applied Energy*, 254: 113604, 2019. ISSN 0306-2619. https://doi.org/10.1016/j.apenergy.2019.113604.

32 G. Marco Meinardi. Solution Comparison for Blockchain Cloud Services From Lead-ing Public Cloud Providers. https://www.gartner.com/en/documents/3906862/solution-comparison-for-blockchain-cloud-services-from-l.

33 C. Gorenflo, S. Lee, L. Golab, and S. Keshav. FastFabric: scaling Hyperledger Fabric to 20,000 transactions per second. In *2019 IEEE International Conference on Blockchain and Cryptocurrency (ICBC)*, pages 455–463, May 2019. https://doi.org/10.1109/BLOC .2019.8751452.

34 Linux Foundation. Hyperledger Blockchain Performance Metrics. *Whitepaper*, 2018.

35 Performance and Scale Working Group. https://wiki.hyperledger.org/x/ToIk.

36 P. W. D. Charles. A Blockchain Benchmark Framework. https://github.com/hyperledger/ caliper, 2019.

37 Prometheus - monitoring system & time series database. https://prometheus.io/.

6

Blockchain for Spectrum Management in 6G Networks

Asuquo A. Okon[1], Olusegun S. Sholiyi[2], Jaafar M. H. Elmirghani[3], and Kumudu Munasinghe[1]

[1]*Faculty of Science and Technology, University of Canberra, Canberra, ACT 2601, Australia*
[2]*National Space Research and Development Agency, Obasanjo Space Centre, Umaru Musa Yar'adua Expressway, P.M.B. 437, Garki, Abuja, Nigeria*
[3]*School of Electronic and Electrical Engineering, University of Leeds, Leeds LS2 9JT, UK*

6.1 Introduction

The relentless drive toward reliable data connectivity, which is crucially important for the actualization of an intelligent, automated, and ubiquitous digital world, is pushing the boundaries of wireless networks. Over the past decade, mobile access technology has evolved rapidly in response to the capacity demands and exponential growth in the number of wireless devices [1]. According to forecasts from Cisco Systems, it is estimated that about 28.5 billion mobile devices will be shipped by 2022, with smartphones expected to account for 50% of the global mobile traffic based on a compound annual growth rate (CAGR) of 30% [2]. While this trend is expected to increase year on year, studies have shown that about 80–90% of this traffic will be generated indoors [3]. With current fourth-generation (4G) long-term evolution (LTE) networks being stretched to their theoretical limits, mobile network operators (MNOs) have already commenced commercial deployment of fifth-generation (5G) networks in a bid to provide better network performance to new applications and services. Some of the key performance indicators (KPIs) of 5G include 1000 times higher wireless area capacity, support for much higher data rates in excess of 10 Gbps, ultra-low latency of 1 ms, very high reliability, and improved spectral and energy efficiency [1, 4, 5].

The sixth-generation (6G) networks also known as International Mobile Telecommunications (IMT 2030) are expected to push these KPIs even further by delivering peak data rates of at least 1 Terabyte per second (Tb/s), user-experienced data rate of 1 Gigabyte per second (Gb/s), over-the-air latency of 10–100 µs, 10× connectivity density of 5G, as well as 10× to 100× energy efficiency and 5× to 10× spectrum efficiency over those of 5G [6]. Promising technologies for the 6G ecosystem include terahertz communication, very-large-scale antenna arrays (i.e. supermassive (SM) multiple-input, multiple-output (MIMO)), large intelligent surfaces and holographic beamforming, orbital angular momentum multiplexing, laser and visible-light communications (VLC), blockchain-based spectrum

Wireless Blockchain: Principles, Technologies and Applications, First Edition.
Edited by Bin Cao, Lei Zhang, Mugen Peng and Muhammad Ali Imran.
© 2022 John Wiley & Sons Ltd. Published 2022 by John Wiley & Sons Ltd.

sharing, quantum communication and computing, molecular communications, and the Internet of Nano-Things [6, 7]. The goal is to actualize the vision of an intelligent information society characterized by seamless and ubiquitous communication between anybody, anything, anywhere, anytime, and anyhow – a paradigm often referred to as the 5A concept [5].

Network densification as a key strategy toward meeting the capacity requirements of next-generation wireless networks has been widely adopted [3]. This approach entails extreme exploitation of spatial reuse through increased deployment of macro-base station (MBS) per unit area, the aim being the achievement of higher spectral efficiency. Indeed, the authors in Ref. [8] are of the view that contrary to MBS densities of 8–10 BSs/km^2 in 4G cellular networks, the MBS densities of future networks are anticipated to be in the region of 40–50 BSs/km^2, representing a 400% increase in cellular deployments. There is no doubt that increased densification has achieved some level of success in providing ubiquitous connectivity, higher data rates, and low latency communication networks. However, the deployment of more MBS comes with attendant challenges such as increased size and complexity of the network in addition to huge financial costs incurred in rolling out new MBS.

Even as 5G networks are currently being deployed boasting of superior performance (higher data rates, lower latency, and spectral efficiency) over fourth-generation long-term evolution (4G LTE) networks, they still fall short of the demands of a fully connected and intelligent digital society. Consequently, research focus is increasingly shifting toward the next generation of wireless networks – 6G networks, envisaged to actualize the vision of future communications. This new vision will be characterized by advances such as the extreme automation of industrial manufacturing operations; harnessing of artificial intelligence capabilities within the cloud and fog networks; full integration of people, machines, and computing resources; and integration of space, air, ground, and underwater networks to provide ubiquitous and unlimited wireless connectivity [6]. The realization of these future networks will see the evolution of a new ecosystem that will place extreme demands for innovative network architectures that will be natively intelligence-driven.

Two technologies – software-defined networks (SDN) [9, 10] and blockchain [11, 12] – have emerged as very promising candidates that will enable the uptake of 6G networks. Different from traditional networks where the control and data planes reside in the router and switching devices [13], SDN decouples the control plane from the data plane to allow for simpler programmable network elements while moving complex control logic to an external controller, which is a software platform running on a commodity server technology [14]. Blockchain is a distributed ledger technology (DLT), for recording a growing list of digital actions or transactions, chained to each other and distributed across nodes [12, 15]. Indeed, there have been growing interests by several authors in employing the smart contract feature of blockchain for sharing spectrum assets among operators in a secure and trusted manner [11, 15, 16]. This chapter will focus on how 6G networks can actualize technical cooperation and business agreements between MNOs that can enable seamless roaming of mobile user across different operator networks.

6.2   Background

It is a widely held view in the research community that 5G networks and beyond will not just be an incremental improvement over their predecessors but rather a revolutionary leap forward in terms of data rates, latency, massive connectivity, network reliability, and energy efficiency [1]. The need to support the exponential increase in the number of wireless devices and data traffic characteristics of 5G networks has made the deployment of small cells (femto cells, pico cells, and micro-cells) a very crucial requirement especially in meeting capacity and coverage demands as well as providing quality of service (QoS) guarantees for vertical industries. These small cells are low-cost and low-power base stations that have shorter coverage area relative to macro-base stations and are often employed for offloading macro-traffic or to complement macro-cells in dense networks [17]. The migration to higher frequencies in 6G networks will witness an even more extreme densification of these small cells in a bid to dramatically enhance network capacity as well as provide ubiquitous coverage. This is particularly crucial in indoor environments in order to address coverage holes brought about by the attenuation of signals from macro-base stations arising from building penetration losses (BPL).

6.2.1   Rise of Micro-operators

Recently, the emergence of micro-operators (µOs) [17] has received a lot of research interest and has defined a concept whereby industry/business stakeholders can exploit their domain-specific knowledge to establish local small-cell networks where context-related services and content can be offered with quality of service guarantees in high-demand areas such as shopping malls, hospitals, campuses, sport arenas, and enterprises [17, 18]. Two good examples of successful commercial deployments of small-cell networks include Cloudberry, the first small-cell operator in the world [19, 20], and Denseair [21]. These µOs through their small-cell deployments are responsible for providing the indoor radio access network (RAN) in collaboration with facility owners and service providers such as MNOs and vendors under a neutral host arrangement. These µOs can thus be regarded as a carrier for carriers.

Two main benefits of this approach to the MNOs are (i) massive cost savings in terms of capital expenditure (CAPEX) and operating expenditure (OPEX) through the reduction in the cost of rolling out and managing new MBS deployments, which in turn reduces time to market and (ii) extension of the MNO geographic coverage and network footprint, thus allowing their subscriber to access network services and applications over a neutral operator's network. This has the potential of providing additional revenue streams to the MNO and µO. Therefore, in addition to providing the RAN for MNOs, µOs play a strategic role in managing business agreements between the various actors through the deployment of a single shared infrastructure.

Several works in the literature have been conducted on small-cell deployment for multi-operator support [11, 17, 18, 20]. In Ref. [17], the authors propose a framework for indoor small-cell deployment managed by micro-operators that leverages on network slicing to provide customized services. The authors identified insufficient coverage between local area networks (LANs) and wide area networks (WANs) as a motivation for the deployment of small cells. In their framework, the micro-operator is responsible for virtualizing the small cell into network slices realized using SDN and NFV, and these slices can form part of the product/service offering of a micro-operator to MNOs, service providers, or end users. While it is envisaged that through this framework MNOs can improve their indoor coverage without much expenses, it also opens up additional revenue stream for the micro-operator as they can charge the MNOs for accessing its local network. The network slices to be used will depend on the end-to-end (E2E) requirements such as bandwidth, data rate, latency, number of users, etc. The assumption that network slicing should be operated by the MNO while third-party service providers buy the network slices from MNOs has been challenged [18]. In their view, new stakeholders such as micro-operators could deploy ultra-dense small-cell RAN for tailored service delivery to various infrastructure providers including MNOs by employing network slices and spectrum sharing techniques. The different service models proposed include cloud-based Platform as a Service (PaaS), Software as a Service (SaaS), Infrastructure as a Service (IaaS), and Network as a Service (NaaS) as an extension of all the aforementioned models. With network slicing, operators could meet the diverse use case requirements and exploit the benefits of a common network infrastructure by abstracting the slice functionality to expose the capabilities of third-party service providers through open Application Programming Interfaces (APIs). However, network slices require radio resources for meeting agreed service quality; hence, the authors are of the view that efficient deployment of ultra-dense small-cell RAN calls for novel spectrum micro-licensing models.

6.2.2 Case for Novel Spectrum Sharing Models

Two spectrum sharing models recently approved by regulators include licensed shared access (LSA) in Europe and the citizens broadband radio services (CBRS) in the United States [22], where both models introduce mobile communications on bands shared with incumbent users. In their proposed high-level architecture, two entities are introduced – the first entity called the spectrum manager has the responsibility of controlling spectrum access according to spectrum availability information while the second entity known as the co-existence manger coordinates between different license holders including micro-licensees with different levels of spectrum access rights.

The problem of small-cell network sharing with a focus on the business model implications for different multi-operator deployment solutions for indoor scenarios is discussed in Ref. [20]. Four business models were identified in their study and these include (i) distributed antenna systems (DAS), (ii) multi-operator small cells using common frequencies, (iii) multi-operator small cells using dedicated frequencies, and (iv) multi-operator access using roaming. The multi-operator small cells are femtocell networks comprising two types of nodes: femtocell access points (FAPs) and femtocell gateways (FeGW). The DAS solution is commonly used to improve indoor coverage for voice services and suffers capacity limitations in supporting the high demands of next-generation networks. The small-cell networks on the other hand can support the very high capacity demands of future networks. From the business case perspective, the indoor deployment models can be further classified as shown in Figure 6.1.

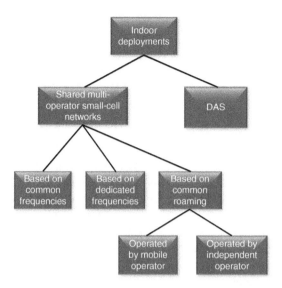

Figure 6.1 Classification of indoor network deployments based on the business model. Source: Okon et al. [23]. Licensed under CC BY 4.0.

In the multi-operator small-cell network based on a common frequency, operators share the same frequency over the radio access network provided by the FAPs while the FeGW connects to and manages the access points over the Internet and thereafter forwards operator traffic to their respective core network. In the dedicated frequency approach, each operator employs a different frequency and the FeGW located in the same premises as the FAP is used to distinguish the various streams of traffic, which are forwarded to different operators over the Internet. In the roaming-based approach, control of the spectrum is key to the business [20] for the independent operator, where the capacity to offload traffic and computationally intensive tasks is a vital service offering.

6.2.2.1 Blockchain for Spectrum Sharing

There has been a growing interest in the use of blockchain for spectrum sharing. A good example of the application of blockchain in the sharing of network resources among operators in small-cell networks can be found in Ref. [11]. Here, the smart contract feature of blockchain is used to manage the attach and detach procedures of a user equipment (UE) to mobile networks, such that the user subscription information and authentication keys are stored in a distributed ledger instead of a centralized home subscriber server (HSS). The distributed ledger can thus be seen as a functional replacement of the HSS. Authentication and security procedures are performed by communicating with this ledger. In the network sharing arrangement, multiple copies of the same ledger are distributed throughout the network, and all nodes (core network elements) retain copies of the entire blockchain. This will facilitate the provision of services to subscribers from other operators based on successful network authentication. Hence, peer-to-peer transactions between different operators can be executed by the smart contract based on the user's consumption and smart contract contents, which could be time variant depending on the geographical location of the small cells [11]. The potential benefits of employing blockchain technology, as a distributed digital ledger, in telecommunications, are well documented. Some of these benefits include interoperability, traceability, reliability, and capability to execute autonomous transactions [16]. Indeed, the authors in Ref. [11] proposed adopting blockchain technology as a facilitator for sharing network resources among operators in a secure and trusted manner.

6.2.2.2 Blockchain in 6G Networks

With a plethora of new applications and services churned out on a daily basis, the demands on wireless networks to increasingly deliver on stringent QoS guarantees has been unprecedented. This has made the need for better spectrum management even more pertinent. Specifically, novel approaches that allow for intelligent configuration and management of spectrum assets (for example, the air interface) on the fly will be required to bring about increased reliability, resilience, and trust to a trustless environment. To this end, blockchain-based spectrum sharing has been identified as a very promising technology for providing a secure, smarter, low cost, and highly efficient decentralized spectrum sharing in 6G networks [6]. This trend will see the evolution of new spectrum sharing models between operators aimed at bringing about seamless roaming access to subscribers in a cost-effective manner. Mobile devices/user equipment would become more network-aware with the capability to automatically switch between mobile network operators (MNOs) based on predefined agreements between MNOs without the intervention of the user. It is envisaged that such developments will significantly improve the QoS and QoE of consumers.

Going by the foregoing review, it is clear that as wireless networks continue to evolve into intelligence-driven and fully ubiquitous infrastructure to support different vertical applications, there is a need for creating business and technical cooperation between operators that will enable subscribers from one MNO to use the services of another MNO in a shared network. Such a framework could enable MNOs to leverage on areas where each MNO has a comparative advantage to provide value-added services to their subscribers in a cost-effective manner, for example, in locations where an MNO may have poor network coverage. To the best of our knowledge, our solution is the first to come up with an architecture for enabling business agreements among MNOs. It will be interesting to develop an integrated framework that can efficiently manage network operations as well as the business agreements between the various actors in a multi-operator small-cell network. This gap provides the motivation for this research work as we propose a blockchain-based approach implemented on top of an SDN infrastructure for supporting multi-operator small-cell deployments. To this end, this chapter will focus on one of the enabling technologies for 6G – blockchain-based spectrum sharing by designing an architecture for managing both the network operations and business agreements of multi-operator small-cell deployments using ablockchain. The initial version of this work can be found in Ref. [23]. Specifically, we intend to employ the smart contract feature of blockchain to realize an autonomous, distributed, secure, and reliable architecture for small-cell networks. An interesting feature of the blockchain-based approach to network management is that it enables autonomous transactions such that contract clauses embedded in the smart contracts are automatically executed once the triggering conditions are met (e.g. breach of SLAs) [16, 24].

6.3   Architecture of an Integrated SDN and Blockchain Model

The architecture of an integrated SDN and blockchain model is shown in Figure 6.2.

Such models would typically be deployed in indoor environment such as shopping malls, hospitals, and campus, with high user traffic and where the radio signal from an MNO's macro-base station may become severely attenuated, owing to BPL. This architecture combines SDN and blockchain functionalities to allow for interoperability between MNOs. The μO is responsible for deploying this infrastructure, which comprises the RAN and the integrated SDN and blockchain platform. From Figure 6.2, it can be seen that our architecture will enable mobile subscribers for example from MNO1 to access slice 2, which has been designated for IoT application from MNO3, subscribers from MNO2 can access mobile broadband services (slice 3) from MNO1, while subscribers from MNO3 can access gaming services from MNO2. Also, the μO can provide to the local customers who are not subscribers to any MNO, access to Internet services over the wireless network. In Sections 6.3.1–6.3.3, we will provide more details on both the design and network operation of our model.

6.3.1   SDN Platform Design

The role of SDN is to decouple the network operations into two (2) planes: the control plane and the data plane. This allows for ease of network management and programmability via

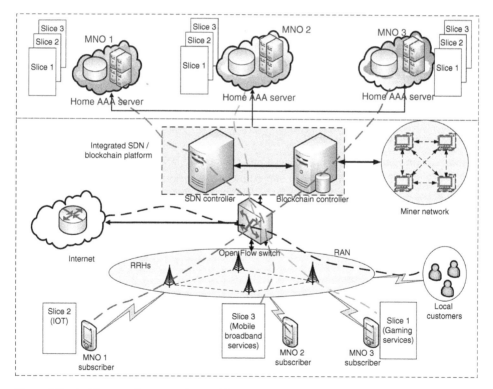

Figure 6.2   Architecture for supporting multi-operator small-cell deployments.

software programs. The control logic is migrated to the SDN controller, which is implemented in software and is responsible for programming of forwarding devices as well as handling the allocation of traffic to the data plane. The SDN controller is designed based on the OpenFlowswitch (OF-Switch) 13 module developed by Ref. [25] using the ns-3 simulator as well as the architectural framework by Ref. [26]. The OF-Switch13 module provides support for OpenFlow protocol version 1.3 by incorporating both a switch device and a controller in its design. However, unlike Ref. [26], where the GPRS tunneling protocol (GTP) is retained, our design eliminates the control signaling overhead incurred from maintenance of GTP tunnels by replacing the serving gateway (S-GW) and packet data network gateway (P-GW) with OF-Switches.

The OF-switch performs packet forwarding based on rules provided by the SDN controller. Thus, the data plane consists of a set eNodeBs (eNBs) connected to the switch ports of the OF-Switch through which the UE traffic is forwarded to its destination. The SDN controller on the other hand handles the control plane functionalities of the mobility management entity (MME) and policy charging and rules function (PCRF). This makes it in charge of executing the control signal logic required for UE authentication, mobility management, network topology, and flow table configuration.

The controller communicates with the OF-Switch in the data plane using the OpenFlow protocol via the southbound interface [14, 27]. Packets arriving at the data plane are checked for a matching rule at the switch after which they are forwarded to their destination through the appropriate switch ports. If no matching rule is found, the OF-Switch forwards the packet to the SDN controller for further processing. Furthermore, flow modification messages from the SDN controller enable the OF-switch to maintain a mapping of port numbers to the media access control (MAC) address of eNBs. In this way, the controller has a global view of the network consisting of UEs, eNBs, and OF-Switches. A key advantage of this programmable platform is that it allows applications to adapt to the network based on real-time information [28].

6.3.2 Blockchain Network Layer Design

The blockchain network (BN) is deployed just above the SDN platform and enables verification of network operations via distributed network authorization. Because of the storage constraints on the blockchain network, not all the information on the network operations can be recorded on the blockchain. As such, mostly control signaling information pertaining to attachment and detachment of the UE from the RAN, mobility management, resource allocation, etc., are recorded as they can be considered as transactions to be represented on a block for subsequent broadcast and verification by all nodes on the network. In this way, the network can maintain an accurate and auditable trail of devices, their locations, and resource utilization. The blockchain network interfaces with the SDN controller via the northbound interface using APIs (such as Python or C++). When exchanging information (e.g. service request, service authorization, spectrum sharing policy, etc.) with the blockchain network, the SDN controller (sender) accomplishes this by using its private key to sign its own transactions that are addressable on the network via the corresponding public key. The blockchain network (receiver) uses the same hash function as the sender to create the hash value of the transaction and compares this value with the hash obtained by decrypting the sender's digital signature with the corresponding public key in a bid to authenticate the sender's digital key. These transactions are written into the blockchain by means of smart contracts.

Smart contracts are computer programs with contractual clauses embedded in them, which are automatically executed once the enforcing or triggering conditions are met. In addition, the smart contract takes over the functions of the HSS as it now stores user subscription information in a digital ledger. The MNOs communicate with the smart contract by writing transactions to the smart contract address and are made part of the private blockchain network to enable network interoperability. The choice of a private blockchain platform over a public version is justified by the fact that the large volume of traffic from external users in the public blockchain introduces extra delays in reaching consensus, which ultimately impacts negatively on network performance and scalability. As a result, private blockchains provide higher throughput in terms of number of transactions per second as access is provided to a limited number of participants/nodes making it possible to reach consensus more quickly when compared to public blockchain (for example, Hyperledger fabric is able to record up to twenty thousand (20,000) transactions per second [29] compared to Bitcoin that can only process a theoretical maximum of seven (7) transactions per second [16]). Also, private blockchains such as Hyperledger fabric and Tendermint employ higher energy efficient consensus mechanisms when compared with the Proof of Work (PoW) scheme used in public blockchains such as Bitcoin [30]. Furthermore, the smaller size of private blockchain allows for greater ease of management of the network. In a private blockchain, all the nodes have to obey the rules of the network issued by a single monolithic body (e.g. spectrum regulator) in order to ensure efficient spectrum utilization while also enforcing penalties against MNOs who violate the transmission standard for the frequency band. Because the source code for the smart contract is publicly visible, the blockchain platform where the code resides guarantees immutability [24]; in that way, all parties to the smart contract are assured that all possible suspicious behaviors will be checked. Once the transactions have been verified as valid, it is appended to the chain as a block using a hash value as shown in Figure 6.3.

The centralization of SDN controllers, while enabling ease of management of the entire network, has an inherent drawback in that it introduces a single point of failure bottleneck in the network [16, 31]. In addition, the lack of consistent records of network data poses difficulties for network management given the heterogeneous nature of the network. An integrated SDN and blockchain architecture using smart contracts would enable a distributed consistent record of SDN data among all nodes in the blockchain, thus allowing for more efficient management of network resources. In addition, by eliminating total network control by any one node, the associated risks of single-point failure and multi-vendor device isolation can be mitigated, thereby enhancing fault recovery [32]. Also, in networks consisting of multiple SDN controller domains, the integration of blockchain and SDN makes it possible to provide network redundancy and faster recovery from network failure in the event any SDN controller fails [33]. This is possible since the network information stored in the blockchain is shared across all the other SDN controllers, making it possible for another SDN controller to manage the affected domain. Furthermore, security concerns in SDN especially with respect to sharing network resources can be addressed using blockchain [33], as it ensures the privacy of network users by preventing

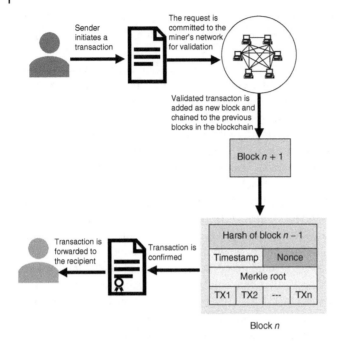

Sender initiates a transaction

The request is committed to the miner's network for validation

Validated transacton is added as new block and chained to the previous blocks in the blockchain

Block *n* + 1

Transaction is forwarded to the recipient

Transaction is confirmed

Harsh of block *n* − 1	
Timestamp	Nonce
Merkle root	
TX1 TX2 --- TXn	

Block *n*

Figure 6.3 Formation of blockchain — each block carries a list of transactions and harsh functions of the previous block.

untrusted members in the SDN network from viewing and modifying shared resources and records.

6.3.3 Network Operation and Spectrum Management

At the heart of the novelty of our design is the integrated SDN and blockchain architecture which serves to facilitate business agreements between MNOs while also managing radio access control in a shared network environment. Our framework seeks to address the coverage hole problem experienced by mobile subscribers while maintaining their existing UEs. To solve this problem, the SDN controller employs control logic to implement policies necessary for data flow control. This flow occurs in two directions. In the downward flow direction, the SDN controller generates packet forwarding rules and interacts with the data plane via the southbound interface based on network operation policies defined by SDN applications. In the upward direction, the controller collects network status and synchronization information from the underlying infrastructure (data plane), which is used to build a global view of the network; this view is presented to SDN applications through the northbound interface. The BN on the other hand handles user registration and maintains a shared billing system of both MNOs and an SLA manager, which is implemented in the blockchain using a smart contract. With this configuration, the smart contract becomes responsible for coordinating and providing billing settlements and agreed levels of service to subscribers. Furthermore, service-level authorization by blockchain ensures that only subscribers who have subscribed to a specific service have access to that service. Figure 6.4 shows the logical flow of information between blockchain and SDN.

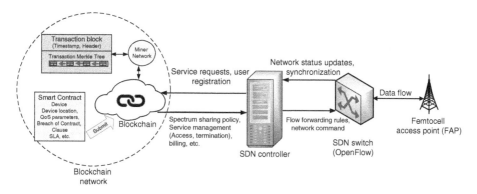

Figure 6.4 Logical flow of information between blockchain and SDN.

The logical centralized control provided by the SDN controller enables it to maintain a global view of the network allowing for efficient monitoring of spectrum usage, proper coordination of spectrum mobility, and effective implementation of spectrum sharing strategies among MNOs. The blockchain network advertises the service profile of the devices on the network to the SDN controller and enforcement of the terms of the contract, which is translated into flow rules and made available to the SDN switches. These flow rules determine which devices should be granted access to the network and what type/class of service these devices can access. Thus, access to the network is controlled at two levels – service level using blockchain and device level using SDN. Furthermore, the blockchain layer facilitates a distributed peer-to-peer network where non-trusting members (MNOs) can interact with each other without a trusted intermediary, thus creating a framework for trusted interaction in a trustless environment.

From the mobile subscriber perspective, a digital transaction is carried out when a phone call is made, text messages are sent, or data are used on the network, leading to a large amount of transactional information that has to be verified to ensure that customers are billed correctly. In heterogeneous networks consisting of diverse devices with different service profiles and service requirements, the lack of consistent billing records of network data could pose serious difficulties for network management and billing settlement. Our integrated SDN and blockchain architecture using smart contracts enables a distributed consistent record of network management data, which ensures accurate billing and effective network management. Specifically, the smart contract helps maintain a record of which devices has access to what resources. Figure 6.5 shows how the smart contract in the blockchain layer can be implemented on top of the SDN layer.

In terms of network operation, Figure 6.6 illustrates the signaling sequence and transactions that occur between the different nodes and how the smart contract is used for access control and network management. First, the UE radio resource control (UE RRC) entity at the UE performs periodic channel state measurements and sends the report to the home eNB (HeNB) RRC entity. As soon as the HeNB receives the measurement reports, it forwards only those related to handover events to the SDN controller, which are employed in implementing handover decisions. The report includes such parameters as reference signal received power (RSRP) and reference signal received quality (RSRQ), which are used as event triggering conditions for handover. Currently, ns-3 supports five (5) event-based

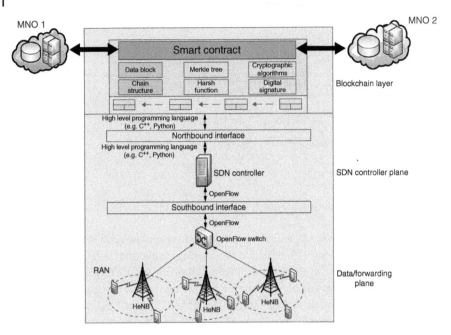

Figure 6.5 Smart contract implemented in blockchain layer as an overlay on SDN.

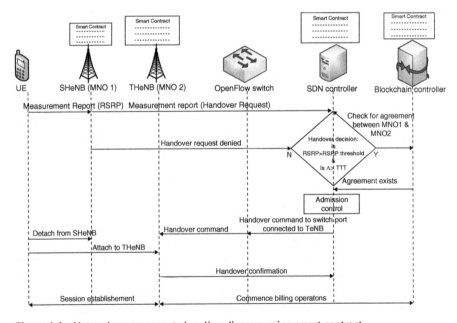

Figure 6.6 Network management signaling diagram using smart contract.

triggering criteria: A1, A2, A3, A4, and A5 [34]. For the purposes of this study, we are interested in the event A3 triggering condition. The event A3 triggering condition is used for implementing the A3RsrpHandoverAlgorithm also known as the strongest cell algorithm. This is motivated by the fact that once the RSRP of the home MNO (MNO1) degrades significantly as may occur when a UE moves outside its coverage area, the next available MNO will have a higher RSRP. In the current LTE standard, the UE performs handover as soon as the RSRP of the target HeNB (THeNB) exceeds the RSRP of its serving HeNB (SHeNB) by a threshold value called the hysteresis [34] or handover margin [26]. The condition for handover is given by

$$RSRP_{THeNB} > RSRP_{SHeNB} + \beta \qquad (6.1)$$

where β is the handover margin. In our model, the SHeNB represents the HeNB of the home MNO while THeNB represents the HeNB of the next available MNO. A further condition known as the time to trigger (TTT) is required before commencing handover procedures and it indicates the time duration over which the THeNB's RSRP must be continuously higher than the SHeNB's RSRP in order to trigger handover. The RSRP threshold values and TTT are implemented at the SDN controller as matching conditions. Note that it is critically important that these two conditions are met before handover can be initiated in order to mitigate the "ping-pong handover problem" [35] – a phenomenon characterized by back and forth signaling storm that results in too frequent handovers. Unlike current LTE implementation where handover is either initiated from the UE or HeNBs, we argue that it will be better if the handover decision is made by the SDN controller since it has a global view of the network and hence has a greater potential of realizing efficient resource allocation. The optimum placement of the SDN controller therefore presents a very interesting and important research problem that has been earmarked for future work. Upon receiving the handover request, the SDN controller compares the received values with the matching conditions given by (1) and TTT values. Once the conditions are met, the controller exchanges messages with the smart contract, which checks if there are any pre-existing agreements between MNO1 and MNO2 for that UE. If such agreements exist, the controller performs admission control procedures to ascertain that MNO2 core network can provide the agreed QoS guarantees as stipulated in the SLA and implements the A3RsrpHandoverAlgorithm by forwarding the handover request to the OF-Switch port connected to the THeNB. At the same time, the UE detaches from SHeNB and attaches to the THeNB. The THeNB thereafter replies with a handover confirmation packet to the controller. Finally, the subscriber establishes a session with MNO2 over the THeNB and billing procedures are automatically initiated.

6.4 Simulation Design

The architecture is modeled using a network simulator 3 (ns-3) and a blockchain platform. The ns-3 simulator is integrated with an SDN platform realized using the Open-Flow switch (OFSwitch) 1.3 module while the blockchain platform consists of an off-chain smart contract, Ethereum private blockchain, and the simulation module. The consensus mechanism employed is the proof of work (PoW). The smart contract enables management of access rights needed to facilitate interoperability between multiple operators. The contracts are designed based on the different vertical services users subscribe to and serve as an interface between multiple operators. The private blockchain is initialized on a test-net using Geth (go-Ethereum) software and thereafter the genesis file is set up. The genesis file is used to define the genesis block which is the first block of the blockchain [16] and provides information about specifications of the block including such variables as gas limit, difficulty level, coinbase, timestamp, and transaction fee.

Table 6.1 Simulation parameters.

Parameter	Value
Network configuration	
Simulation grid	180 m × 100 m
Number of UEs	10
Number of HeNBs	3
Number of MNOs	3
Number of SDN controllers	1
SDN controller type	OFSwitch13InternalHelper
Number of OFSwitches	3
UE speed	2 m/s
UE TX Power	100 mW
eNB TX power	100 mW
Carrier Frequency	3.6 GHz
Pathloss model	Friis
Type of traffic	UDP
Mobility model (eNB)	Constant position
Mobility model (UE)	Constant velocity
Handover algorithm	A3RsrpHandoverAlgorithm
Handover margin (hysteresis)	3 dB
TTT	256 ms
Blockchain configuration	
Blockchain platform	Geth (go-Ethereum)
Consensus mechanism	Proof of Work (PoW)
Number of virtual machines	15
Total gas limit	3 000 000
Cost per transaction	21 000
Number of transactions per block	142
Block time	10
Transactions/sec	14.2

In studying the impact of SDN and blockchain platform in radio access management, we assume that a contractual agreement between the different MNOs is already in place. The SDN controller is designed using the OFSwitch13InternalHelpler function, which is included in the new OFSwitch13 module in ns-3. The parameters used in the simulation are presented in Table 6.1, and the RAN simulation scenario is illustrated in Figure 6.7. Any UE from a given MNO (e.g. MNO1 subscriber) seeking to access services on different MNOs (e.g. Host on MNO2) conducts regular measurement reports to confirm spectrum availability of the intended MNO (by virtue of a higher RSRP) and notifies the SDN controller via the SHeNB. The UEs and HeNBs have a transmit power of 100 mW with the HeNB having a coverage radius of 30 m. The above scenario is modeled in an NS3 simulator to study the performance of the network when users move between three HeNBs as shown in Figure 6.7.

The duration of the simulation takes into consideration the length of the time required to transfer the control signal from a UE moving at a velocity of 2 m/s, between three MNOs beginning from the cell edge of MNO1 to the edge of MNO3, while also allowing extra time for the simulation to settle into a steady state. Based on the above considerations, a simulation duration of 150 seconds is chosen. Once access has been granted by the SDN controller to the RAN, the UE makes a transaction to the smart contract address to initialize and trigger the services specified in the contract. It is important to note that the time it takes for the BN to change the contract from one operator to the next one will affect the network availability for the user. Hence, it is a critical requirement that this contract switching delay be kept to a minimum.

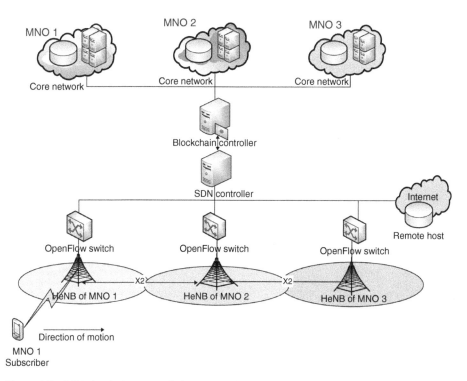

Figure 6.7 RAN simulation scenario in an ns-3 simulator.

6.5 Results and Analysis

6.5.1 Radio Access Network and Throughput

The simulation performance of the SDN controller in handling the handover decisions across the three HeNBs is shown in Figure 6.8. In our simulation, the SDN controller is responsible for making the handover decisions because it maintains a global view of the network. The graph shows the periodic RSRP measurements (in dBm) provided by the UE to the SDN controller as it moves from one HeNB to another. At the beginning of the simulation, the UE is connected to HeNB 1 (the Home MNO) because it has the highest RSRP (strongest cell). This is indicated by the dotted dark gray line. As the UE moves from the edge of HeNB1 along the positive x-axis (UE, 0, 0), the RSRP values increase accordingly and reaches its peak when the UE is at the center of the HeNB1. Further movement of the UE away from the center toward the other edge of HeNB1 sees a decline in the RSRP values reported to the SDN controller. At the same time, it is observed that there is a gradual increase in the RSRP value of HeNB2, which represents the next available MNO (second dotted dark gray line), because of the movement of the UE away from HeNB1 toward the vicinity of HeNB2.

It should be noted that at this point, the UE is still connected to HeNB1, albeit its RSRP value continues to decrease until the RSRP value of HeNB2 exceeds that of HeNB1 by a threshold value and persists in that condition for a certain duration (TTT) – we have set these values in our simulation as 3 dB and 256 ms, respectively. These values that represent

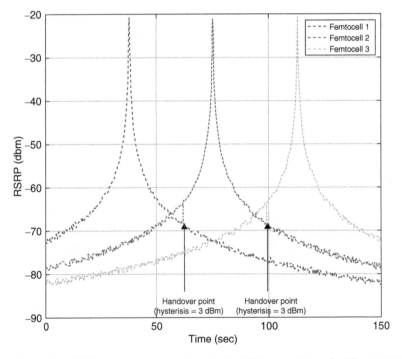

Figure 6.8 RSRP measurements received by the SDN controller as the UE moves across femtocells.

our matching conditions are chosen because a 3 dB difference represents a 50% drop in a signal power level, which is very significant, while a delay of 256 ms is just about adequate to accommodate handover procedures and transactions to the blockchain while also ensuring session continuity for real-time applications such as voice and video. Upon receiving these matching conditions, the SDN controller writes to the smart contract to confirm the presence of a contract or agreement between both MNOs and thereafter automatically initiates the handover procedure for transferring the UE from HeNB1 to HeNB2 (as it is now the cell with the strongest RSRP) by invoking the A3RsrpHandoverAlgorithm. The handover point is indicated by the dotted vertical light gray line. The same procedure is executed when the UE moves away from the coverage area of HeNB2 into the vicinity of HeNB3. The second handover point is indicated by the second dotted vertical light gray line.

Figure 6.9 shows a comparison of the network performance in terms of the average throughput measured across ten (10) UEs with and without blockchain implementation. It can be seen in Figure 6.9a that in the absence of a smart contract agreement, users experience a total break in network connectivity and hence a disruption in services as they move between HeNBs. The absence of a smart contract agreement between MNOs means that there is no control signal available from the next operator to facilitate handover and session transfer between MNOs when users move between operators. In contrast, however, Figure 6.9b shows the performance of the network when a smart contract agreement is in place. In this scenario, the throughput is maintained at near optimum levels (10 Mbps) with very minimal disruption as the user moves between different MNOs because of the enforcement of the smart contract. The is because the smart contract makes a control

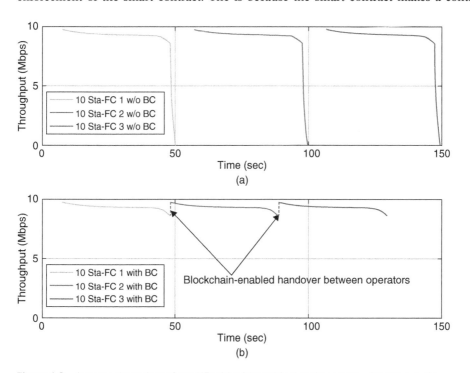

Figure 6.9 Average throughput from UEs: (a) without blockchain and (b) with blockchain.

signal available to serve as a mobility anchor point from other operators who are co-parties to the agreement in order to enable the handover between MNOs as soon as users move outside the coverage area of any MNO. Furthermore, the execution of the smart contract ensures that operators provide the prescribed levels of service as stipulated in the SLA to enable seamless continuity of sessions as illustrated in the graph. The handover delay in moving from one HeNB to the next available HeNB is equivalent to the round trip time in writing a transaction to the smart contract address, which from our simulation is approximately 70 ms. This delay figure is enough to meet the stringent delay requirements required for real-time applications such as voice and video (150–200 ms) [36].

6.5.2 Blockchain Performance

The blockchain performance is evaluated across different instances with 1, 5, 15, and 25 users actively interacting with the smart contract; each instance was simulated for 600 seconds as shown in Figure 6.10.

The blockchain transaction simulation is carried out on fifteen (15) Ethereum virtual machines (VM) in order to provide insights into the frequency of transactions between multiple users and the number of times users access services from the MNOs. The virtual machines run on a Dell Latitude 7400 with Intel(R) Core (TM) i7-8665U CPU@2.11 GHz, 16 GB of RAM, and 1 Gbps ethernet connection. The difficulty level is set to 1 for every block, and the approximate block processing time is set to about 10 seconds using the PoW consensus algorithm. It is observed that the number of transactions increases over time across the different categories of users; however, the number of transactions mined decreases with

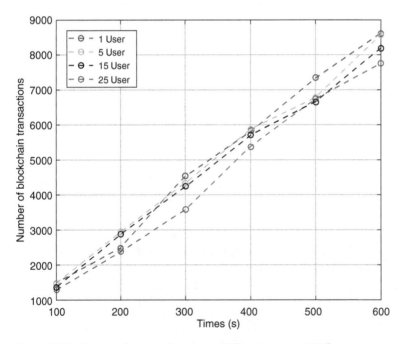

Figure 6.10 Number of transactions across different users on MNO$_2$.

increase in users. This is because a higher number of users making transactions on the network means that more blocks will be generated, which will need to be confirmed. The mechanism of the PoW is designed such that the mining difficulty level is adjusted automatically every two weeks based on the block production rate, in order to maintain steady average production of 1 block every 10 minutes. If the blocks are generated too fast as can arise because of the increasing cumulative hashing (computational) power of miners, the difficulty level increases to ensure that the 10 minute interval between the new blocks is maintained. Conversely, if the time between blocks increases way beyond 10 minutes (translating to lower block generation rate), the difficulty level is reduced accordingly.

6.5.3 Blockchain Scalability Performance

The recording of every single transaction in the blockchain results in a very huge amount of data to be stored in the blockchain, which comes with severe penalties such as increased latency (i.e. time to execute a transaction) and reduced scalability. To cope with latency and scalability issues which constitute a major challenge in blockchain implementations, our solution adopts the strategy of moving most of the transactional data off-chain, leaving only the most recent transaction on the blockchain. To facilitate this without breaking the block's hash, all transactions are hashed in a Merkle tree with only the root included in the block's hash. This is supported by our results in Figure 6.11, which shows an analysis of the number of transactions per second at different instances of the simulation with increasing number of users.

It is observed that the number of transactions executed at each instance of the simulation (100, 300, 400, and 600 seconds) is relatively consistent for each set of users and decreases just marginally with an increasing number of users. From the simulation results, the huge number of successful transactions recorded notwithstanding the increase in number of

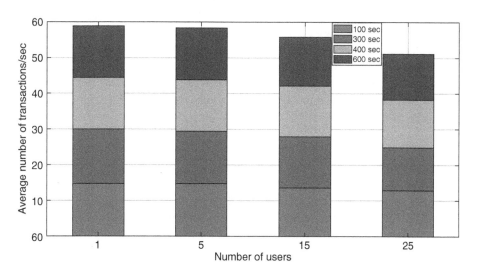

Figure 6.11 Average number of transactions per second at different instances during the simulation with increasing number of users.

users clearly demonstrates the scalability of our solution and how smart contracts can enable interoperability across network operators.

Our proposed multi-operator small-cell solution very much aligns with provisions of the latest 3GPP Release 15 and beyond which allow support for multi-operator core network (MOCN), whereby the RAN can be shared by multiple core networks [37, 38]. This blockchain agreement executed via smart contracts enables spectrum sharing through the shared use of the RAN by multiple core networks. The serving MNO provides the radio access elements, e.g. eNBs/HeNBs, which are shared to provide access to both home and visiting subscribers. The MOCN used in our scenario is justified because the MNOs each maintain their core networks and shared access is provided over the RAN. However, our approach uses smart contracts to enable RAN sharing so that subscribers from other operators can have access to the shared radio elements provided by the serving MNO (sharing operator). It is important to note that the approach we have introduced here is generic and can apply to any part of the spectrum that is currently in use by cellular systems or other parts that may be included by 3GPP in future.

6.6 Conclusion

In this chapter, we considered an integrated blockchain and SDN architecture for multi-operator support in 6G networks. This solution employs the smart contract feature of blockchain to enable the creation of business and technical agreements between MNOs for intelligent and efficient management of spectrum assets (i.e. the radio access network). The disruptive power of blockchain technologies provides a unique opportunity to re-think and re-invent long established principles of wireless communications, bringing with it a very promising prospect of identifying areas that are ripe for innovation. Specifically, smart contracts have the potential to remove trust barriers through a cost-effective implementation of simple business agreements in the form of SLAs between MNOs. The ledger function provided by blockchain by means of the smart contract enables end-to-end full visibility in the business network, making transactions between all participating parties transparent and auditable. Thus, by integrating blockchain and SDN, the foundation for creating trusted interactions in a trustless environment can be established. This is evident in our simulation results where mobile subscribers can enjoy seamless roaming access whenever they traverse between operators.

Acknowledgments

This chapter is an adaptation of Blockchain and SDN Architecture for Spectrum Management in Cellular Networks (https://ieeexplore.ieee.org/document/9094658) by A. A. Okon, I. Elgendi, O. S. Sholiyi, J. M. H. Elmirghani, A. Jamalipour, and K. Munasinghe and is used under a Creative Commons Attribution 4.0 License.

References

1 M. Shafi, A. F. Molisch, P. J. Smith et al. 5G: a tutorial overview of standards, trials, challenges, deployment, and practice. *IEEE Journal on Selected Areas in Communications,* 35(6): 1201–1221, 2017.

2 Cisco. Cisco Visual Networking Index: Global Mobile Data Traffic Forecast Update, 2017–2022 White Paper, 2020. https://www.cisco.com/c/en/us/solutions/collateral/service-provider/visual-networking-index-vni/white-paper-c11-738429.html#Trend2 (accessed 14 January 2020).

3 F. Hu. *Opportunities in 5G Networks: A Research and Development Perspective*, 1st ed. CRC Press Taylor & Francis Group, 2016.

4 M. Lauridsen, L. C. Gimenez, I. Rodriguez, et al. From LTE to 5G for connected mobility. *IEEE Communications Magazine,* 55(3)*:* 156–162, 2017, doi: https://doi.org/10.1109/MCOM.2017.1600778CM.

5 Z. S. Bojkovic, R. B. Bakmaz, and B. M. Bakmaz. Research challenges for 5G cellular architecture. In *Presented at the 12th International Conference on Telecommunication in Modern Satellite*, Cable and Broadcasting Services (TELSIKS), 2015.

6 Z. Zhang, Y. Xiao, Z. Ma, et al. 6G wireless networks: vision, requirements, architecture, and key technologies. *IEEE Vehicular Technology Magazine,* 14(3): 28–41, 2019, doi: 10.1109/MVT.2019.2921208.

7 H. Viswanathan and P. Mogensen. Communications in the 6G Era. *IEEE Access,* 8: 57063–57074, 2020, doi:https://doi.org/10.1109/ACCESS.2020.2981745.

8 X. Ge, S. Tu, G. Mao, et al. 5G ultra-dense cellular networks. *IEEE Wireless Communications,* 23(1)*:* 72–79, 2016, doi: https://doi.org/10.1109/MWC.2016.7422408.

9 P. Ameigeiras, J. J. Ramos-Munoz, L. Schumacher, et al. Link-level access cloud architecture design based on SDN for 5G networks. *IEEE Network,* 29(2)*:* 24–31, 2015, doi: https://doi.org/10.1109/MNET.2015.7064899.

10 I. Elgendi, K. S. Munasinghe, and A. Jamalipour. A three-tier SDN based distributed mobility management architecture for DenseNets. In *Presented at the IEEE ICC 2016 – Mobile and Wireless Networking Symposium*, 2016.

11 B. Mafakheri, T. Subramanya, L. Goratti, and R. Riggio. Blockchain-based infrastructure sharing in 5G small cell networks. In *Presented at the 14th International Conference on Network and Service Management (CNSM 2018)*, 2018.

12 X. Ling, J. Wang, T. Bouchoucha, et al. Blockchain radio access network (B-RAN): towards decentralized secure radio access paradigm. *IEEE Access,* 7: 9714–9723, 2019, doi: https://doi.org/10.1109/ACCESS.2018.2890557.

13 P. Göransson and C. Black. *Software Defined Networks A Comprehensive Approach*. Elsevier Inc., 2014.

14 H. Venkatarman and R. Trestian. *5G Radio Access Networks: Centralized RAN, Cloud-RAN, and Virtualization of Small Cells*. Taylor & Francis Group, 2017.

15 M. Saravanan, S. Behera, and V. Iyer. Smart contracts in mobile telecom networks. In *Presented at the 23RD Annual International Conference in Advanced Computing and Communications (ADCOM)*, 2017.

16 H. Dai, Z. Zheng, and Y. Zhang. Blockchain for Internet of Things: a survey. *arXiv:1906.00245v2[cs.NI]*, pp. 1–19, 2019.

17 J. Walia, H. Hämmäinen, and M. Matinmikko. 5G micro-operators for the future campus: a techno-economic study. In *Presented at the Internet of Things Business Models, Users, and Networks*, 2017.

18 M. Matinmikko-Blue, S. Yrjölä, and M. Latva-aho. Micro operators for ultra-dense network deployment with network slicing and spectrum micro licensing. In *Presented at the IEEE 87th Vehicular Technology Conference (VTC Spring)*, 2018.

19 Cloudberry. Cloudberry mobile. http://www.cloudberrymobile.com (accessed August 9, 2019).

20 J. Markendahl and A. Ghanbari. Shared smallcell networks – multi-operator or third party solutions – or both?. In *Presented at the Fourth International Workshop on Indoor and Outdoor Small Cells 2013*, 2013.

21 Denseair. Dense air. http://denseair.net/ (accessed August 9, 2019).

22 M. Matinmikko, M. Latva-aho, P. Ahokangas, et al. Micro operators to boost local service delivery in 5G. *Wireless Personal Communication*, 95: 69–82, 2017, doi:https://doi.org/10.1007/s11277-017-4427-5.

23 A. A. Okon, I. Elgendi, O. S. Sholiyi, et al. Blockchain and SDN architecture for spectrum management in cellular networks. *IEEE Access*, 8: 94415–94428, 2020, doi:https://doi.org/10.1109/ACCESS.2020.2995188.

24 E. Pascale, J. McMenamy, I. Macaluso, and L. Doyle. Smart contract SLAs for dense small-cell-as-a-service. *arXiv:1703.04502v1 [cs.NI]*, pp. 1–2, 2017.

25 L. J. Chaves, I. C. Garcia, and E. R. M. Madeira. OFSwitch13: enhancing ns-3 with OpenFlow 1.3 support. In *Presented at the WNS3 '16: Proceedings of the Workshop on ns-3*, Seattle, Washington, USA, 2016.

26 F. H. Khan and M. Portmann. A system-level architecture for software-defined LTE networks. In *Presented at the 2016 10th International Conference on Signal Processing and Communication Systems (ICSPCS)*, Gold Coast, QLD, Australia, 2016.

27 V. Nguyen, A. Brunstrom, K. Grinnemo, and J. Taheri. SDN/NFV-based mobile packet core network architectures: a survey. *IEEE Communication Surveys and Tutorials*, 19(3): 1567–1602, 2017, doi:https://doi.org/10.1109/COMST.2017.2690823.

28 P. Sharma, S. Singh, Y. Jeong, and J. H. Park. DistBlockNet: a distributed blockchains-based secure SDN architecture for IoT networks. *IEEE Communications Magazine*, 55(9): 78–85, 2017, doi:https://doi.org/10.1109/MCOM.2017.1700041.

29 C. Gorenflo, S. Lee, L. Golab, and S. Keshav. FastFabric: scaling hyperledger fabric to 20,000 transactions per second. In Presented at the arXiv:1901.00910v2 [cs.DC], 4 March 2019, 2019.

30 A. A. Monrat, O. Schelén, and K. Andersson. A survey of blockchain from the perspectives of applications, challenges, and opportunities. *IEEE Access*, 7: 117134–117151, 2019, doi:https://doi.org/10.1109/ACCESS.2019.2936094.

31 J. Weng, J. Weng, J. Liu, and Y. Zhang. Secure software-defined networking based on blockchain. *arXiv:1906.04342v1 [cs.CR]*, pp. 1–19, 2019.

32 C. Xu, N. Xu, and B. Yin. Research on key technologies of software-defined network based on blockchain. In *Presented at the IEEE International Conference on Service-Oriented System Engineering (SOSE)*, 2019.

33 A. Yazdinejad, R. Parizi, A. Dehghantanha, and K. Choo. Blockchain-enabled authentication handover with efficient privacy protection in SDN-based 5G networks. *arXiv:1905.03193v1 [cs.NI]*, pp. 1–12, 2019.

34 NS3. ns-3 Model Library Release ns-3.30. NS3 Project, August 2019 21, 2019.

35 T. Bilen, B. Canberk, and K. R. Chowdhury. Handover Management in Software-Defined Ultra-Dense 5G Networks.. *IEEE Network*, 31(4): 49–55, 2017, doi:https://doi.org/10.1109/MNET.2017.1600301.

36 Y. Chen, T. Farley, and N. Ye. QoS requirements of network applications on the internet. *Journal of Information, Knowledge, Systems Management,* 4: 55–76, 2004.

37 3GPP. (3GPP) Evolved Universal Terrestial Radio Access X2 application protocol (X2AP) (Rel 14), 3GPP Std. TS 36.423. In *(E-UTRA)*, vol. 14.1.0, 2017.

38 ETSI. 5G: system architecture for the 5G system (Rel 15), Version. 15.3.0, document 3GPP TS 23.501, 2018.

7

Integration of MEC and Blockchain

Bin Cao, Weikang Liu, and Mugen Peng

State Key Laboratory of Networking and Switching Technology, Beijing University of Posts and Telecommunications, Beijing, 100876, China

7.1 Introduction

In the face of the explosive growth of Internet of things (IoT) and 5G services, such as high-definition video, automatic drive, and augmented reality, etc., the cloud computing model gradually shows their shortcomings in the real-time process, network constraints, resource overhead, and privacy protection [1]. To enable massive and complex computing services, mobile edge computing (MEC) systems were proposed to support the emerging services [2–5]. Because of the shortening of transmission links, edge computing can respond to service requests quickly and efficiently on the data side. In addition, edge computing can also provide basic services offline, which reduces the dependence of services for online real-time network access [2, 6]. In MEC systems, in addition to the need for a single server to quickly complete computing tasks on the edge side, there also exist situations, multiple MEC servers need to collaborate to integrate the resources they own to complete large-scale computing tasks or data migration [7, 8]. However, multiserver collaboration typically involves multiple servers belonging to different domains whose trust is isolated from each other as they are managed by different organizations. Therefore, there are trust and security issues when MEC servers of different domains need to collaborate with other servers [7, 9, 10].

To address these issues, blockchain that is a tamper-proof transaction database shared by all nodes participating in a network based on a consensus protocol is introduced. Features such as security, transparency, and decentralization allow it to be a distributed peer-to-peer network where non-trusting members can interact with each other in a verifiable manner without a trusted intermediary. To ensure data security in mobile commerce between mobile devices, blockchain has been integrated as an efficient security solution for establishing trust between mobile devices in a decentralized network. The development for integration of blockchain in IoT mobile applications is hindered by a major challenge brought by heavy computational process [11]. PoW-based public blockchain networks (PBN's) security relies on a proof-of-work (PoW) procedure called mining, which is a difficult mathematical problem making blockchain almost impossible to be tampered with [3]. However, mobile devices are restricted in key areas related to communication and computation such

Wireless Blockchain: Principles, Technologies and Applications, First Edition.
Edited by Bin Cao, Lei Zhang, Mugen Peng and Muhammad Ali Imran.
© 2022 John Wiley & Sons Ltd. Published 2022 by John Wiley & Sons Ltd.

as memory, battery, and processing because of design choices that guarantee their mobility [5, 9], so that they fall short in their capacity to afford the high computing resources to find the value in mining process [10]. To support mining tasks execution in them, offloading mining tasks to MEC servers is suggested and a lot of research has been done in this aspect.

In summary, blockchain can be employed to ensure the reliability and irreversibility of data in MEC systems, and in turn, MEC can also solve the major challenge in the development of blockchain in IoT applications, which is brought by heavy computational process.

7.2 Typical Framework

7.2.1 Blockchain-Enabled MEC

7.2.1.1 Background

Many efforts have been made on computation offloading and resource allocation of MEC systems [12–16]. However, the above existing methods are not suitable for some specific environments because of the following challenges.

1. *Security and privacy issues*: Most of the existing studies [17, 18] pay little attention to the security and privacy of MEC. The interaction between heterogeneous edge nodes and the migration of service across edge nodes are likely to challenge its security and privacy. To address these issues, blockchain has been envisioned as a promising approach [19]. Different from traditional digital ledger approaches, which depend on a trusted central authority, blockchain employs community verification to synchronize the decentralized ledgers that are replicated across multiple nodes [20]. Blockchain can facilitate the establishment of trusted, secure, and decentralized MEC systems. In blockchain-enabled MEC systems, MEC servers not only handle their tasks but also deal with the task (e.g. generate blocks and perform consensus process) from blockchain systems, which makes the design of the system more complex. Therefore, the design and optimization of blockchain and MEC should be implemented simultaneously.

2. *Cooperative computation offloading*: This approach has been only considered by a few researchers in the previous works. Most existing computation offloading schemes [17, 18] assume that computing tasks can be directly offloaded to MEC servers via wireless communications. However, a mobile device may be experiencing weak or intermittent connectivity and thus cannot directly offload computing tasks to MEC servers. If computing tasks are forced to offload to MEC servers directly, the computation offloading performance of mobile devices may be affected because of signal loss. A mobile device must offload computing tasks to MEC servers with the help of neighboring nodes. Therefore, it is necessary to study cooperative computation offloading. Furthermore, if there exist malicious nearby nodes, the data security and privacy of mobile devices will be susceptible to attacks. Therefore, the trust model needs to be considered in cooperative computation offloading.

7.2.1.2 Framework Description

As shown in Figure 7.1, the blockchain-enabled MEC system is composed of an MEC system and a blockchain system. In the MEC system, a single macro-cell base station (MBS) is

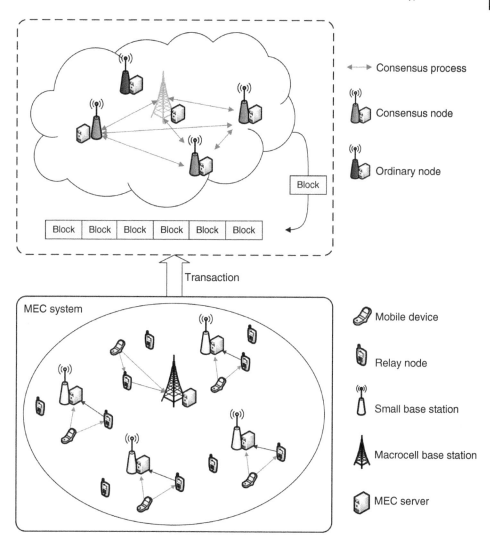

Figure 7.1 Architecture for blockchain-enabled MEC system. Source: Feng et al. [12] / with permission of IEEE.

located in the center of the coverage area. Several small base stations (SBSs) are distributed around the MBS, and all BSs are connected by wire links, each of which is integrated with an MEC server. Each BS serves multiple mobile devices and each mobile device is running independent and fine-grained tasks.

Because mobile devices have relatively weak computing capability, the computing tasks of mobile devices need to be completed with the help of MEC servers to improve the quality of the user's computation experience. Cooperative communications are adopted to offload tasks in the MEC system, i.e. offloading computation tasks with the help of relay nodes, to meet the computation requirements of mobile devices that are far from MEC servers. When offloading computation tasks by relaying, there may be selfish and malicious nodes. Therefore, security plays an important role in realizing cooperative communications [21].

In the blockchain system, the blockchain nodes consist of all BSs. These nodes have two types of roles: (i) ordinary nodes and (ii) consensus nodes. The blockchain system mainly deals with transactions, i.e. offloading data records, from the MEC network. To handle the transactions, the blockchain system needs to complete two steps. One is the block generation and the other is the consensus process. Ordinary nodes only transfer and accept ledger data, while consensus nodes produce blocks and perform the consensus process. However, there may be a security issue during the block generation process, i.e. malicious consensus nodes may tamper with transaction data. Therefore, the trust value of each candidate should be considered when voting for consensus nodes. Consensus nodes with high trust value are likely to ensure a secure and reliable block generation process and consensus process [22]. Meanwhile, the blockchain system can also store some parameters for calculating the trust value of the interactive nodes (i.e. relay nodes and consensus nodes), such as network status, resource availability, and trustworthiness of interactive nodes [23].

7.2.2 MEC-Based Blockchain

7.2.2.1 Background
The most famous consensus protocol that can be adopted in the permissionless chain network is PoW, which was proposed by Nakamoto in Bitcoin [24]. In the PoW-based permissionless chain network, participants can be divided into two roles: (i) blockchain user and (ii) miner. Blockchain users generate transactions, and miners try to mine valid blocks. Blockchain miners can be users. The PoW consensus algorithm regulates that the miners should change different nonce with the previous block content to generate a new valid block through a specific hash function, which is also called the mining process. To incentive more miners to participate in the mining process, the first solver will be rewarded certain tokens from the system and blockchain user [25].

The permissionless chain network is especially suitable to be deployed among IMDs for constructing the decentralized autonomous framework in wireless communication networks. With the help of blockchain, service providers of IMDs only need to invest a small operating fee to keep the self-organized system operating normally. However, the mining process is considered to be a computationally intensive process, which obstructs those lightweight resource-limited IMDs from being directly participating in the consensus process. MEC has been introduced to leverage available computation resources in mobile environments. With the help of MEC, IMDs can acquire more computation and communication resources, while decreasing the latency and enhancing quality of service (QoS) compared with the traditional centralized system, which are key requirements in the fifth-generation (5G) networks [26].

7.2.2.2 Framework Description
As illustrated in Figure 7.2, there are three entities in the system that are edge computing service providers (ESPs), IMDs, and MEC server manager. More specifically, IMDs are distinguished according to its mining strategy. For ease of expression, the IMD that adopts solo mining strategy is denoted by sIMD, and the IMD that adopts collusive mining strategy is denoted by cIMD. sIMD mines alone and it can only rent computation resources from an

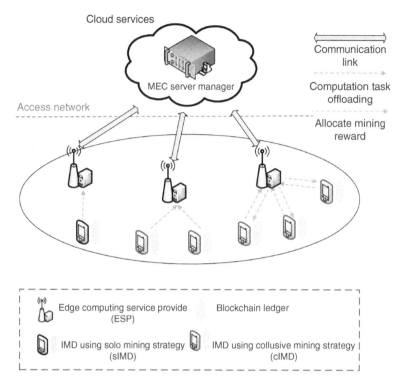

Figure 7.2 Architecture for MEC-based blockchain system. Source: Zhao et al. [1].

ESP. Once the sIMD mines a valid block, it obtains the overall mining reward alone. cIMDs rent computation resources from ESPs and they pool their computation resources and mine collusively. Contrary to sIMD, once cIMDs mined a valid block, they share the common mining reward. The sharing of mining reward process is led by the corresponding ESP. An ESP is constituted by two parts, a communication unit and a server. The communication unit is responsible for communicating with IMDs and managing them within its communication range, while sharing the common mining reward of cIMDs.

Some research works have discussed the access management of IMDs using blockchain technology, like [27]. The constrained application protocol (CoAP) protocol [28] can be adopted in the communication link between ESP and IMDs, while also providing secure communication channels through Datagram Transport Layer Security (DTLS) [29]. Actually, ESP can provide an interface to translate the CoAP protocol into JSON format for the ease of incorporating into Ethernet and extending application scenarios in the future.

To provide the computation task offloading function for IMDs, ESP should equip a server with strong processing ability. Additionally, ESPs are managed by an MEC server manager in the cloud area. The underlying IMDs constitute a P2P network; they run blockchain clients and can communicate with each other. An IMD can be a blockchain user or a miner. The difference between the blockchain user and miner is whether they participate in the mining process. Blockchain user only generates transactions, and the consensus is accomplished by miners. The PoW consensus algorithm is deployed among blockchain miners to reach consensus on the newly generated transactions from blockchain user. Considering the limited storage ability among IMDs, each IMD can store partial blockchain ledger, i.e. the latest several blocks. Meanwhile, as discussed in [30], considering the sleep state among IMDs, they can acquire the latest blockchain ledger from neighbor IMDs once booting.

7.3 Use Cases

In addition to the framework design, there have been many combination cases for specific scenarios. In Section 7.3.1, the convergence of blockchain and federated learning is studied to achieve a secure training process. In Section 7.3.2, the blockchain-assisted device authentication is proposed to carry out the decentralized information exchange among multiple domains for industrial IoT.

7.3.1 Security Federated Learning via MEC-Enabled Blockchain Network

7.3.1.1 Background

Recently, the ubiquitous proliferation of end devices and popular applications brings a huge burden on existing networks. The upcoming fully decentralized and intelligent communications and networking systems for providing innovative secure services trigger the investigation of prospective technologies, and federated learning [31, 32] is a promising approach for privacy-preserved edge intelligence in such distributed scenarios. The local training is executed by users on their data, which usually adopts the gradient descent optimization algorithm. Then, users keep their data with themselves but send the parameters to the server for aggregation. Thus, federated learning achieves edge intelligence by learning from distributed data in a privacy-preserved manner. However, the authors of [33] demonstrated that gradient updates might leak significant information about customers' training data. Attackers can recover data from gradients uploaded by customers. Besides, the federated approach for training the model is susceptible to model poisoning attacks. Moreover, information leakage risks exist in the third party's mobile edge computing server [34].

To address the issues mentioned above, a series of works have studied leveraging blockchain for secure federated learning. Kim et al. [35] designed BlockFL with multiple miners to coordinate federated learning (FL) tasks and maintain the global model. Zhou et al. [36] proposed using blockchain to maintain the global model within a community to reach a consensus and leveraging all reduce protocol to transmit and update the model among multiple communities. The global model is updated continuously and promoted by various communities. Weng et al. [37] proposed a DeepChain architecture to solve the problems of privacy and audibility by storing the local machine learning model in blockchain and regarding these models as transactions. In order to store sensitive and personal data in a diskless environment, Lu et al. [38] adopt blockchain to form a data-sharing platform that is beneficial for data privacy and security. Besides, Kang et al. [39] leverage the blockchain to achieve secure reputation management for workers with non-repudiation and tamper resistance properties in a decentralized manner. The differences between architectures of the traditional FL and the blockchained FL are shown in Figure 7.3.

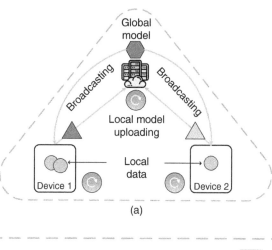

(a)

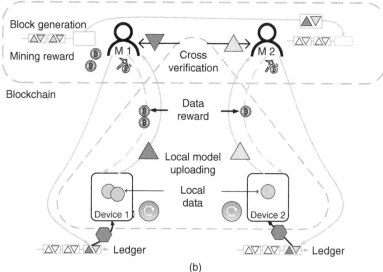

(b)

Figure 7.3 (a) The structure of traditional FL and (b) the structure of the proposed BlockFL. Source: Kim et al. [35] / with permission of IEEE.

7.3.1.2 Blockchain-Driven Federated Learning

As shown in Figure 7.3, the logical structure of BlockFL consists of devices and miners. The miners can physically be either randomly selected devices or separate nodes such as network edges (i.e. base stations in cellular networks), which are relatively free from energy constraints in the mining process. The one-epoch operation in BlockFL is presented in Figure 7.4.

The BlockFL operation of the device D_i at the 1th epoch is described by the following seven steps [35].

1. *Local model update*: The device D_i executes local training with the number N_i of iterations.

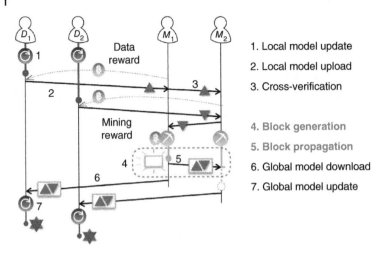

Figure 7.4 The one-epoch operation of BlockFL with and without forking. Source: Kim et al. [35] / with permission of IEEE.

2. *Local model upload*: The device D_i uniformly randomly associates with the miner M_i; if $M = D$, then M_i is selected from $M \backslash D_i$. The device uploads the local model updates and the corresponding local computation time to the associated miner.

3. *Cross-verification*: Miners broadcast the obtained local model updates. At the same time, the miners verify the received local model updates from their associated devices or the other miners. The verified local model updates are recorded in the miner's candidate block until it reaches the block size or the maximum waiting time T_{wait}.

4. *Block generation*: Each miner starts running the PoW until either it finds the nonce or it receives a generated block.

5. *Block propagation*: Denoting as $M_\wedge \in M$ the miner who first finds the nonce. Its candidate block is generated as a new block and broadcast to all miners. In order to avoid forking, an ACK, irrespective of whether forking occurs or not, is transmitted once each miner receives the new block. If a forking event occurs, the operation restarts from Step 1. A miner that generates a new block waits until a predefined maximum block ACK waiting time $T_{a,wait}$.

6. *Global model download*: The device D_i downloads the generated block from its associated miner.

7. *Global model update*: The device D_i locally computes the global model update by using the aggregate local model updates in the generated block.

7.3.1.3 Experimental Results

In this part, the experimental results of BlockFL are provided to verify the efficiency and robustness of the training process. By default, the number of devices and the number of miners are set as 10, and $N_i : Uni(10, 50), \forall D_i \in D$. Other simulation parameters are given as $T_{wait=50\ ms}, T_{a,wait=500\ ms}$.

As shown in Figure 7.5, the BlockFL and the traditional FL achieve almost the same accuracy for an identical N_D. On the other hand, the learning completion latency of our BlockFL is lower than that of the vanilla $FL(N_M = 1)$ as in Figure 7.6, which shows the scalability in

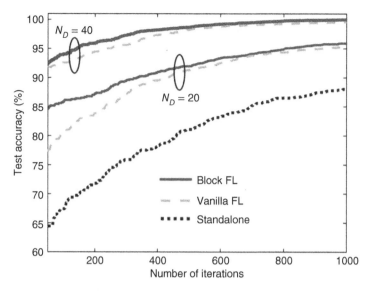

Figure 7.5 Best accuracy of BlockFL, traditional FL, and standalone without federation. Source: Kim et al. [35] / with permission of IEEE.

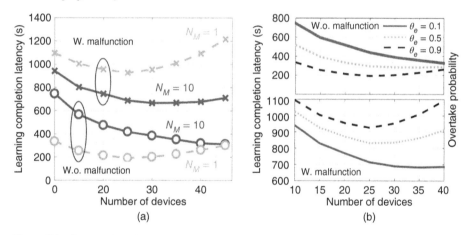

Figure 7.6 Average learning completion latency versus the number of devices (a) under the miners' with/without malfunction and (b) for different energy constraints θ_e. Source: Kim et al. [35] / with permission of IEEE.

terms of N_M and N_D representing numbers of miners and devices, respectively. The average learning completion latency is computed for $N_M = 1$ and $N_M = 10$ with or without the miners' malfunction. The malfunction is captured by adding Gaussian noise $N : (-0.1, 0.01)$ to each miner's aggregate local model updates with a probability of 0.05. Without malfunction, a larger N_M increases the latency because of the increase in their cross-verification and block propagation delays. In BlockFL, each miner's malfunction only distorts its associated device's global model update. Such distortion can be restored by federating with other devices that associate with the miners operating normally. Hence, a larger N_M may achieve a shorter latency for $N_M = 10$ with the malfunction.

Figure 7.6a shows that there exists a latency-optimal number N_D of devices. A larger N_D enables to utilize a larger amount of data samples, whereas it increases each block size and blocks exchange delays, resulting in the convex-shaped latency. In Figure 7.6b, it is assumed that some miners cannot participate if their battery level is lower than a predefined threshold value θ_e, normalized battery level, and $\theta_e \in [0, 1]$. Without the malfunction of miner nodes, the learning completion latency becomes larger for a lower θ_e because of an increase in cross-verification and block propagation delays. On the contrary, when the malfunction occurs, a lower θ_e achieves a shorter latency because more miners federate with leading to robust global model updates.

7.3.2 Blockchain-Assisted Secure Authentication for Cross-Domain Industrial IoT

7.3.2.1 Background

With the proposal of Industry 4.0 [40] and other similar concepts [41] mentioned frequently in recent years, the Industrial Internet of Things (IIoT) is deemed as one of the crucial enabling technologies [42] to put these concepts into practice. On the other hand, it has become a trend that devices inside an administrative domain (e.g. factory) using IIoT technologies connect to automate manufacturing tasks, which can significantly improve productivity and reduce management cost. However, it is hard to have a complete product manufactured in a standalone domain as manufacturing is getting more sophisticated. The entire production process is more likely to span across several domains that have a cooperation relationship. Thus, the blockchain-assisted secure authentication (BASA) mechanism for cross-domain industrial IoT is proposed to improve the security during the process. The identity-based signature (IBS) techniques are employed and extended for cross-domain authentication without introducing any trusted third party. In such a condition, public key certificates are no longer needed, which reduces the heavy work of digital certificate issuing, maintaining, and revoking. Besides, a flexible identity management mechanism, which can efficiently revoke the identities (public keys) of IIoT devices while preserving their privacy, is designed.

7.3.2.2 Blockchain-Driven Cross-Domain Authentication

Several roles exist in the proposed mechanism, which includes IIoT devices, Key Generation Center (KGC), Blockhain Agent Server (BAS), and Authentication Agent Server (AAS). Specifically, KGC is unique in an administrative domain that is responsible for the management of private keys of IIoT devices in that domain. BAS and AAS are two task-specific server introduced for agent missions. Every KGC in an administrative domain needs to build a node to maintain the global ledger of a consortium blockchain. The node server only receives domain-specific information from KGC and writes it into the blockchain. The authentication process can be divided into two key operations, i.e. signature generation and verification. Thus, AAS is introduced to run signature generation and verification operations on behalf of the requesting devices. These two operations can be finished under the coordination of KGC and BAS. According to their functionalities, these roles are grouped into different layers, which are illustrated in Figure 7.7.

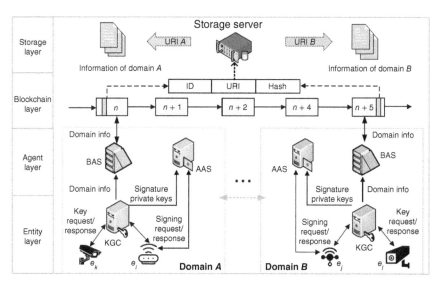

Figure 7.7 Layered architecture of the proposed cross-domain authentication mechanism. Source: Shen et al. [43] / with permission of IEEE.

IIoT devices and KGC are included on the entity layer as they are the least roles in identity-based cryptography (IBC) systems. BAS and AAS are two task-specific servers introduced for agent missions. Besides, two more layers are introduced, which include the blockchain layer and storage layer. The blockchain layer can be treated as a common secure channel for domain-specific information sharing. Blockchain only stores the least information, i.e. domain identifier and its binding values consisting of a uniform resource identifier (URI) and a hash value computed upon the real domain-specific data. URI points to the actual storage file location on the Internet, where real domain-specific data are stored.

Every KGC in an administrative domain needs to build a node to maintain the global ledger of a consortium blockchain. The consortium blockchain node encapsulates the domain-specific information into transactions and writes them into blocks. This domain-specific information will be acquired by other domains for authentication. It would be overloaded if the ledger maintaining work is afforded by KGC. Thus, it is better to split the consortium blockchain node into a separate server. The node server only receives domain-specific information from KGC and writes it into the blockchain. There is no doubt that such a node server can be treated as an agent of KGC. In addition to KGC, BAS also cooperates with AAS to complete cross-domain authentication.

The blockchain layer here represents the consortium blockchain used underneath. It is a global distributed ledger composed of blocks encapsulating numbers of transactions, which carry domain-specific information related to different administrative domains. This information is shared by each KGC in a domain and will be used when cross-domain authentication happens. The global distributed ledger is maintained by a set of preselected nodes, each representing a KGC of an administrative domain.

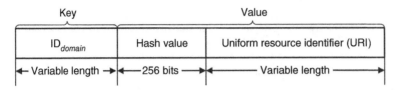

Figure 7.8 Data fields indicating domain-specific information encapsulated into transactions. Source: Shen et al. [43] / with permission of IEEE.

1. *Domain-specific information formation*: In fact, the domain-specific information may contain quite a lot of bytes. Considering the transaction latency and throughput of blockchain, it is better to write minimal information as possible. The domain-specific information written to the global ledger formats is shown in Figure 7.8, where ID_{domain} denotes the unique identifier distinguishing a domain from others.
2. *Uniform resource identifier* (URI): URI is a universal naming and routing method to locate a piece of resource on the Internet, e.g. URI can be a uniform resource locator (URL). URI here points to a file hosted on a cloud service, where the details of domain-specific information are stored.
3. *Hash value*: Hash value is computed on the domain-specific information file. Because the file is stored off-blockchain and hosted on the cloud service governed by a third party, it could be potentially altered by a cyber adversary or a cloud service provider. The hash value on the blockchain is used to verify the authenticity of the real domain-specific information of a domain.

Real domain-specific information is stored off-chain, which includes domain name, domain master public key, domain system parameters, and a public key list of entities in that domain. The real domain-specific information is stored in a single file (e.g. a JSON file) hosted in a cloud service, e.g. AliCloud and Microsoft Azure. To protect data from being modified by malicious adversaries, the whole file is hashed and the hash value is further written into the blockchain. Therefore, the authenticity of the data can be easily verified through the newest hash value maintained on the blockchain compared with the recomputed one upon the actual file.

In the proposed authentication mechanism, entities are authenticated under the coordination of the three main components, i.e. KGC, AAS, and BAS in each domain. The authentication process of entity e_i^A in Domain A cross-authenticated by entity e_j^B in Domain B is illustrated in Figure 7.9.

7.3.2.3 Experimental Results

In this part, the experiments to evaluate the performance of BASA in terms of several critical metrics, including computational overhead, communication overhead, write latency, and query latency, are conducted. All machines are inter-connected in a local network. The operations of KGC and AAS in each domain are executed in a single desktop with AMD Ryzen 5 2600X CPU @3.6 GHz and 16.00 GB memory. The operations of BAS are executed in a virtual machine where 4 GB memory is set, hosted on the desktop using VMware Workstation 15 Pro. The operations of IIoT devices are executed in a laptop with Intel(R) Core(TM) i5-4200 CPU @1.7 GHz and 4 GB memory, whose computing power is similar to the widely

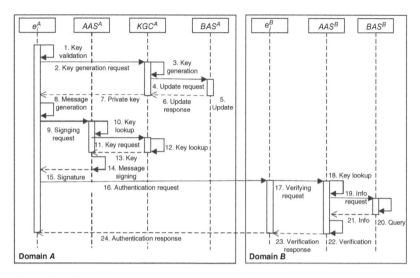

Figure 7.9 Overview of cross-domain authentication process [43]. Entity e_i^A in Domain A is authenticated by entity e_j^B in Domain B under the coordination of KGC, AAS, and BAS in each domain. Source: Shen et al. [43] / with permission of IEEE.

used smartphone. Windows 10 Pro 64-bit operating system and JDK 11.02 are installed on each machine.

As Figure 7.10a shows, the computation cost of all the mechanisms grows linearly as the number of users increases because the computational overhead is approximate for each user. It is obvious that BASA outperforms the other compared mechanisms. Figure 7.10b shows the computational overhead on the server-side. Similar to the user-side, the execution time for BASA on the server-side is lower than the other compared mechanisms. From Figure 7.10a, b, we find that BASA costs approximate overhead on both user and server-side because it uses a digital signature-based technique to realize authentication and unilateral authentication is a symmetrical process. Thus, it is not surprising that the user and server-side consume almost the same computation power.

Figures 7.10c, d show time cost raised by chaincode. Specifically, latency happens when querying data from or writing data into the blockchain ledger. The simulated querying or writing operations are concurrent within 90 m imitating a practical environment where a public key of a device is invoked. As Figure 7.10c shows, the time cost of querying data stays at a low level, which is about 75-90 ms. It is noted that the number of domains does not affect the query time because every time data are queried from the blockchain ledger, the chaincode retrieves data from the local copy of the ledger. As Figure 7.10d shows, the time cost first stays at a low level no matter how many domains are included. However, as the concurrent writes increase, the time cost increases sharply. Interestingly, the more domains are included, the earlier the sharp increase point appears. It is reasonable that consensus time increases as the endorsing and validating nodes increase. It is noted that the write and query latency can be further reduced by exploiting some optimizing techniques.

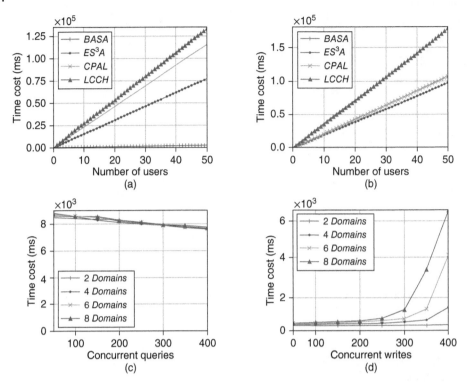

Figure 7.10 Time consumption of BASA with varying parameters (AU: Authentication and KN: Key Negotiation). (a) Cost on users for AU and KN, (b) cost on Server for AU and KN, (c) time cost on querying data by chaincode, and (d) time cost on writing data by chaincode. Source: Shen et al. [43] / with permission of IEEE.

7.4 Conclusion

In this chapter, the integration of blockchain and MEC has been investigated and discussed. On the one hand, the incorporation of blockchain into edge computing enhances security, privacy, and automatic resource usage. On the other hand, the incorporation of edge computing into blockchain brings a powerful decentralized network and rich computation and storage resources to the network edge. For instance, the integration of blockchain and federated learning via MEC brings a huge improvement to security during the training process. Moreover, the security and efficiency of cross-domain device authentication are improved with the assistance of blockchain.

References

1 N. Zhao, H. Wu, and Y. Chen. Coalition game-based computation resource allocation for wireless blockchain networks. *IEEE Internet of Things Journal*, 6 (5): 8507–8518, 2019.

2 Y. C. Hu, M. Patel, D. Sabella, N. Sprecher, and V. Young. Mobile edge computing-a key technology towards 5G. *ETSI White Paper*, 11 (11): 1–16, 2015.

3 L. Lei, Z. Zhong, K. Zheng, J. Chen, and H. Meng. Challenges on wireless heterogeneous networks for mobile cloud computing. *IEEE Wireless Communications*, 20 (3): 34–44, 2013.

4 T. Zhao, S. Zhou, X. Guo, Y. Zhao, and Z. Niu. A cooperative scheduling scheme of local cloud and internet cloud for delay-aware mobile cloud computing. In *2015 IEEE Globecom Workshops (GC Wkshps)*, pages 1–6. IEEE, 2015.

5 F. Wang, J. Xu, X. Wang, and S. Cui. Joint offloading and computing optimization in wireless powered mobile-edge computing systems. *IEEE Transactions on Wireless Communications*, 17 (3): 1784–1797, 2017.

6 M. Satyanarayanan. The emergence of edge computing. *Computer*, 50 (1): 30–39, 2017.

7 H. Yang, Y. Wu, J. Zhang, H. Zheng, Y. Ji, and Y. Lee. Blockonet: blockchain-based trusted cloud radio over optical fiber network for 5G fronthaul. In *Optical Fiber Communication Conference*, pages W2A–25. Optical Society of America, 2018.

8 H. Yang, Y. Liang, J. Yuan, Q. Yao, A. Yu, and J. Zhang. Distributed blockchain-based trusted multi-domain collaboration for mobile edge computing in 5G and beyond. *IEEE Transactions on Industrial Informatics* 16 (11): 7094–7104, 2020.

9 W. Yu, H. Zhang, Y. Wu, D. Griffith, and N. Golmie. A framework to enable multiple coexisting internet of things applications. In *2018 International Conference on Computing, Networking and Communications (ICNC)*, pages 637–641. IEEE, 2018.

10 R. Yu, J. Ding, S. Maharjan, S. Gjessing, Y. Zhang, and D. H. K. Tsang. Decentralized and optimal resource cooperation in geo-distributed mobile cloud computing. *IEEE Transactions on Emerging Topics in Computing*, 6 (1): 72–84, 2015.

11 H. Arasteh, V. Hosseinnezhad, V. Loia, A. Tommasetti, O. Troisi, M. Shafie-khah, and P. Siano. IoT-based smart cities: a survey. In *2016 IEEE 16th International Conference on Environment and Electrical Engineering (EEEIC)*, pages 1–6. IEEE, 2016.

12 J. Feng, F. R. Yu, Q. Pei, X. Chu, J. Du, and L. Zhu. Cooperative computation offloading and resource allocation for blockchain-enabled mobile edge computing: a deep reinforcement learning approach. *IEEE Internet of Things Journal* 7 (7): 6214–6228, 2019.

13 J. Kwak, Y. Kim, J. Lee, and S. Chong. Dream: dynamic resource and task allocation for energy minimization in mobile cloud systems. *IEEE Journal on Selected Areas in Communications*, 33 (12): 2510–2523, 2015.

14 J. Du, L. Zhao, X. Chu, F. R. Yu, J. Feng, and I. Chih-Lin. Enabling low-latency applications in LTE-A based mixed fog/cloud computing systems. *IEEE Transactions on Vehicular Technology*, 68 (2): 1757–1771, 2018.

15 J. Feng, Q. Pei, F. R. Yu, X. Chu, and B. Shang. Computation offloading and resource allocation for wireless powered mobile edge computing with latency constraint. *IEEE Wireless Communications Letters*, 8 (5): 1320–1323, 2019.

16 Y. Mao, J. Zhang, S. H. Song, and K. B. Letaief. Stochastic joint radio and computational resource management for multi-user mobile-edge computing systems. *IEEE Transactions on Wireless Communications*, 16 (9): 5994–6009, 2017.

17 J. Zhao, Q. Li, Y. Gong, and K. Zhang. Computation offloading and resource allocation for cloud assisted mobile edge computing in vehicular networks. *IEEE Transactions on Vehicular Technology*, 68 (8): 7944–7956, 2019.

18 F. Zhou, Y. Wu, R. Q. Hu, and Y. Qian. Computation rate maximization in UAV-enabled wireless-powered mobile-edge computing systems. *IEEE Journal on Selected Areas in Communications*, 36 (9): 1927–1941, 2018.

19 D. Miller. Blockchain and the internet of things in the industrial sector. *IT Professional*, 20 (3): 15–18, 2018.

20 Z. Xiong, Y. Zhang, D. Niyato, P. Wang, and Z. Han. When mobile blockchain meets edge computing. *IEEE Communications Magazine*, 56 (8): 33–39, 2018.

21 G. Han, J. Jiang, L. Shu, and M. Guizani. An attack-resistant trust model based on multidimensional trust metrics in underwater acoustic sensor network. *IEEE Transactions on Mobile Computing*, 14 (12): 2447–2459, 2015.

22 J. Kang, Z. Xiong, D. Niyato, D. Ye, D. I. Kim, and J. Zhao. Toward secure blockchain-enabled internet of vehicles: optimizing consensus management using reputation and contract theory. *IEEE Transactions on Vehicular Technology*, 68 (3): 2906–2920, 2019.

23 Z. Yao, D. Kim, and Y. Doh. Plus: parameterized and localized trust management scheme for sensor networks security. In *2006 IEEE International Conference on Mobile Ad Hoc and Sensor Systems*, pages 437–446. IEEE, 2006.

24 S. Nakamoto. Bitcoin: a peer-to-peer electronic cash system. Cryptography Mailing list at https://metzdowd.com, 03 2009.

25 M. Liu, F. R. Yu, Y. Teng, V. C. M. Leung, and M. Song. Joint computation offloading and content caching for wireless blockchain networks. In *IEEE INFOCOM 2018-IEEE Conference on Computer Communications Workshops (INFOCOM WKSHPS)*, pages 517–522. IEEE, 2018.

26 V. W. S. Wong. *Key Technologies for 5G Wireless Systems*. Cambridge University Press, 2017.

27 O. Novo. Blockchain meets IoT: an architecture for scalable access management in IoT. *IEEE Internet of Things Journal*, 5 (2): 1184–1195, 2018.

28 Z. Shelby, K. Hartke, and C. Bormann. The constrained application protocol (COAP), 2014.

29 E. Rescorla and N. Modadugu. Datagram transport layer security version 1.2., 2012.

30 C. Xu, K. Wang, P. Li, S. Guo, J. Luo, B. Ye, and M. Guo. Making big data open in edges: a resource-efficient blockchain-based approach. *IEEE Transactions on Parallel and Distributed Systems*, 30 (4): 870–882, 2018.

31 B. McMahan, E. Moore, D. Ramage, S. Hampson, and B. A. y. Arcas. Communication-efficient learning of deep networks from decentralized data. In *Artificial Intelligence and Statistics*, pages 1273–1282. PMLR, 2017.

32 K. Bonawitz, H. Eichner, W. Grieskamp, D. Huba, A. Ingerman, V. Ivanov, C. Kiddon, J. Konečný, S. Mazzocchi, H. B. McMahan, et al. Towards federated learning at scale: system design. *arXiv preprint arXiv:1902.01046*, 2019.

33 L. Melis, C. Song, E. De Cristofaro, and V. Shmatikov. Exploiting unintended feature leakage in collaborative learning. In *2019 IEEE Symposium on Security and Privacy (SP)*, pages 691–706. IEEE, 2019.

34 Y. Zhang, T. Gu, and X. Zhang. MDLdroid: a chainSGD-reduce approach to mobile deep learning for personal mobile sensing. In *2020 19th ACM/IEEE International Conference on Information Processing in Sensor Networks (IPSN)*, pages 73–84. IEEE, 2020.

35 H. Kim, J. Park, M. Bennis, and S.-L. Kim. Blockchained on-device federated learning. *IEEE Communications Letters*, 24 (6): 1279–1283, 2019.

36 S. Zhou, H. Huang, W. Chen, P. Zhou, Z. Zheng, and S. Guo. PIRATE: a blockchain-based secure framework of distributed machine learning in 5G networks. *IEEE Network*, 34 (6): 84–91, 2020.

37 J. Weng, J. Weng, J. Zhang, M. Li, Y. Zhang, and W. Luo. DeepChain: auditable and privacy-preserving deep learning with blockchain-based incentive. *IEEE Transactions on Dependable and Secure Computing*, 1-1, 2019 (Early access).

38 Y. Lu, X. Huang, Y. Dai, S. Maharjan, and Y. Zhang. Blockchain and federated learning for privacy-preserved data sharing in industrial IoT. *IEEE Transactions on Industrial Informatics*, 16 (6): 4177–4186, 2019.

39 J. Kang, Z. Xiong, D. Niyato, S. Xie, and J. Zhang. Incentive mechanism for reliable federated learning: a joint optimization approach to combining reputation and contract theory. *IEEE Internet of Things Journal*, 6 (6): 10700–10714, 2019.

40 H. Lasi, P. Fettke, H.-G. Kemper, T. Feld, and M. Hoffmann. Industry 4.0. *Business & Information Systems Engineering*, 6 (4): 239–242, 2014.

41 M. Shen, X. Tang, L. Zhu, X. Du, and M. Guizani. Privacy-preserving support vector machine training over blockchain-based encrypted IoT data in smart cities. *IEEE Internet of Things Journal*, 6 (5): 7702–7712, 2019.

42 X. Du, M. Zhang, K. E. Nygard, S. Guizani, and H.-H. Chen. Self-healing sensor networks with distributed decision making. *International Journal of Sensor Networks*, 2 (5–6): 289–298, 2007.

43 M. Shen, H. Liu, L. Zhu, K. Xu, H. Yu, X. Du, and M. Guizani. Blockchain-assisted secure device authentication for cross-domain industrial IoT. *IEEE Journal on Selected Areas in Communications*, 38 (5): 942–954, 2020.

8

Performance Analysis on Wireless Blockchain IoT System

Yao Sun[1], Lei Zhang[1], Paulo Klaine[1], Bin Cao[2], and Muhammad Ali Imran[1]

[1]*James Watt School of Engineering, University of Glasgow, G12 8QQ, Glasgow, United Kingdom*
[2]*Institute of Network Technology, Beijing University of Posts and Telecommunications, Beijing, 100876, China*

8.1 Introduction

The Internet of Things (IoT) is envisioned as one of the most promising technologies for constructing a global network with machines and devices that will gradually cover almost all aspects of human life [1, 2]. With such an important role it plays in our life, there is a consensus that the security problem present in IoT devices is of the first priority to be addressed [3, 4] because of the devices' ease of access as well as its hardware/software constraints. However, despite its importance, today's methods to tackle security in IoT are inefficient. For example, whenever vulnerability is found, there needs to be a tremendous amount of effort and cost invested in a short amount of time in order to update IoT devices. Moreover, in current centralized IoT systems, a cloud server is necessary for the identification, authorization, and communication among low-end devices, resulting in huge operational costs on the construction and maintenance of servers. Finally, because of the involvement of third parties in centralized IoT systems, the high agent cost makes smart contracts [5] (such as autonomous micropayments and information exchange) among devices unattractive and thus poses a bottleneck on the prosperity of IoT ecosystems, while also limiting interoperability between devices.

In order to effectively address the aforementioned problems, intensive work has been done. As a revolution in systems of record, blockchain has been regarded as a promising technology to address IoT's trust and security concerns, as well as its high-maintenance costs [5–7]. Specifically, blockchain enables secure and reliable transactions/communications between two smart devices without the need of a central authority, which improves the settlement time from days to almost instantaneous [8], while also saving agent fees.

Although good prospects can be expected, there are some problems associated with the combination of blockchain and IoT. As pointed by Christidis and Devetsikiotis [5] and Xu et al. [9], it is computationally intensive and power consuming to implement current blockchain consensus mechanisms, especially proof-of-work (PoW) based ones, as most IoT devices may not bear the necessary computing capability, memory, as well as power needed to run and store the blockchain information. Other relevant state-of-the-art works

Wireless Blockchain: Principles, Technologies and Applications, First Edition.
Edited by Bin Cao, Lei Zhang, Mugen Peng and Muhammad Ali Imran.

focus on blockchain protocol design in terms of its consensus mechanism [5 10–12], trust, and privacy [13, 14]. However, despite their importance, these works do not consider an efficient network architecture to support low-cost blockchain-enabled wireless IoT systems. Only the authors in [15] propose a blockchain-enabled IoT network architecture with a central management system, which is more suitable for a small local network (such as smart homes, the use-case considered by the authors).

From the communication perspective, in traditional blockchain systems, a common assumption is that communications among the nodes are perfect without any throughput and delay constraints. However, considering the unstable wireless channel quality, interference, limited resources, and different network topologies, it is necessary to investigate the impact of wireless communication on blockchain-enabled IoT systems [16]. In particular, fundamental communication metrics such as SINR and throughput should be analyzed to determine how the wireless communication quality may affect or constrain a blockchain-enabled IoT network deployment (e.g. node distribution), protocols (e.g. size of block and frequency of transactions), and transaction consensus delay. On the other hand, given a transaction throughput bound in blockchain (e.g. one block in every 10 minutes, as defined in Bitcoin [17]), it is valuable to know how to deploy IoT nodes in order to optimally meet this bound. In general, the performance of blockchain-enabled IoT systems in terms of transaction throughput (the number of confirmed transactions in a unit time), communication throughput (the amount of transmitted data in a unit time), and transaction successful rate should be associated with both blockchain transaction arrival rate in time domain and the nodes geographic distribution in the spatial domain [18]. Therefore, a fundamental challenge is in establishing a valid framework of analytical models considering the two-dimensional randomness to evaluate the performance of blockchain-enabled IoT system.

Although there has been tremendous research work on wireless network modeling, the existing performance analysis focusing on traditional wireless networks cannot be directly applied to blockchain-enabled IoT system because these works do not consider blockchains. In [19–21], stochastic geometry is exploited to evaluate the network performance in terms of association probability, throughput, and outage probability in traditional cellular networks [19], heterogeneous cellular networks [20], and millimeter wave networks [21], respectively. These investigations focus on the spatial domain without considering the time domain, whose characteristics are much different for blockchains. Most existing works that simultaneously consider spatial and time domains, such as [22–24], combine queuing theory and stochastic geometry to evaluate delay performance. However, considering the characteristics of blockchain (e.g. low transaction arrival rate and limited transaction throughput), the performance evaluations could be much different.

In addition to communication performance analysis for blockchain-enabled IoT systems, security performance analysis is also necessary when some attacks exist in this system. Considering the vulnerability of wireless links, the types of attacks could be various. To the best of the authors' knowledge, there are no such analytical models dedicated to blockchain-enabled wireless IoT systems. Based on that, in this chapter, we first present a new network model for blockchain-enabled IoT systems. Then, we theoretically analyze the performance of blockchain-enabled IoT systems and propose an optimal blockchain full function node[1] deployment based on the analysis. Numerical results show that the difference between analytical and simulation results is as low as 4%, which clearly validate the accuracy of our theoretical analysis.

8.2 System Model

In this section, we first present the blockchain-enabled IoT network model and then describe the wireless communication model by considering the spatio-temporal characteristics of this network.

8.2.1 Blockchain-Enabled IoT Network Model

Consider a blockchain-enabled IoT network, as shown in Figure 8.1, which consists of two main elements: IoT transaction nodes (TNs) and full function nodes (FNs). TNs can be basically seen as traditional low-cost, low-power IoT devices supported by the blockchain system. At any time point, the TNs can be classified as two types: active TNs, with data being transmitted, and idle TNs, with no transmissions, with nodes being able to be in a single state at a time and switching between states in different time periods. The detailed time domain characteristics will be further discussed in Section 8.2.2. FNs are nodes in the blockchain with high computing and storage power. They have full functionalities to support blockchain protocols; thus, they are capable of performing transaction confirmation, data storage, and building new blocks. To guarantee the security and improve the effectiveness in blockchain, FNs are connected with each other through high data rate links via an independent interface. In addition, all FNs are connected with TNs through wireless communications. In this context, transactions are considered to be any valuable information transmitted between TNs. Because these nodes have low computing power, the transactions are processed and confirmed by FNs, which will then store them in blocks, forming the blockchain.

The processing of a transaction can be described as follows: Once a transaction arrives at a specific TN, the TN should broadcast the information to FNs by using wireless IoT networks. This information should be received by as many FNs as possible in order to enhance the security level. However, in this chapter, and as a starting point, we assume that each transaction successfully received by a single FN is secure[2]. The received information by an FN through the wireless channel will be shared within all FNs via the dedicated connections (e.g. wired links). Then, the FN that has the right of building a block will insert this transaction into the chain, and all FNs need to update their own ledger accordingly. Note that our focus is in the uplink transmission from TN to FN; even though the downlink performance can be performed in a similar manner, it is out of the scope of this chapter.

It is also considered that the association relationship between TNs and FNs is not fixed. More specifically, Figure 8.1 presents an instantaneous network state at a certain time. Note

---

1 Full function node has high computing and storage power with full functionalities to support blockchain protocols, which will be explained in detail in the system model.
2 Nevertheless, our framework can be generalized to the scenario where each transaction should be successfully received by multiple FNs.

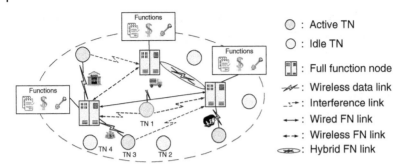

Figure 8.1 Blockchain-enabled IoT network model.

that the network state including the active TNs, the active FNs, the association relationship between TNs and FNs, etc., can be dynamic along with time. In other words, TNs can choose a suitable FN to be served according to the current network state (in this work, we assume that TNs choose the nearest trustable FN). Also note that the FN set and some nodes in the TN set can dynamically switch depending on the network needs. In the case that an FN is shut down or attacked, the TNs that are associated with this FN can be associated with other nearby trustable FNs; therefore, the centralization problem is avoided in our system.

We assume that some malicious FN nodes exist in the network, and the percentage of malicious FN nodes does not exceed half the number of total FNs; thus, it does not violate the security requirement of blockchain systems. Also, malicious FNs can make Denial of Service (DoS) attacks (such as rejecting to forward information). In this case, it is equivalent to the FNs being shut down, and the TNs associated with them will seek for other nearby FNs. However, the malicious FNs cannot change the transaction because it is authenticated by digital signature [25]. If any part of the transaction is modified by the malicious FN, the signature would be invalidated; thus, this transaction cannot be inserted into a block. As a result, the TN will seek for other nearby FNs when the transaction is not confirmed within a predefined duration (which is also equivalent to the malicious FN being shut down).

We assume that the connection between FNs can be of several types including wired link, wireless point-to-point link, and wired and wireless hybrid relay link, as shown in Figure 8.1. The connection type between two specific FNs is determined by the environment around the two FNs. For example, traditional wireless links can be used when the distance between the two FNs is short, while wired links should be used in a more complicated or difficult to access environment. This type of hybrid wired-wireless connections could be more practical and cost-effective in large-scale IoT networks, while also guaranteeing the transmission security between FNs by using some wireless physical layer security techniques, such as physical layer authentication [26].

Before moving to the wireless communication model, we need to clarify two definitions in blockchain-enabled IoT networks: transaction throughput and communication throughput. Transaction throughput is defined as the number of transactions confirmed in a unit time by the blockchain system; usually, we use transactions per second (TPS) as its metric. In communication unlimited case, without delay and throughput constraints, TPS is typically limited by the blockchain's block size, consensus mechanism, as well as the transaction arrival rate. For example, the maximum TPS of Bitcoin and Ethereum is of 7 and 20, respectively [27]. Communication throughput, on the other hand, is defined as the amount of transmitted data in a unit time for this network; usually, we use bits per second (bps) as its unit, and it is related to the wireless channel condition, radio resource allocation, as well as radio access technology (e.g. LTE and WiFi). In wireless blockchain networks, the required communication throughput R can be calculated based on the transaction throughput C_T as follows:

$$R \geq L \cdot C_T, \tag{8.1}$$

where L is the packet length for a transaction. As the limitation of C_T, there is a maximum required communication throughput.

8.2.2 Wireless Communication Model

Here, we present the wireless communication model with respect to the spatio-temporal domain characteristics of blockchain-enabled IoT networks. Let us start by first describing the network spatial distribution. Let A be the considered two-dimensional area where TNs and FNs are assumed to be distributed as a homogeneous PPP with density λ_d and λ_f, respectively. It is practical to assume that the minimum distance between a TN and an FN is set as d_{min}. The association rule between TNs and FNs is based on distance, i.e. a TN is associated with the nearest FN.

Once a transaction arrives at a specific TN, the TN should be in active mode to broadcast the information to FNs. It is considered that the transaction packet length L is usually very short. As per 3GPP standards, the data packet size (i.e. active time of TN) can be modeled as a Pareto distribution with shape parameter $\alpha = 2.5$ and minimum data packet size of 20 bytes [28]. Therefore, the expectation of the data packet size in IoT networks is around 33.3 bytes, which can be transmitted in a single Transmission Time Interval (TTI) in 1 ms by using long-term evolution (LTE) system [29]. Moreover, the traffic arrival rate is as low as from half an hour to several hours [28], implying that the traffic of IoT nodes is not very active. Hence, based on the above, it is reasonable to assume that a TN's active time is a small constant with $t \ll T$, and thus, the number of arrived transactions M in time T can be assumed to be a Poisson distribution with parameter $\lambda_a T$.

Regarding the wireless channel model, in this work, only the uplink transmission is considered. Given a specific TN served by an FN, the desired signals experience path loss $g(d)$, where d is the distance between the TN and the FN. Moreover, other TNs can also generate interference to the considered TN. From the time domain perspective, only the active TNs are counted as interference TNs, whereas from the spatial domain, we assume that only the TNs within a certain circular area, where the serving FN is located as the center with radius D_0, as shown in Figure 8.2, could contribute to the interference. We set the transmit power of all TNs as P; therefore, the received SINR can be expressed as

$$SINR(D_1, N_I, \mathbf{D}_2) = \frac{Pg(D_1)}{\sum_{i=1}^{N_I} Pg\left(D_2^{(i)}\right) + \sigma}, \tag{8.2}$$

where D_1 is the distance between the desired TN and the serving FN, $\mathbf{D}_2 = \left[D_2^{(1)}, D_2^{(2)}, \ldots, D_2^{(N_I)}\right]$ is the distance vector for all interference TNs, N_I is the number of interference TNs, and σ is the noise power. For convenience, the frequently used notations are summarized in Table 8.1.

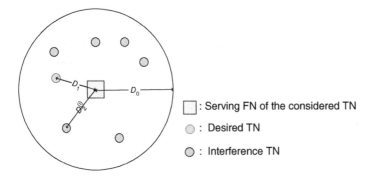

Figure 8.2 Interference area for a specific TN.

Table 8.1 Frequently used notations.

Notation	Definition
A	The whole considered area
D_0	Radius of interference area
D_1	Distance between desired TN and the serving FN
$\mathbf{D}_2$	Distance vector of all interference UEs (a vector)
$D_2^{(i)}$	Distance between interference TN i and the FN
λ_d	TN density
λ_f	FN density
λ_a	Blockchain transaction arrival rate
M_i	Number of arrival transactions in time T for TN i
N_0	The number of total TNs
K	The number of TNs in interference area
N_I	The number of interference TNs
t	Radio transmission time for a transaction
T	The total considered time
L	The packet length of each blockchain transaction
P	TN transmit power
$g(d)$	Channel path loss model (a function of distance)
f_X	Probability density function of random variable X

8.3 Performance Analysis in Blockchain-Enabled Wireless IoT Networks

In this section, we analyze the theoretical transmission performance in terms of SINR, transaction data packet (TDP) transmission successful rate, as well as overall communication throughput by considering the blockchain characteristics, such as short transaction packet length, low TN active rate, and limited transaction throughput. We first derive the probability density function (PDF) of SINR according to the spatio-temporal domain modeling and then calculate the blockchain TDP transmission successful rate based on the PDF of the SINR. Finally, we obtain the overall communication throughput PDF in a close-form expression under the constraints of blockchain transaction throughput and the TDP transmission successful rate.

8.3.1 Probability Density Function of SINR

Let us start with the SINR distribution analysis. Given that both TNs and FNs are geographically distributed as homogeneous PPP, it is reasonable to investigate the SINR performance of any arbitrary TN. To derive the PDF of SINR in (8.2), the desired signal power and interference need to be studied separately.

For a specific TN, the desired signal power S is a random variable written as $S = Pg(D_1)$, where D_1 is the distance between the TN and the serving FN. As the transmit power, P, is fixed, S is only related to D_1. Proposition 8.1 gives the PDF of D_1. For convenience, we use capital letters to denote random variables, and the corresponding lowercases to the value of random variables.

Proposition 8.1   *The PDF of the distance D_1 between a specific TN and the serving FN is*

$$f_{D_1}(d_1) = 2\pi\lambda_f d_1 \exp\{-\lambda_f\pi(d_1)^2\}. \tag{8.3}$$

According to the association rules presented in Section 8.2, $D_1 > d_1$ is the event that there is no FN distributed in the circular area with radius d_1. Thus, $\Pr(D_1 > d_1) = \exp\{-\lambda_f\pi(d_1)^2\}$. Hence, the PDF of D_1 is $f_{D_1}(d_1) = \frac{d(F_{D_1}(d_1))}{d(d_1)} = 2\pi\lambda_f d_1 \exp\{-\lambda_f\pi(d_1)^2\}$.

Therefore, based on Proposition 8.1, we obtain that the PDF of desired signal power as

$$f_S\left(S = Pg(d_1)\right) = f_{D_1}(d_1) = 2\pi\lambda_f d_1 \exp\{-\lambda_f\pi(d_1)^2\}. \tag{8.4}$$

We now study the distribution of the received interference. We start from the number of interference TNs N_I. As stated in Section 8.2, only active TNs located in the interference area are counted for interference contribution[3]. The number of TNs $K(K \geq N_I)$ within the interference area is still a variable of Poisson distribution with density parameter $\pi(D_0)^2\lambda_d$. On the other hand, as the transmission time for TN is t, the TNs that are active during time period $[-t, t]$ can bring interference. For a TN, the number of arrived transactions is distributed as a PPP with parameter $2t\lambda_a$. Therefore, the probability of a TN during time period $[-t, t]$ of being active is

$$\Pr(active) = 1 - \exp\{-2t\lambda_a\}. \tag{8.5}$$

The probability of the number of interference TNs $N_I = n_I$ given $K = k$ is

$$\Pr(N_I = n_I | K = k) = C_k^{n_I}(1 - \exp\{-2t\lambda_a\})^{n_I}, \tag{8.6}$$

---

3 This approximation is removed when the area considered is infinite.

where $C_k^{n_I}$ is the combinatorial number. Therefore, the PDF of N_I is

$$
\begin{aligned}
f_{N_I}(n_I) &= \Pr(N_I = n_I) \\
&= \sum_{k=n_I}^{N_0} \Pr(N_I = n_I | K = k) \Pr(K = k),
\end{aligned}
\tag{8.7}
$$

where

$$
\Pr(K = k) = \frac{\left(\pi (D_0)^2 \lambda_d\right)^k}{k!} \exp\{-\pi (D_0)^2 \lambda_d\}.
\tag{8.8}
$$

Next, we investigate the distance $D_2^{(i)}$ between an interference TN i and the FN. Proposition 8.2 gives the PDF of $D_2^{(i)}$.

Proposition 8.2 *The PDF of the distance D_2 is*

$$
f_{D_2^{(i)}}\left(d_2^{(i)}\right) = \frac{2d_2^{(i)}}{(D_0)^2},
\tag{8.9}
$$

where D_0 is the radius of the interference area.

The interference TNs are distributed as a PPP with density $\pi (D_0)^2$. Therefore, for TN i, the PDF of location (X, Y) is

$$
f_{(X,Y)} = \frac{1}{\pi (D_0)^2}.
\tag{8.10}
$$

The CDF of distance $D_2^{(i)}$ can be calculated as

$$
\begin{aligned}
F_{D_2^{(i)}}(d_2^{(i)}) &= \iint_{X^2+Y^2 \le (d_2^{(i)})^2} \frac{1}{\pi (D_0)^2} \sqrt{X^2 + Y^2} dX dY \\
&\stackrel{(a)}{=} \int_0^{d_2^{(i)}} \int_0^{2\pi} \frac{r}{\pi (D_0)^2} d\theta dr \\
&= \left(\frac{d_2^{(i)}}{D_0}\right)^2,
\end{aligned}
\tag{8.11}
$$

where (a) is obtained by using the following replacement

$$
\begin{cases} X = r\cos\theta, \\ Y = r\sin\theta. \end{cases}
\tag{8.12}
$$

Hence, based on (8.12), the PDF of $D_2^{(i)}$ is

$$
f_{D_2^{(i)}}\left(d_2^{(i)}\right) = F'_{D_2^{(i)}}\left(d_2^{(i)}\right) = \frac{2d_2^{(i)}}{(D_0)^2}.
\tag{8.13}
$$

From Proposition 8.2, we can express the PDF of interference I_i generated by TN i as

$$
f_{I_i}\left(I_i = Pg(d_2^{(i)})\right) = f_{D_2^{(i)}}(d_2^{(i)}) = \frac{2d_2^{(i)}}{(D_0)^2}.
\tag{8.14}
$$

The total interference, denoted by $I(N_I, \mathbf{D}_2)$, is related to the number of interference TNs N_I and the distance $\mathbf{D}_2$ of these interference TNs. From (8.7) and (8.15), we have the PDF of $I(N_I, \mathbf{D}_2)$

$$f_I(N_I = n_I, \mathbf{D}_2 = \mathbf{d}_2) = f_{N_I}(n_I) \Pr(\mathbf{D}_2 = \mathbf{d}_2 | N_I = n_I)$$

$$= f_{N_I}(n_I) \left(\frac{2}{(D_0)^2} \right)^{n_I} \prod_{n=1}^{n_I} d_2^{(n)}. \tag{8.15}$$

As SINR expressed in (8.2) is related to D_1, N_I, and $\mathbf{D}_2$, the PDF of SINR can be expressed as

$$f_{SINR}(D_1 = d_1, N_I = n_I, \mathbf{D}_2 = \mathbf{d}_2)$$
$$= f_{D_1}(D_1 = d_1) f_I(N_I = n_I, \mathbf{D}_2 = \mathbf{d}_2), \tag{8.16}$$

where f_{D_1} and f_I are given in (8.7) and (8.17), respectively.

8.3.2 TDP Transmission Successful Rate

When the received SINR is greater than β (given as the SINR threshold that FNs can successfully decode the received information bits), a blockchain transaction transmission is successful. Therefore, we need to calculate the probability $\Pr(SINR > \beta)$ that can be expressed as

$$\Pr(SINR > \beta) = \iiint_\Omega f_{SINR} d\Omega, \tag{8.17}$$

where Ω is the area of $(D_1, N_I, \mathbf{D}_2)$ that satisfies $SINR(D_1, N_I, \mathbf{D}_2) > \beta$. As f_{SINR} is obtained in (8.19), we only need to find the satisfied area Ω.

For the distance D_1 between the desired TN and the serving FN, it is safe to say that the SINR cannot be greater than β when $D_1 > D_0$. Thus, the satisfied range of D_1 is $[0, D_0]$. Then, for a given $D_1 = d_1$, we need to determine the satisfied number of interference TNs N_I as well as the locations of these TNs, $\mathbf{D}_2$. To obtain the close-form expression of $\Pr(SINR > \beta)$, we use the following approximation:

The number of interference TNs $N_I \cong E(N_I)$, $\tag{8.18}$

where $E(N_I)$ is the expectation of random variable N_I. As the total number of TNs, N_0, in IoT networks is usually quite large, this approximation can be considered accurate. In addition, simulation results, as shown in Section 8.6, indicate that the approximation is very effective in all considered scenarios. Based on the TN distribution and transaction arrival models, we have that

$$E(N_I) = E(K) \cdot \Pr(active)$$
$$= \pi (D_0)^2 \lambda_d (1 - \exp\{-2t\lambda_a\}) \triangleq \bar{n}_I, \tag{8.19}$$

where $\Pr(active)$ is defined as (8.5). Thus, the SINR is only related to D_1, and $\mathbf{D}_2$, and it can be rewritten as

$$SINR(D_1, \mathbf{D}_2) = \frac{Pg(d_1)}{\sum_{i=1}^{\bar{n}_I} I_i + \sigma}. \tag{8.20}$$

For a given $D_1 = d_1$, we have

$$\Pr\left(SINR(D_1 = d_1, \mathbf{D}_2) > \beta\right) = \Pr\left(\sum_{i=1}^{\bar{n}_I} I_i < \frac{Pg(d_1) - \sigma}{\beta}\right). \tag{8.21}$$

Because of the high TN density in blockchain networks and the large radius of the interference area, D_0, $\bar{n}_I$ can be considered as a large number. Moreover, $I_i(i = 1, 2, \ldots, \bar{n}_I)$ is a set of random variables with independent identically distribution, and thus, $\sum_{i=1}^{n_I} I_i$ can be seen as a normal distribution $N(\mu_I, \delta_I{}^2)$, where $\mu_I = \bar{n}_I E(I_i)$ and $\delta_I = \sqrt{\bar{n}_I} D(I_i)$, where $E(I_i)$ is the expectation and $D(I_i)$ is the variance of variable I_i, respectively [30].

In the following, we present the derivations of μ_I and δ_I, respectively.

$$\mu_I = \bar{n}_I E(I_i)$$
$$= \bar{n}_I \int_{d_2^{(i)} = d_{min}}^{D_0} Pg\left(d_2^{(i)}\right) \Pr\left(D_2^{(i)} = d_2^{(i)}\right)\left(d_2^{(i)}\right)$$
$$= \bar{n}_I \int_{d_2^{(i)} = d_{min}}^{D_0} Pg\left(d_2^{(i)}\right) \frac{2d_2^{(i)}}{(D_0)^2} d(d_2^{(i)})$$
$$= \frac{2P\bar{n}_I}{(D_0)^2}\left(d_2^{(i)}[G(D_0) - G(d_{min})] - \overline{G}(D_0) - \overline{G}(d_{min})\right), \tag{8.22}$$

where G is the primitive function of path loss model $g(d)$ and $\overline{G}$ is the primitive function of G.

$$\delta_I = \sqrt{\bar{n}_I} D(I_i)$$
$$= \sqrt{\bar{n}_I}\left(E\left(I_i^2\right) - E^2\left(I_i\right)\right)$$
$$= \sqrt{\bar{n}_I}\left[\int_{d_2^{(i)} = d_{min}}^{D_0} P^2 g^2\left(d_2^{(i)}\right)\frac{d_2^{(i)}}{2(D_0)^2} d\left(d_2^{(i)}\right) - \left(\frac{\mu_I}{\bar{n}_I}\right)^2\right]. \tag{8.23}$$

Denote $I = \sum_{i=1}^{\bar{n}_I} I_i$, and $I \sim N\left(\mu_I, \delta_I{}^2\right)$. Let $Y = \frac{I - \mu_I}{\delta_I}$, and thus $Y \sim N(0, 1)$. Therefore,

$$\Pr\left(\sum_{i=1}^{\bar{n}_I} I_i < \frac{Pg(d_1) - \sigma}{\beta}\right) = \Pr\left(Y < \frac{\frac{Pg(d_1) - \sigma}{\beta} - \mu_I}{\delta_I}\right)$$
$$= \Phi\left(\xi\left(d_1\right)\right), \tag{8.24}$$

where $\xi\left(d_1\right) = \frac{\frac{Pg(d_1) - \sigma}{\beta} - \mu_I}{\delta_I}$, Φ is the cumulative density function of the standard normal distribution. Therefore, we have

$$\Pr\left(SINR > \beta\right) = \iiint_\Omega f_{SINR} d\Omega$$
$$= \int_{d_1 = d_{min}}^{D_0} f_{D_1}\left(d_1\right) \Phi\left(\xi\left(d_1\right)\right) d\left(d_1\right). \tag{8.25}$$

Note that (8.37) is actually the close-form expression of $\Pr\left(SINR > \beta\right)$, which can be calculated analytically when function f_{D_1}, the value of μ_I, δ_I as well as the parameters β, and σ, are given.

8.3.3 Overall Communication Throughput

Denote by R the overall required communication throughput, which can be expressed as

$$R = L \cdot \Pr\,(SINR > \beta) \left(\sum_{i=1}^{N_0} M_i \right), \quad 0 \leq R \leq W, \tag{8.26}$$

where N_0 is the total number of TNs, M_i is the number of arrived transactions for TN i, and W is the communication throughput when the transaction throughput reaches the maximum value. For a given λ_d and λ_f, the variables N_0, L and $\Pr\,(SINR > \beta)$ are constant, while M_i is a set of independent identically PPP-distributed random variables with parameter $\lambda_a \cdot T$. Let $M = \sum_{i=1}^{N_0} M_i$. Because N_0 is a large number, M is a random variable with normal distribution $N(\mu_M, \delta_M{}^2)$, where $\mu_M = N_0 E\,(M_i)$ and $\delta_M = \sqrt{N_0} D\,(M_i)$ [30]. As M_i is distributed as a PPP, $E\,(M_i) = \lambda_a T$ and $D\,(M_i) = \lambda_a T$. Therefore, we have

$$\mu_M = N_0 \lambda_a T, \tag{8.27}$$

$$\delta_M = \sqrt{N_0} \lambda_a T. \tag{8.28}$$

As mentioned in Section 8.2, the maximum required communication throughput by the blockchain system in time T is W. For a given λ_d and λ_f, the PDF of overall required communication throughput R is

$$f_R\,(r = mLPr\,(SINR > \beta))$$
$$= \begin{cases} f_M\,(m) = N\left(\mu_M, \delta_M{}^2\right), & r < W \\ 1 - \Phi\,(m^*), & r = W \end{cases}, \tag{8.29}$$

where Φ is the cumulative density function of standard normal distribution, and

$$m^* = \frac{\frac{W}{L\,\Pr\,(SINR>\beta)} - \mu_M}{\delta_M}. \tag{8.30}$$

For a given λ_d and λ_f, the variables L and $\Pr\,(SINR > \beta)$ are constant. If the overall communication throughput $r < W$, we have $f_R\,(r) = f_M\,(m) = N\left(\mu_M, \delta_M{}^2\right)$. If $r = W$, $mL\,\Pr\,(SINR > \beta) = W$, and thus $m = \frac{W}{L\,\Pr\,(SINR>\beta)} \triangleq \tilde{m}$. Because of the maximum transaction throughput constraint, we have $f_R\,(r = W) = \Pr\,(M \geq \tilde{m})$. Let $Y = \frac{M - \mu_M}{\delta_M}$, and thus $Y \sim N\,(0,1)$ as $M \sim N\left(\mu_M, \delta_M{}^2\right)$. Thus, $\Pr\,(M \geq \tilde{m}) = \Pr\left(Y \geq \frac{\tilde{m} - \mu_M}{\delta_M}\right) = 1 - \Phi(\frac{\tilde{m} - \mu_M}{\delta_M})$, where Φ is the cumulative density function of the standard normal distribution.

8.4 Optimal FN Deployment

For a given TN deployment, we can increase the communication throughput by deploying more FNs and thus support higher blockchain transaction throughput. However, as mentioned in Section 8.2.1, once the transaction throughput reaches its maximum value, increasing communication throughput cannot improve the transaction throughput anymore. Thus, for the sake of saving cost, it is worth to minimize the FN density subject to the blockchain transaction throughput constraint.

From 8.39, we know that R is a function of both λ_f and M, given λ_d and λ_a. To explore the relationship between FN density λ_f and TN density λ_d, we use $E(M) = \mu_M$ to calculate the conditional expectation of R as

$$\ddot{E}(R) = E\left(R \,|\, M = \mu_M\right)$$
$$= \min\left\{L \Pr\left(SINR > \beta\right) \mu_M, W\right\}. \tag{8.31}$$

As it can be seen, (8.45) shows that the overall communication throughput should increase with the number of transactions at the start and stay unchanged when it reaches the maximum value W (i.e. the blockchain transactions saturated). According to (8.45), we find that $\ddot{E}(R)$ can exactly depict this relationship. Therefore, the following Definition 8.1 states the optimal FN deployment.

Definition 8.1 For given λ_d and λ_a, the FN deployment Θ is optimal if the Poisson distribution density $\lambda_f{}^*$ satisfies $\lambda_f{}^* = \arg\min_{\lambda_f}\left(\ddot{E}(R) = W\right)$.

According to Definition 8.1, the optimal FN deployment problem can be formulated as:

$$\min \lambda_f$$
$$s.t. \ddot{E}(R) = W. \tag{8.32}$$

Because of the complexity of $\Pr(SINR > \beta)$, we cannot find the optimal $\lambda_f{}^*$ in a closed-form solution. Fortunately, as $\Pr(SINR > \beta)$ is a monotonic increasing function with FN density λ_f, we design Algorithm 8.1 to find the optimal value of FN density $\lambda_f{}^*$ for a given λ_d and λ_a. In Algorithm 8.1, we first calculate the value of $\overline{n}_I$, μ_I, δ_I and $\xi\left(d_1\right)$ and then determine the searching region of FN density; finally, we find the optimal FN density in the searching region. The detailed steps are stated in Algorithm 8.1.

We find that the major computational complexity of our optimal FN deployment algorithm, Algorithm 8.1, lies on the search stage, where we should determine the optimal FN density from the range $\left[\frac{\lambda_f{}^0}{2}, \lambda_f{}^0\right]$. In the worst case, we need to search $\log_2\lceil\frac{\lambda_f{}^0}{2\epsilon}\rceil$ rounds, where ϵ is a termination parameter. In each round, we should calculate $\ddot{E}(R)$ and compare it with the maximum value W, thus determining the new search area for the next round. The computational complexity of these operations in a round can be seen as a constant, denoted as $O(1)$. Therefore, the total computational complexity of Algorithm 8.1 is $O(\log_2\lceil\frac{\lambda_f{}^0}{2\epsilon}\rceil)$.

8.5 Security Performance Analysis

In this section, we analyze the security performance of the proposed system. Specifically, our analysis focuses on three typical types of malicious attacks: (i) eclipse attack, (ii) random link attack, and (iii) random FN attack.

8.5.1 Eclipse Attacks

As stated in [31], an eclipse attack is defined as the one in which the attacker monopolizes all the downlink and uplink connections of a TN (denoted as the victim), thus isolating the

---

Algorithm 8.1 : Algorithm of Optimal FN Deployment.

Require: all the parameters (except λ_f) and the termination parameter $\epsilon > 0$.
Ensure: optimal FN density λ_f^*.

    Initialization:

1: calculate $\bar{n}_I$, μ_I, δ_I and $\xi\left(d_1\right)$.

    Find searching region:

2: set λ_f^0 as the initial value

3: calculate $\Pr\left(SINR > \beta\right)$ and $\ddot{E}\left(R\right)$ based on λ_f^0

4: **if** $\ddot{E}\left(R\right) < W$ **then**

5:    $\lambda_f^0 = 2\lambda_f^0$ and go back to line 3

6: **else**

7:    break

8: **end if**

    Search stage:

9: set $a = \frac{\lambda_f^0}{2}$, $b = \lambda_f^0$

10: **while** $|b - a| > \epsilon$ **do**

11:    set $\lambda_f^* = \frac{a+b}{2}$

12:    calculate $\Pr\left(SINR > \beta\right)$ and $\ddot{E}\left(R\right)$ based on λ_f^*

13:    **if** $\ddot{E}\left(R\right) < W$ **then**

14:      set $a = \lambda_f^*$

15:    **else**

16:      set $b = \lambda_f^*$

17:    **end if**

18: **end while**

19: **output** λ_f^*

---

victim from the rest of the network. The attacker in this way can filter the victim's view of the blockchain and conduct some activities for his own purposes, such as disrupting the blockchain network, wasting the computer power, etc. [31]. To make eclipse attacks more difficult, several countermeasures have been proposed in [31], including deterministic random eviction, random selection, test before evict, etc. Besides the countermeasures proposed in [31], we can also exploit wireless channel characteristics as well as blockchain protocols to further address eclipse attacks.

Eclipse attacks can be addressed from two aspects: physical layer security aiming at protecting wireless links from attacks and blockchain network protocols to avoid information modifications. Let us elaborate them respectively. Physical (PHY) layer security can be used to avoid the wireless links being attacked and thus are able to address DoS attacks. PHY-layer authentication techniques exploit the spatial decorrelation property of the PHY-layer information, such as received signal strength indicators, received signal strength, channel phase response, channel impulse responses, and channel state information to distinguish radio transmitters and thus detect spoofing attacks with low overhead [26].

From the blockchain network protocol perspective, private key technology should be used to avoid transaction information modifications. Private key consists of a string of characters, which is only known by the TN itself. Each TN owns a pair of private and public keys. The private key is used to sign the transactions [25], which can only be made with private and public keys. Thus, even if the links of the victim TN are monopolized, the transaction information cannot be modified by the attacker because of the lack of information about the private key.

Therefore, by using the technologies of physical layer security and blockchain network protocol (authentication encryption), eclipse attacks could be dealt effectively. Moreover, as discussed in Section 8.2.1, when the attacked TNs are uniformly distributed in the network, the proposed framework is still valid by reducing the TN density accordingly.

8.5.2 Random Link Attacks

A random link attack consists of some links being randomly attacked or blocked because of the instability of the wireless channel. One alternative to mitigate this type of attack is by considered PHY-layer security technology as well. Besides PHY-layer security technology, other methods can also be considered for addressing this type of attack. By sending some acknowledgment signaling (e.g. Hybrid Automatic Repeat Request (HARQ) in LTE systems), TNs can judge whether the current wireless link is blocked. If the link is blocked, the corresponding TN will connect to another nearby FN via a non-attacked wireless link. Because of the increase of the distance between the TN and its serving FN, the TDP (transaction data packet) transmission successful probability (i.e. $\Pr(\text{SINR} > \beta)$) will be decreased, which can be derived from Eq. (8.37). Therefore, the transaction throughput could be decreased with the same FN density. However, as all FNs have a copy of the whole blocks in the blockchain, the transaction security can also be guaranteed in this case. Similar to the eclipse attack, simulation results will be presented in Section 8.6 to show the validity of our model in the presence of this attack.

8.5.3 Random FN Attacks

A random FN attack is defined as the one in which an attacker randomly monopolizes some FNs; thus, the FNs cannot communicate with any TNs. In other words, whenever a specific FN is attacked, it cannot serve any TNs. Based on Proposition 8.1, Proposition 8.2, and Eq. (8.37), we know that the system performance in terms of SINR, TDP transmission successful rate, transaction throughput, etc., could be degraded. However, similar to the above two cases, when FNs are uniformly/randomly attacked, the framework proposed by this chapter is still valid by reducing the FN density accordingly.

8.6 Numerical Results and Discussion

In this section, we first validate the accuracy of our theoretical analysis by comparing the theoretical results with the simulation results in different scenarios. Then, we evaluate the relationship between blockchain transaction throughput and wireless communication

Table 8.2 Simulation parameters.

Parameter	Value	
Radius of the considered area	150 m	
Radius of the interference area, D_0	50 m	
TN transmit power, P	20 dBm	
Path loss mode, l g(d)	$g(d) = d^{-2.5}$	[20]
Total time, T	10 000 s	
Transaction packet length, L	256 bits	[33]
Transaction arrival density, λ_a	$\frac{1}{1800}$s^{-1}	[28]
Noise power, σ	−104 dBm	[32]

throughput and demonstrate how the latter can cause a bottleneck for the former. Next, we give the optimal FN deployment under different TN densities. Finally, we evaluate the performance of the proposed system under the three typical attacks considered.

8.6.1 Simulation Settings

We consider an IoT network that is composed of multiple TNs operating blockchain transactions and multiple FNs supporting blockchain service. The network coverage is set as a circular area with a radius of 150 m, whereas the radius of the interference area is $D_0 = 50$ m. The transmit power of TNs is given as 20 dBm, and the noise power is −104 dBm [32]. The transaction packet length consists of 256 bits [33], and the transaction arrival rate is $\frac{1}{1800}$ s^{-1} [28]. The total considered time in the simulation is of 10 000 s. For convenience, all the parameters are summarized in Table 8.2.

8.6.2 Performance Evaluation without Attacks

In this first experiment, we examine the TDP transmission successful rate, i.e. the probability Pr $(SINR > \beta)$, with fixed FN density of 320 per km^2 and varying TN density. The analytical results are computed from Eq. (8.37). In detail, we first calculate f_{D_1} (the PDF of D_1) based on Proposition 8.1. We then obtain the value of μ_I and δ_I by using Eqs. (8.28) and (8.32), respectively, and thus, the value of $\xi(d_1)$. Substituting them in (8.37), we get Pr $(SINR > \beta)$. For simulation purposes, it is considered that if the received SINR for a transaction transmission is greater than β, this transaction is transmitted successfully; otherwise, it counts as a failure. Figure 8.3 shows the probability Pr $(SINR > \beta)$ for both analytical and simulation results with different TN densities for the SINR threshold parameter $\beta = -15$ dB and $\beta = -9$ dB. From this figure, we can see that the curves of analytical results for both values of β match closely to those of the simulations (closed-form expressions). For example, when the TN density equals to 1.0×10^5 and $\beta = -15$ dB, the successful rate for analytical results and simulations is 76% and 77%, respectively, implying that the difference between the analytical results and simulations is trivial. Moreover, under both

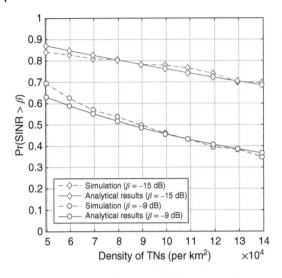

Figure 8.3 Comparisons of Pr (*SINR* > β) vs. TN density (FN density is per 320 km^2).

$\beta = -15$ dB and $\beta = -9$ dB scenarios, the probability Pr (*SINR* > β) decreases with the TN density because of the increasing interference, as expected. We also find that Pr (*SINR* > β) is much lower for $\beta = -9$ dB than that of $\beta = -15$ dB with the same TN density due to stringent SINR requirements.

In the second experiment, we compare the TDP transmission successful rate with fixed TN density 1.0×10^5 per km^2 but varying FN densities, with both analytical and simulation results obtained in the same way as in the first experiment. Figure 8.4 shows the probability Pr (*SINR* > β) for both analytical and simulation results with different FN densities under the SINR threshold parameter β. From this figure, we can again find that the differences between the analytical and simulation results are quite negligible (for example, 4% for $\beta = -15$ dB and 3% for $\beta = -9$ dB when FN density is 200 per km^2). These numerical results clearly validate the accuracy of the proposed model and show the effectiveness of the approximation in (8.23). Moreover, Pr (*SINR* > β) increases with the FN density because of the decreasing distance between the desired TN and the serving FN, as expected.

Next, we evaluate the performance of the overall communication throughput as a function of TN density. Considering the characteristics of the new block generation in the blockchain system (e.g. a new block with a size of 1 MB transaction data is generated every 10 minutes in Bitcoin [17]), the overall communication throughput in this chapter is calculated as the total data volume that is successfully transmitted in every 10 minutes for all TNs. Because of the limitation of the maximum transaction throughput (MTT) in blockchain, the overall required communication throughput will remain unchanged once the transaction throughput achieves the MTT. Figure 8.5 shows the overall throughput with varying TN density from 1.0×10^5 to 1.0×10^6 per km^2 under different parameters β and MTT. The FN density is fixed to 5000 per km^2. From Figure 8.5, we can see that the communication throughput for all four scenarios is increased when the TN density is low. With TN density increasing, the curve with parameters $\beta = -15$ dB, MTT = 7TPS, $\beta = -9$ dB, MTT = 7TPS, and $\beta = -15$ dB, MTT = 14TPS arrives to the maximum communication throughput sequentially. The curve with parameters $\beta = -9$ dB, MTT = 14TPS

Figure 8.4 Comparisons of Pr ($SINR > \beta$) vs. FN density (TN density is 1.0×10^5 per km²).

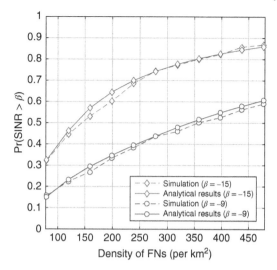

Figure 8.5 Comparisons of overall throughput vs. TN density (the FN density is 5000 per km²).

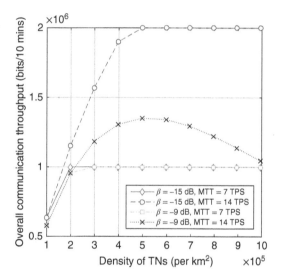

cannot achieve the maximum throughput under any TN density scenario. Note that the throughput is not the maximum value when the TN density is 5×10^5 per km², as it does not remain unchanged after that. When the TN density is greater than 5×10^5 per km², the overall communication throughput for the parameter pair $\beta = -9$ dB, MTT = 14TPS is decreased because of the high interference. This provides a valid theoretical guidance for the design of future blockchain-enabled IoT systems.

In the next experiment, we investigate the relationship between the overall communication throughput and FN density with fixed TN density 4.0×10^5 per km². Intuitively, under a given TN density, the more the FNs are deployed, the greater the SINR would be received, and thus, the higher overall communication throughput can be achieved. Figure 8.6 shows the overall communication throughput with varying FN density from 1000 to 8000 per km²

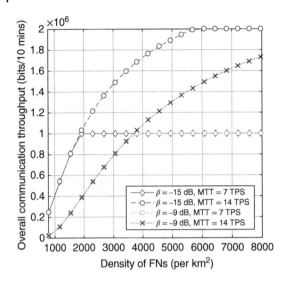

Figure 8.6 Comparisons of overall throughput vs. FN density (the TN density is 40 000 per km^2).

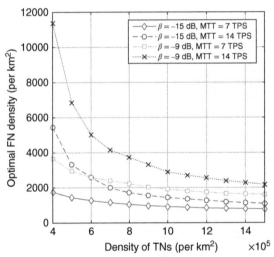

Figure 8.7 Comparisons of optimal FN density vs. TN density.

under different parameters β and MTT. From this figure, we can see that the communication throughput for all the four scenarios is increased when the FN density is low. When the throughput reaches its maximum value, it is unchanged, which means that the transaction throughput achieves the MTT. For example, when $\beta = -15$dB, MTT = 7TPS, the communication throughput is unchanged for a FN density higher than around 2000 per km^2, which means that under the TN density 4.0×10^5 per km^2, the optimal FN density is about 2000 per km^2. Similarly, we can find that the optimal FN density with fixed TN density 4.0×10^5 per km^2 for $\beta = -9$ dB, MTT = 7TPS, $\beta = -15$ dB, MTT = 14TPS, and $\beta = -9$ dB, MTT = 14TPS is of about 4000 per km^2, 5800 per km^2 and larger than 8000 per km^2, respectively.

Next, we investigate the optimal FN deployment for different TN densities. Figure 8.7 shows the optimal FN density with varying TN density from 4.0×10^5 to 1.5×10^6 per km^2

under different parameters β and MTT. From this figure, we can find that when the TN density is lower than 8.0×10^5 per km^2, the optimal FN density decreases rapidly. The rationale behind this fact is that that the more TNs are deployed, the more transactions happen in time T; thus, it is easier to achieve the maximum throughput, and, consequentially, the less number of FNs are needed. However, when the TN density is higher than 8.0×10^5 per km^2, the optimal FN density decreases slowly. This is because of high interference, which is introduced, resulting in lower TDP transmission successful rate. Therefore, although the number of transactions is increased, the overall throughput decreases slowly, and thus, the change in the optimal FN density is also slow.

8.7 Chapter Summary

In this chapter, we investigated the performance of blockchain-enabled IoT networks. We first theoretically analyzed SINR, TDP transmission successful rate, as well as the overall communication throughput by considering the characteristics of blockchain in the spatio-temporal domain. Then, based on this performance analysis, we designed an optimal blockchain full function node deployment scheme to achieve the maximum transaction, and communication throughput with minimum full function node density. Finally, we analyzed the security performance in the system under three typical kinds of attacks, where we have proposed to adopt approaches such as physical layer security algorithms to mitigate their effects. Numerical results validated the accuracy of our theoretical analysis, indicating that the difference between the proposed closed-form expressions simulated and analytical results is usually less than 4%.

The work in this chapter provides a framework for the blockchain-enabled wireless IoT system design through a detailed spatio-temporal model. It can be served as a foundation for future research on system performance analysis, protocols, and algorithm design. For instance, one potential research topic is to use this model to develop new and optimized communication protocols by considering the broadcasting natural in blockchain systems. In addition, by adopting physical layer security techniques, secure wireless blockchain system design against active attacks can be a promising research topic.

References

1 I. Lee and K. Lee. The Internet of Things (IoT): applications, investments, and challenges for enterprises. *Business Horizons*, 58 (4): 431–440, 2015.

2 A. Bassi and G. Horn. Internet of Things in 2020: a roadmap for the future. *European Commission: Information Society and Media*, 22: 97–114, 2008.

3 M. U. Farooq, M. Waseem, A. Khairi, and S. Mazhar. A critical analysis on the security concerns of Internet of Things (IoT). *International Journal of Computer Applications*, 111 (7): 1–6, 2015.

4 T. Xu, J. B. Wendt, and M. Potkonjak. Security of IoT systems: design challenges and opportunities. In *Proceedings of the 2014 IEEE/ACM International Conference on Computer-Aided Design*. IEEE Press, 2014, pp. 417–423.

5 K. Christidis and M. Devetsikiotis. Blockchains and smart contracts for the Internet of Things. *IEEE Access*, 4: 2292–2303, 2016.

6 Z. Zheng, S. Xie, H.-N. Dai, and H. Wang. Blockchain challenges and opportunities: a survey. *International Journal of Web and Grid Services*, 14 (4): 352–375, 2016.

7 B. Cao, Y. Li, L. Zhang, L. Zhang, S. Mumtaz, Z. Zhou, and M. Peng. When Internet of Things meets blockchain: challenges in distributed consensus. *IEEE Network*, 33 (6): 133–139, 2019.

8 i-SCOOPBlockchain and the Internet of Things: the IoT Blockchain Opportunity and Challenge, 2018. https://www.i-scoop.eu/blockchain-distributed-ledger-technology/blockchain-iot/, Accessed: September 21.

9 H. Xu, L. Zhang, Y. Liu, and B. Cao. Raft based wireless blockchain networks in the presence of malicious jamming. *IEEE Wireless Communications Letters*, 9 (6): 817–821, 2020.

10 G. BitFury. Proof of Stake versus Proof of Work. https://bitfury.com/content/downloads/pos-vs-pow-1.0.2.pdf, Sep. 2015, [Online; accessed 28-September-2018].

11 S. Popov. The Tangle. https://www.iota.org/research/academic-papers, Apr. 2018, [Online; accessed 28-September-2018].

12 W. Li, C. Feng, L. Zhang, H. Xu, B. Cao, and M. Imran. A scalable multi-layer PBFT consensus for blockchain. *IEEE Transactions on Parallel and Distributed Systems*, 32 (5): 1146–1160 2020.

13 A. Narayanan, J. Bonneau, E. Felten, A. Miller, and S. Goldfeder. *Bitcoin and Cryptocurrency Technologies: A Comprehensive Introduction*. Princeton University Press, 2016.

14 M. Iansiti and K. R. Lakhani. The Truth about Blockchain. *Harvard Business Review*, 95 (1): 118–127, 2017.

15 A. Dorri, S. S. Kanhere, and R. Jurdak. Blockchain in Internet of Things: Challenges and Solutions. [Online] *arXiv preprint arXiv:1608.05187*, 2016.

16 H. Xu, P. V. Klaine, O. Onireti, B. Cao, M. Imran, and L. Zhang. Blockchain-enabled resource management and sharing for 6G communications. *Digital Communications and Networks* 6 (3): 261–269, 2020.

17 I. Eyal, A. E. Gencer, E. G. Sirer, and R. van Renesse. Bitcoin-NG: a scalable blockchain protocol. in *USENIX Symposium on Networked Systems Design and Implementation*, 2016, pages 45–59. [Online]. Available: http://arxiv.org/abs/1510.02037.

18 Y. Sun, L. Zhang, G. Feng, B. Yang, B. Cao, and M.A. Imran. Blockchain-enabled wireless Internet of Things: performance analysis and optimal communication node deployment. *IEEE Internet of Things Journal* 6 (3): 5791–5802, 2019.

19 M. Haenggi, J. G. Andrews, F. Baccelli, O. Dousse, and M. Franceschetti. Stochastic geometry and random graphs for the analysis and design of wireless networks. *IEEE Journal on Selected Areas in Communications*, 27 (7): 1029–1046, 2009.

20 K. Smiljkovikj, P. Popovski, and L. Gavrilovska. Analysis of the decoupled access for downlink and uplink in wireless heterogeneous networks. *IEEE Wireless Communication Letters*, 4 (2): 173–176, 2015.

21 T. Bai and R. W. Heath. Coverage and rate analysis for millimeter wave cellular networks. *IEEE Transactions on Communications*, 14 (2): 1100–1114, 2015.

22 Y. Zhong, T. Q. S. Quek, and X. Ge. Heterogeneous cellular networks with spatio-temporal traffic: delay analysis and scheduling. *IEEE Journal on Selected Areas in Communications*, 35 (6): 1373–1386, 2017.

23 N. Sapountzis, T. Spyropoulos, N. Nikaein, and U. Salim. An analytical framework for optimal downlink-uplink user association in HetNets with traffic differentiation. in *IEEE Global Communications Conference (GLOBECOM)*, 2015, pages 1–7.

24 Y. Zhong, G. Wang, R. Li, and T. Q. S. Quek. Effect of Spatial and Temporal Traffic Statistics on the Performance of Wireless Networks. [Online] *arXiv preprint arXiv:1804.06754*, 2018.

25 Z. Zheng, S. Xie, H.-N. Dai, X. Chen, and H. Wang. Blockchain challenges and opportunities: a survey. *International Journal of Web and Grid Services*, 14 (4): 352–375, 2018.

26 L. Xiao, X. Wan, and Z. Han. Phy-layer authentication with multiple landmarks with reduced overhead. *IEEE Transactions on Wireless Communications*, 17 (3): 1676–1687, 2018.

27 T. Huimin, S. Yong, and D. Peiwu. Public blockchain evaluation using entropy and TOPSIS. *Expert Systems With Applications*, 117: 204–210, 2018.

28 3GPP TR 45.820 v13.10. Cellular system support for ultra low complexity and low throughput internet of things (CIoT). 2015.

29 3GPP TS 36.201 Release 8. LTE Physical Layer - General Description. 2007.

30 P. L. Hsu and H. Robbins. Complete convergence and the law of large numbers. *Proceedings of the National Academy of Sciences of the United States of America*, 33 (2): 25–31, 1947.

31 E. Heilman, A. Kendler, A. Zohar, and S. Goldberg. Eclipse attacks on bitcoin's peer-to-peer network. In *USENIX Security Symposium*, 2015, pages 129–144.

32 Y. Sun, G. Feng, S. Qin, and S. Sun. Cell association with user behavior awareness in heterogeneous cellular networks. *IEEE Transactions on Vehicular Technology*, 67 (5): 4589–4601, 2018.

33 S. Chen, J. Zhang, R. Shi, J. Yan, and Q. Ke. A comparative testing on performance of blockchain and relational database: foundation for applying smart technology into current business systems. In *International Conference on Distributed, Ambient, and Pervasive Interactions*, 2018, pages 21–34. [Online]. Available: http://link.springer.com/10.1007/978-3-642-39351-8.

9

Utilizing Blockchain as a Citizen-Utility for Future Smart Grids

Samuel Karumba[1], Volkan Dedeoglu[3], Ali Dorri[2], Raja Jurdak[2,3], and Salil S. Kanhere[1]

[1] *School of Computer Science and Engineering, UNSW, 2052, Sydney, NSW, Australia*
[2] *School of Computer Science, QUT, 4000, Brisbane, Australia*
[3] *Data61, CSIRO, 4006, Brisbane, Australia*

9.1 Introduction

To date, the conventional architecture of electricity markets is hierarchical, highly dependent on centralized generation, and inflexible because of the limited pool of sellers (wholesalers) and buyers (retailers) [1]. This structure is not efficient in several ways: (i) it has restricted demand and supply matching options imposed by its unidirectional energy and information flow; (ii) it has inevitably centrally controlled trading algorithms that hinder demand flexibility, making energy more expensive; lastly (iii) it is much less adapted to distributed generation, fronted by the growing number of distributed energy resources (DERs). DERs range from solar panels, windmills, turbines, batteries, to electric vehicles used to generate, store, and transport energy from renewable energy sources (RES). To counter these problems, a growing consensus now views the transition to distributed energy trading (DET) markets as a key strategy [2].

One of the most significant impacts of DET markets has been the rise of prosumers: entities become both producers and consumers of energy [3]. The authors in [1] consider a prosumer as a residential renewable energy producer equipped with a small-scale generation unit such as a solar panel or a windmill. Additionally, the authors suggest that the prosumer could invest in a battery to store excess generation and a smart meter to simultaneously monitor energy generation, consumption, and supply levels within the distribution network or on the consumer side of the network. The shift toward a prosumer-driven economy implies that there is a decentralization of value creation, often made possible by the adoption of peer-to-peer (P2P) energy exchanges and feed-in tariff (FiT) rates [4]. FiT is the price received by prosumers for injecting their excess electric power to the grid. Furthermore, DET also brings various benefits such as increasing the demand flexibility, grid efficiency, and reduces the energy prices on the consumer side [5].

Although prosumerism is accelerating, P2P DET is arguably still nascent. A full realization of the P2P DET infrastructure will depend on the availability of several other important aspects such as trust management, security and privacy, traceability, and bidirectional communication mechanism. Therefore, the rise of smart grids [6], powered by information and

Wireless Blockchain: Principles, Technologies and Applications, First Edition.
Edited by Bin Cao, Lei Zhang, Mugen Peng and Muhammad Ali Imran.
© 2022 John Wiley & Sons Ltd. Published 2022 by John Wiley & Sons Ltd.

communication technologies, might prove to be an essential ingredient in scaled participation of residential homes, commercial buildings, electric vehicles, and data centers as active nodes in a smarter and resilient distribution network.

The common information and communication technologies adopted in smart grids include the Internet-of-Things (IoT) and artificial intelligence (AI). On the one hand, IoT features high connectivity among devices and participants in a smart grid, aimed at collecting a huge amount of information for monitoring physical electrical systems that enable demand side management (DSM) [7]. DSM is a real-time response to supply and demand from the end-user. On the other hand, AI is used to provide an analysis of the generation and consumption patterns based on the collected IoT data for supply and demand response [8]. While AI can be used to derive knowledge from data acquired by IoT devices, prosumers in P2P energy markets may be reluctant to cooperate because of data-related issues (e.g. privacy, ownership, and integrity). In practice, third party auditor's (TPA) frameworks are used to manage data challenges and conflict resolution [9]. However, it could be difficult to find a TPA that is trusted by all users. Also, TPAs risk privacy leakage because of their centralized nature.

Recently, blockchain technology, first applied in Bitcoin [10] — a P2P cryptocurrency system, promises to address the issue of trust in the P2P energy markets. Blockchain is a distributed ledger of chronologically timestamped transactions, immutably secured by hashing them into an ongoing chain of verifiable blocks. Each block is added through consensus, and no one has full control over the consensus process [9]. Control is pegged on its salient features of trust through distributed consensus, security, and privacy through cryptography and traceability through immutable provenance [11]. In recent years, blockchain has gained traction for both development and utilization in the energy sector. For instance, the authors in [12] reviewed 140 blockchain-based initiatives in the energy sector and classified them according to their applications as illustrated in Figure 9.1.

Currently, DET is by far the most prominent research topic as indicated in Figure 9.1, with various P2P energy trading platforms being trialed in several locations globally. Examples include WePower[1] from Lithuana, Powerledger[2] from Australia, and LO3 Energy[3] from the United States. In this context, the authors in [4] coined the term "citizen-utilities" to describe the process of managing P2P DET moderated by the blockchain technology. Collectively, citizen-utilities are driving the transition from hierarchical centralized energy markets to distributed energy markets by eliminating the need for trusted third parties.

The objectives of this chapter are to walk through the fundamentals of "citizen-utilities," primarily assessing their impact on efforts to manage distributed generation, storage, and consumption on the consumer side of the distribution network, while intelligently coordinating DET without relying on trusted third parties. Additionally, the chapter highlights some of the open research challenges including scalability, interoperability, and privacy that hinder the mainstream adoption of "citizen-utilities" in the energy sector. Then, to address these research challenges, we propose a scalable citizen-utility that supports interoperability and a privacy-preserving data clearing house (PDCH). Both platforms are based on the hypergraph-based adaptive consortium blockchain (HARB) framework proposed in [11].

---

1 https://wepower.com/
2 https://www.powerledger.io/
3 https://lo3energy.com/

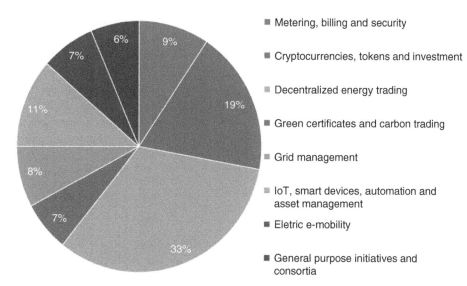

Figure 9.1 Classification of blockchain applications in the energy sector. Source: Andoni et al. [12]. Licensed under CC BY-4.0.

The HARB framework addresses the issues of scalability, interoperability, and transaction privacy for blockchain-based DET applications. The main contributions for the scalable HARB-based citizen utility include the following:

i) A data verification module that extends the HARB framework's clustering module to include an identity manager and a provenance manager.
ii) An adaptive blockchain module that organizes the heterogeneous IoT nodes into different blockchain service modules.
iii) An atomic meta-transaction workflow engine for cross-chain interactions.

The PDCH framework is a blockchain-based provenance tool that addresses the issue of data privacy for on-ledger and off-ledger transaction storage. The framework can be used to verify the authenticity, trustworthiness, and traceability of IoT data produced by buildings during energy generation and consumption operations such as heating, ventilation, cooling, and air conditioning (HVAC), within the DET ecosystem. Its main contributions include the following:

i) A privacy-preserving batch verification scheme for authentic data collection from IoT networks. Our scheme provides certificate-less public key signatures to protect the identity privacy with lower calculation cost suitable for resource-constrained IoT devices where blockchain cannot be installed.
ii) A distributed privacy-preserving multi-tenanted cloud storage mechanism for trust-less and cost-effective data storage. Our mechanism addresses the availability issue by replacing the centralized cloud storage with a distributed one.
iii) Lastly, a privacy-preserving integrity provenance query model based on secure multi-party computation technique. The model ensures the efficiency, integrity, and correctness of the query results regardless of the data volume.

The rest of the chapter is organized as follows: In Section 9.2, we explore the proliferation of citizen-utilities in the emerging prosumer communities such as microgrids, vehicular energy networks, and virtual power plants, illustrated through case studies. Additionally, we review how the existing citizen-utilities are used to reconcile the DSM and the dynamic P2P energy marketplaces through real-time supply and demand. Moreover, we review the challenges and limitations of using blockchain technology in designing citizen-utilities. Then, in Section 9.3, we summarize the HARB and PDCH frameworks that address some of the reviewed blockchain challenges and limitations. Finally, Section 9.4 gives our concluding remarks and outlines future research directions.

9.2 DET Using Citizen-Utilities

In this section, we lay out the groundwork for the rest of the chapter by providing a more detailed overview of the expected purpose, shape, and architecture of the citizen-utility for future smart grids, illustrated in Figure 9.2. We structure the chapter in terms of three questions that we believe are central to this discussion: (i) Who will use citizen-utilities? (ii) How will citizen-utilities be used for energy optimization? and (iii) What problems must be solved to make citizen-utilities commonplace? We provide an overview of these three issues, illustrating with case studies presented by researchers in academia and energy industry.

9.2.1 Prosumer Community Groups

Who will use the citizen-utilities? The study in [13] outlines five main elements of a smart grid that enables seamless energy sharing. These include smart grid infrastructure, bidirectional communication, advanced management system, standards and legislation, and sustainable integration with the prosumers. The prosumer is the fundamental stakeholder in a DET marketplace as illustrated in Figure 9.2. Therefore, this chapter revolves around sustainable integration with the prosumers.

In [14], Rathnayaka introduced the concept of prosumer community groups (PCG) referring to a community of prosumers generating and sharing energy. In other literature, the PCG concept is described using new terms such as electric prosumer community, integrated energy systems, and clean energy communities [13]. Outside the energy sector, the sharing economy is more mainstream with Airbnb[4] and Uber[5] as common examples [15]. Similarly, PCGs are developed around the same "sharing economy" principles and aim to unlock the various diverse sharing objectives in DET.

Examples of PCG include microgrids, virtual power plants (VPP), and electric vehicle networks (EVN), illustrated in Figure 9.3 and further discussed in Sections 9.2.1.1–9.2.1.3. We pick these three prosumer communities as they are the most common sharing networks, where prosumers with relatively similar energy sharing objectives and interest come together to make an effort to pursue a mutual goal or jointly compete in the energy marketplace.

---

4 https://www.airbnb.com
5 https://www.uber.com/

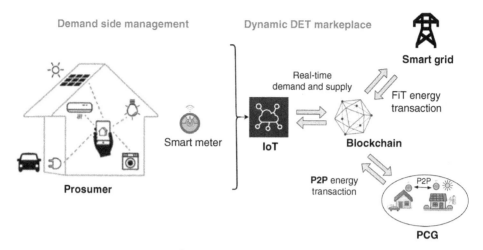

Figure 9.2 Blockchain-based DET system (citizen-utility).

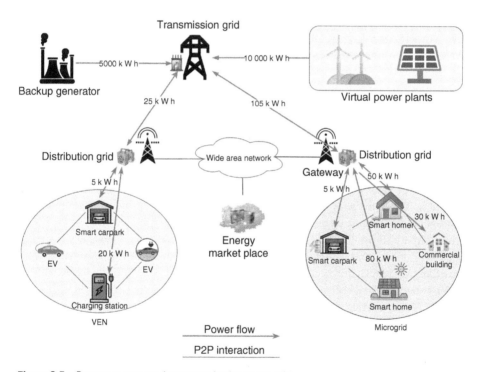

Figure 9.3 Prosumer community groups in the smart grid.

9.2.1.1 Microgrids

The microgrid PCG can be described as a cluster of loads (e.g. home appliances, heating, and cooling systems), DERs, and energy storage systems (ESS) (e.g. battery and flywheel), which are operated in coordination to supply energy reliably and share costly resources (e.g. DERs and ESS) in local communities [16]. Typically, microgrids are centrally controlled, which requires expensive central control devices to collect and process IoT data; also, it acts as a single point of failure [16]. Contrary, in a decentralized control microgrid, prosumers control themselves independently using the advancing technologies such as smart homes. A smart home is a technology that can monitor and control many devices in a home or building by connecting them through a communication network [7]. However, selfish decisions of prosumers may affect the microgrids in a decentralized control system.

Through the Brooklyn Microgrid case study, the authors in [17] demonstrated that blockchain is an eligible technology to operate decentralized microgrids. The Brooklyn Microgrid project run by LO3 Energy in Brooklyn, New York, is based on Ethereum [18] blockchain and smart contracts. The platform enables P2P energy trading, control of DERs for grid balancing, demand response, emergency management, and other use. This provides incentives for investments in local renewable generation and for locally balancing supply and demand in smart grids. Their vision is to create a blockchain-based microgrid intelligent system. Moreover, the authors in [16] highlighted other projects using the decentralized approach for control and business processes in a microgrid. These projects include WePower [19], PowerLedger [20], LO3Energy [21], NRGcoin [22], among others.

9.2.1.2 Virtual Power Plants (VPP)

The VPP prosumer communities are cloud-based services that aggregate DERs including ESS and present them as a single commercial entity to the wholesale electricity market. Individually, prosumer with small-scale DERs has little to no impact in the wholesale markets [23]. Therefore, VPP create a shared value to both prosumers and the grid by efficiently utilizing the existing DERs in a P2P network to contribute to demand balancing in smart grids without the need for new asset investment. However, the traditional VPP that adopt a centralized management model, have the problem of information and data security, unfair benefits distribution, and cumbersome transaction management [24].

Recently, blockchain-based VPP implemented a distributed energy supply to wholesale markets in a P2P framework, which solves the problems of trust, security, and ensures transparency of transactions without the involvement of trusted third parties [24–26]. More specifically, the authors in [25] integrated their proposed blockchain VPP system with an actual VPP pilot project in Dubai with a total aggregated size of 1.8 megawatts (MW) composed of DERs, ESS, and controllable loads. The setup includes a utility-scale battery (1.2 MW), two residential units with solar PV and battery (38 kW), and microgrid with solar PV (270 kW) and batteries (330 kW). Ethereum blockchain protocol was implemented to test and validate the concept of registering VPP transactions and generating DigiWatts (financial incentives) using a smart contract. The case study showed that integration of blockchain and smart contracts in the automation of VPP control, triggered by pre-defined agreements between VPP and DERs owners (i.e. prosumers) can be used to handle market settlements in a secure, transparent, and trust-less manner.

9.2.1.3 Vehicular Energy Networks (VEN)

In a recent report published by International Energy Agency [27], renewable electricity use in transport is anticipated to increase 70% with greater use of electrified rail as well as electric vehicles (EV). Nonetheless, the uncoordinated load of a large number of EVs charging during peak hours poses critical challenges to the grid in terms of overloading and frequency instability. Alternatively, EVs could be valuable as shiftable demand or dispatchable load in the grid in a movement connectivity network referred to as the VEN [28].

To understand how VENs will impact the power distribution grids, the authors in [29] studied vehicle-to-vehicle (V2V) matching communities for charge sharing using blockchain. In most cases, matching of EVs is done by centralized trusted third parties. While this approach can yield an optimal matching, the cost of running a centralized algorithm could be high (e.g. Hungarian Algorithm with a complexity of $O(n^3)$). Additionally, private information (e.g. time and location) about the participant's whereabouts could be leaked. To address the privacy challenge, the authors proposed a privacy-preserving mechanism, which uses blockchain and partial homomorphic schemes, to hide the location of participants. Lastly, the authors proposed a distributed dynamic matching algorithm to address the issue of matching.

Similarly, the authors in [30] studied vehicle-to-grid (V2G), which offers a promising alternative to demand-supply matching in smart grids. For example, EVs can be used to meet the temporally demand of critical loads such as pop-up hospitals built during the Covid-19 pandemic. However, a major challenge of V2G concept is that it requires a robust energy management system that enables increased distributed influence and provides a trusted environment for scheduling and matching. To address the above problem, the authors proposed a novel framework that monitors the DER-based EV charging station along with conventional grid and maintains the energy in-out flow information as well as accounting information over blockchain. Additionally, a time-based pricing policy based on smart contracts is also proposed to incentivize EVs to share their energy.

9.2.2 Demand Side Management

This section answers the question, how will citizen-utilities be used for energy optimization? Historically, FiT market schemes have been the main support instrument to promote RES production, notably solar and wind power, by offering remuneration for each kilowatt-hour of renewable energy produced [31]. However, according to [3], high penetration of DERs has led to the duck curve problem, which does not match with the increasing energy load (e.g. EVs, high-voltage electrical appliances, heating, and cooling systems) as illustrated in Figure 9.4. Intermittent generation from DERs such as solar PV leads to high generation during the day and rapidly subsidizes in the evening. In contrast, residential demand is low during the day while in the evening demand peaks as people get back from work. This challenge raises doubt about the continuation of FiTs and has started the debate about the method that should be used to deal with this problem.

In this context, DSM has been proposed as a solution that can still promote the adoption of RES production while balancing demand and supply to improve the reliability of the smart grid infrastructure [32, 33]. DSM is the ability to influence the use of electricity or to alter the pattern and magnitude of demand by end-users, which have a significant impact on energy cost and demand supply balancing [33]. The authors in [32] presented a taxonomy of DSM models such as demand response, energy efficiency, and spinning reserve, illustrated in Figure 9.5. These models utilize load balancing, scheduling, and other P2P energy

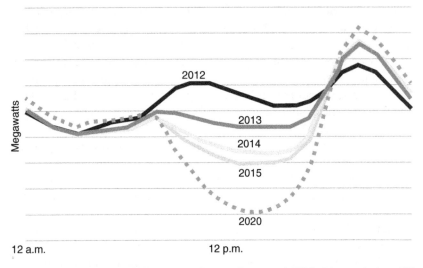

Figure 9.4 Illustration of a Duck curve. Source: Green et al. [3] / with permission of Elsevier.

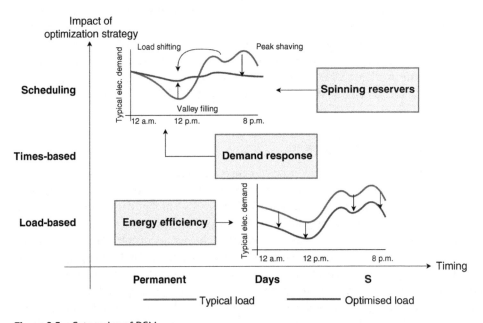

Figure 9.5 Categories of DSM.

trading strategies for DERs optimization. A detailed discussion for each model is provided in Sections 9.2.2.1–9.2.2.3.

9.2.2.1 Energy Efficiency

Energy Efficiency (EE) involves the voluntary reduction of consumers' energy use through energy conservation incentives [34]. According to the International Energy Agency report of 2019 [27], improving energy efficiency would reduce energy bills by more than 500 billion dollars each year. Specifically, the report stated that buildings consume about 40% of the world's energy use, making them the largest single energy consumers. Improving the energy efficiency of residential buildings or industrial sites starts with information and insights into the processes involved [32]. Therefore, IoT-based digital concepts such as smart buildings, smart homes, and smart metering are acting as key enablers.

Although these digital concepts give enormous power to energy consumers to understand, produce, monitor, and control their energy requirements, smart IoT devices and their intelligent applications raise serious security and privacy concerns. For example, reporting on too fine-grained user consumption can reveal whether a house is occupied or not or when the occupants are away by analyzing their behavioral patterns. These problems of privacy and more generally data protection and Cybersecurity are shaping the discussion on how energy efficiency data could be encrypted and shared over a blockchain platform to improve DET transparency, the security of information, and reliability of services.

In a case study presented in [35], the Energy Company Obligation (ECO) scheme in the United Kingdom requires energy suppliers with more than 250 thousands domestic consumers to have a supply threshold of 500 GWh of electricity. The intention is to reduce the amount of energy required to heat the smart homes through the installation of energy efficiency measures such as smart heating controls. The targets for each obligated energy supplier are set by the regulator, Ofgem, who is also responsible for auditing the scheme, preventing fraudulent compliance claims, and reporting progress to the UK government. Additionally, suppliers have been allowed to trade their obligations with each other. Previously, all stakeholders in the value chain including smart homes, government, and suppliers have had to trust Ofgem, a centralized third party system to keep their data secure. However, blockchain can be used to create a decentralized public digital ledger to securely and transparently record all transactions relating to ECO. Additionally, smart contracts developed using blockchain technology can be used to enforce installation and consumption compliance measures.

9.2.2.2 Demand Response

Demand Response (DR) involves alteration of energy consumption levels and/or patterns of consumers and prosumers in response to dynamically changing prices and incentives [34]. Frequently used DR strategies aim at matching energy demand with production by motivating the prosumers to shave or shift their energy production to deal with peak load periods as illustrated in Figure 9.5. To manage this process, the authors in [36] discuss the challenges of introducing a trusted third-party entity referred to as a distributed system operator (DSO), who defines a DR program for coordinating distributed energy prosumers (DEPs) as illustrated in Figure 9.6.

Typically, the DSO sends DR events to all prosumers with a request to modify consumption and the associated incentive. Then, the prosumers send a bid with the amount of energy they are willing to reduce or increase. While the DSO is responsible for accepting the bids and checking if the energy balance on the grid is met, this process is slow and DSO acts as a single point of failure. Moreover, DEPs assume that the DSO can be trusted for fair distribution of incentives and management of their private information.

In their solution, the authors proposed a blockchain-based approach with the implementation of near-real-time automated DR event program, near-real-time financial settlement

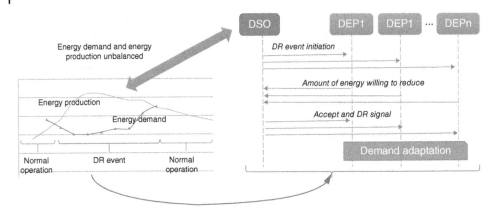

Figure 9.6 Centralized management of demand response. Source: Pop et al. [36]. Licensed under CC BY-4.0.

and event validation, secure energy transactions, and scalability of distributed generation within the global energy mix. Similarly, the authors in [37] leveraged blockchain to implement near-real-time DR for EVs charging and discharging for VEN communities in smart cities using intelligent energy scheduling and trading. The proposed consortium blockchain-enabled secure energy framework for EVs implemented incentive mechanisms tailored for unique characteristics of EVs to incentivize more EVs to participate in DR.

9.2.2.3 Spinning Reserves

Spinning Reserve (SR) is the market concept related to the amount of the energy capacity available in an online reserve, normally used to compensate for power shortage or frequency drop [38]. ESS such as lithium-ion batteries, flywheels, compressed air storage, pumped hydro, and superconducting magnetic sources are some of the most fast-responsive types of SR [39]. Alternatively, EVs with the ability to perform V2G, grid-to-vehicle (G2V), and V2V modes are likely to become a key component for real-time SR applications. According to [40], aggregated EVs can provide large-scale battery storage to offer load frequency control. The need for EV aggregation and relevant offerings of the aggregator models (V2G, G2V, and V2V) is analyzed in [41, 42]. However, key challenges in aggregating distributed EV, for EV-assisted frequency regulation system, are to provide information security, offer user privacy, and how to achieve data immutability, availability, tracking, and trust among stakeholders.

To address these challenges, blockchain has grown in a variety of applications [40, 43]. For instance, [40] have proposed a novel framework for monitoring solar PV-based EV community charging stations and load frequency regulation using blockchain. Additionally, they proposed a smart energy management controller (SEMC) that coordinated the overall operations of the charging stations. SEMC optimally allocates the energy load to EVs by deciding which EVs will operate in either charging or discharging mode. The novelty of this framework is that it offers the V2V, V2G, and G2V power transfer capabilities. Similarly, the authors in [43] proposed a blockchain-based billing system for EV aggregation models. The system enables both the EV and the charging station to mutually authenticate each other before connecting to a charging station and after charging/discharging is complete by recording their transactions on the blockchain to prevent transaction modification. The mutual authentication between the EV and the charging station is based on individual transaction's terms and permits to meet the interest of each participant.

9.2.3 Open Research Challenges

What problems must be solved to enable the production level deployments of citizen-utilities? We studied the blockchain-based DET case studies provided in Sections 9.2.1 and 9.2.2 based on the core concept of Cyberphysical systems [44] and summarized our findings in Table 9.1.

Despite all the advanced merits of blockchain, there are still limitations and issues that are hindering production deployment of citizen-utilities in smart grids. The specific issues that need to be addressed include lack of standardization, privacy leakage, IoT overheads, and blockchain-specific limitations [45].

9.2.3.1 Scalability and IoT Overhead Issues

Most of the case studies reviewed in Table 9.1 implement their citizen-utility using the Ethereum blockchain protocol. Ethereum is a public blockchain protocol that adopts the proof-of-work (PoW) consensus mechanism that was first proposed in [10]. PoW imposes

Table 9.1 A summary review of proposed systems and framework based on blockchains Cyberphysical systems' core concepts.

Case study	Objective	Trust	Transparency	Security	Privacy	Scalability	Interoperability
[17]	P2P energy trading in Microgrids	✓	✓a)	✓	✗	✗	✗
[24]	P2P energy trading in VPP	✓	✓a)	✓	✗	✗	✗
[25]	P2P energy trading in VPP	✓	✓a)	✓	✗	✗	✗
[26]	P2P energy trading in VPP	✓	✓a)	✓	✗	✓	✗
[29]	P2P energy trading in VEN	✓	✓a)	✓	✓	✓	✗
[30]	P2P energy trading in VEN	✓	✓a)	✓	✗	✗	✗
[37]	DSM through DR	✓	✓a)	✓	✓	✓	✗
[36]	DSM through DR	✓	✓a)	✓	✓	✓	✗
[35]	DSM through EE	✓	✓a)	✓	✗	✗	✗
[40]	DSM through SR	✓	✓a)	✓	✓	✗	✗
[43]	DSM through SR	✓	✓b)	✓	✗	✗	✗

a) Fully transparent if implementation is based on public blockchain (Ethereum).
b) Partially transparent if implementation is based on a consortium or private blockchain.

high computation and communication requirements for validating and disseminating blocks in the network so that no single node can take over the network to create a fork [18]. However, this becomes an increasingly challenging issue in smart grids, where a large number of IoT devices may not have the required computational power (e.g. storage, processing, or bandwidth) to utilize the PoW-based blockchain capabilities [46]. Moreover, PoW mechanisms have high energy footprint counter to energy efficiency objective, which hinders the deployment of citizen-utilities in production scale [47].

In an attempt to address the PoW's scalability limitations, the author in [48] adopted the Byzantine Fault Tolerant (BFT) protocol as a pluggable consensus mechanism in their modular Hyperledger Fabric (HLF) blockchain framework. HLF separates the roles of nodes into endorsers for transaction execution and validation, orderers for generating blocks, and committers to store transactions adequately increasing the blockchain's throughput. While this is desirable, execution and validation of transactions in HLF are managed by a consortium, which sacrifices transaction visibility and transparency. Therefore, the design of a better blockchain framework that ensures scalability, trust, transparency, security, and reduces energy footprint is desired.

9.2.3.2 Privacy Leakage Issues

Blockchain systems store data as part of the transactions, which is replicated across the network nodes to provide visibility and transparency [2]. In the DET context, transaction data can be used to provide valuable services such as DSM, price prediction, optimization, and pattern recognition (e.g. for behavior modeling). However, because blockchain ledgers are immutable, permanently recording payment and energy usage records may leak sensitive information including identities, locations, and energy production and consumption patterns, raising privacy concerns especially in local prosumer communities such as microgrids. While advanced privacy-preserving techniques such as Zero-Knowledge Proofs and Elliptic Curve Digital Signature Algorithm can be incorporated to enable privacy, these techniques increase computational complexity and consequently have a higher energy footprint. Therefore, research topics such as secure off-ledger [2] data storage techniques should be explored further to figure out how data privacy could be preserved by off-ledger storage without losing the benefits of blockchain.

9.2.3.3 Standardization and Interoperability Issues

Interoperability refers to the ability of citizen-utilities to exchange and make use of information in the DET marketplace [11]. Because of market deregulation, today, there exists a wide spectrum of blockchain protocols and platforms for building citizen-utilities, which has led to multiple energy trading systems with different methods of data and consensus management. Further, these blockchain-based systems do not support interoperability even when their implementation is based on the same blockchain platform (e.g. Ethereum or HLF), which in turn has led to data and information silos [49]. Given that citizen-utilities may have common DSM goals, the development of standards for interoperability is crucial to achieving the full benefits of DET in smart grids. In an attempt to provide interoperability to blockchain-based systems, ad hoc platforms such as centralized digital notaries have been proposed [50]. A centralized digital notary is a trusted third party (TTP) that verifies or authenticates blockchain transactions and participants to ensure trust between inter-operating networks for value transfer [51], as illustrated in Figure 9.7. Therefore, there is a need for an interoperable blockchain framework design that provides better transaction visibility and transparency, distributed trust, and value transfer between citizen-utilities.

9.3 Improved Citizen-Utilities

In this section, we present our previous and current blockchain-based DET applications for smart grids. Further, we highlight the key contributions of each framework and discuss how our approach addresses the aforementioned research challenges that hinder the mainstream adoption of blockchain in the energy sector.

9.3.1 Toward Scalable Citizen-Utilities

In [11], we propose a Hypergraph-based Adaptive Consortium Blockchain for DET known as the HARB framework. The HARB framework addresses the issues of scalability, interoperability, and transaction privacy for blockchain-based DET applications. While the implementation is focused on DET, the solution can be adapted for other use cases. In this case, therefore, we use the HARB framework to design scalable citizen-utilities. To understand the nature of our solution, first, we give an overview of the addressed challenges. Then, we discuss the proposed scalable citizen-utility based on the HARB framework.

9.3.1.1 Challenges

First, we considered the two PCG at the distribution grid level in Figure 9.3, namely, VEN and microgrid. We assume that their citizen-utility platform is based on different blockchain protocols: Ethereum and Bitcoin blockchains. Recall that smart grids are inherently IoT environments with heterogeneous nodes, some are resource-constrained and others are high-powered computational devices. From the reviewed case studies [52, 53], multiple IoT devices were found to be collapsed into one single node connected in a Local Area Network (LAN) to increase the scalability as illustrated in Figure 9.7. For example, consider smart home 1's LAN owned by a participant with Id: *User_1* in the microgrid platform as shown in Figure 9.7. A device with high computing resources within the LAN (e.g. the EV in Smart Home 1 LAN) is chosen to represent the other resource-constrained IoT devices (i.e. sensor network, smart meter, and solar PV) on the blockchain network. The EV becomes the blockchain node, such that all transactions from other devices in LAN are sent to the blockchain node for validation on the blockchain network. However, the credibility of transactions from other non-blockchain nodes in the LAN cannot be guaranteed, as they could be compromised before they get recorded on the blockchain network. Therefore, we require a suitable blockchain platform to design a scalable citizen-utility.

Second, we consider a participant who interacts in both the VEN and the microgrid network. For example, *User_1* could charge his/her EV at home and then on his/her way to work they are incentivized to discharge the surplus EV energy at a charging station for some incentives (e.g. crypto tokens) from the VEN. In this case, the EV will need to run both the VEN's Ethereum blockchain and the microgrid's Bitcoin blockchain network. However, the current blockchain protocols do not support interoperability between the implemented

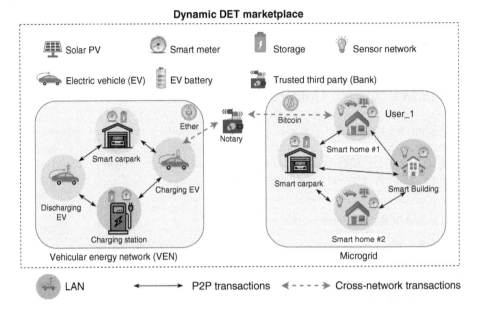

Figure 9.7 Overview scenario for two inter-operating citizen-utilities in a DET marketplace.

citizen-utilities, unless it is through a digital notary, as illustrated in Figure 9.7. A digital notary is a trusted entity that authenticates both the network participants and the energy assets through a centralized lookup registry, as agreed upon by both in the DET marketplace. Consequently, the centralized nature of the digital notary system acts a single point of failure; therefore, there is a need for a distributed energy notary system.

Lastly, we noted that the digital notary is managed by a trusted third party outside the blockchain network. While sensitive transaction data in the blockchain network is protected through pseudonyms and cryptography, the centralized nature of the digital notary's lookup registry risks privacy leakage because of various inference and linking attacks [54]. Such privacy leakage can lead to breaching the confidentiality of transaction data. Moreover, transactions in the blockchain network are replicate across the distributed nodes where data privacy is protected through pseudonyms and cryptography. However, current research has shown that with recent technology, pseudonyms can be anonymized using linking attacks.

9.3.1.2 HARB Framework-Based Citizen-Utility

Fundamentally, P2P network structures use regular graph theory to model complex distributed systems, which is limited by simple pairwise relationships. Pairwise relationships in P2P networks do not represent the complex real-world systems under investigation. As a result, such systems are hard to scale and consequently manifests other challenges such as lack of scalability, interoperability, and privacy as discussed above. To address the P2P network design limitations, the HARB framework is designed based on the hypergraph theory, which is a generalization of graphs [55]. Hypergraphs use high-order relationships to accurately model complex distributed systems by completely describing the nodes' contextual

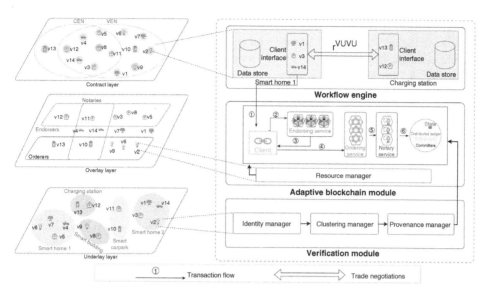

Figure 9.8 High-level overview of the HARB framework. On the left, we have the three network layers and on the right we have an expanded view of the framework components.

information (e.g. spatial, temporal, and identity) and their multi-dimensional interactions (e.g. human–thing, thing–thing, thing–thing–human, etc.).

Therefore, we implement the citizen-utilities based on the three-layered HARB framework's network architecture to provide scalability, interoperability, and privacy to the DET ecosystem. Figure 9.8 illustrates the conceptual overview of the framework with the three network layers: underlay, overlay, and application. Each layer describes various complex interactions between nodes at different levels of granularity backed by hypergraph concepts presented by authors in [55].

The underlay layer is the physical network, where IoT devices are interconnected either through the Wide Area Network (WAN) or LAN. Recall that not all IoT devices have the computation capability to participate as a blockchain node in a LAN. Therefore, to ensure the credibility of the non-blockchain nodes in the LAN, we propose a data verification module that extends the adaptive blockchain module in HARB with a device identity manager. The identity manager is a Certificate Authority (CA) that uses the Public Key Infrastructure [56] to assign identities to the physical devices. Each LAN has its own CA to privately manage its nodes' identities through an on-boarding process. Once the identities are assigned to all network nodes, the clustering manager uses the community discovery mechanism in HARB to group nodes into context, location, or time-oriented clusters or communities. Each cluster becomes a sub-network (sub-chain) in a scalable energy network. Lastly, to model the environment under which the IoT devices interact in, we propose a provenance manager-based HARB framework's complex high-order relationship rules. The provenance module records all historical interaction data on blockchain for traceability of data to its original source.

The overlay layer is the blockchain network layer (on-ledger) that adapts the HARB's adaptive blockchain module (ABM) to include a resource manager component. The resource manager considers the dynamic nature of the IoT devices in the network as nodes join or leave the network, their heterogeneous functionality, and their varying computing capabilities to assign network roles. All historical interaction data and resource level is captured by the provenance manager to determine how to assign the blockchain service roles of endorsing, ordering, or committing. A more detailed description of the roles assignment process is described in [11]. Lastly, we address the issue of interoperability to support value transfer between the different sub-chains using the distributed digital notary module proposed in the HARB framework.

Finally, the contract layer represents the application's network, where end-users interact to exchange value through P2P DET. For user and data privacy, we used the data tagging and anonymization module (dTAM) proposed in HARB to develop an atomic cross-chain workflow engine. dTAM uses the hypergraph coloring concept to tag transactions and zero-knowledge-proofs (ZKP) [57] coupled with linkable ring signatures (LRSig) [58] for user and data anonymization. The resulting data digest is stored on the blockchain while the actual data is stored in off-ledger storage to preserve privacy.

9.3.2 Toward Privacy-Preserving Citizen-Utilities

Historically, the DSM services outlined in Section 9.2.2 have been delivered through the multi-tenanted cloud systems. A multi-tenanted cloud is a centralized cloud computing architecture that allows customers (tenants) to share computing resources. It is often managed by a trusted third party (service provider) to ensure that each tenant's data are isolated and remain invisible to other tenants. Figure 9.9 illustrates the high-level overview of the multi-tenancy cloud architecture. First, the data producers in the sensor network layer

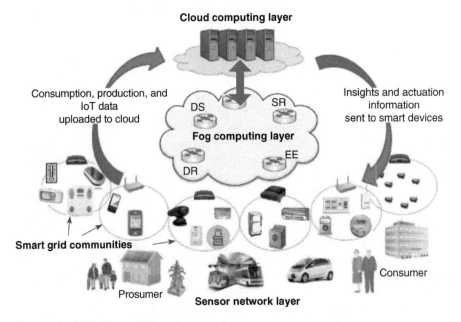

Figure 9.9 DSM citizen-utility system overview.

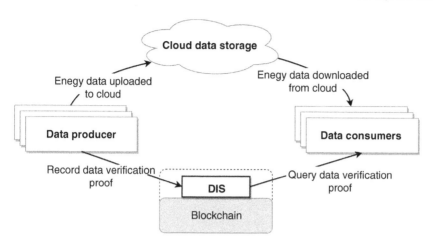

Figure 9.10 Privacy-preserving citizen-utility system overview.

upload their energy data (generation, trading, and consumption) to their cloud instance. Then, the data consumer in the fog computing layer uses the data uploaded to the cloud to provide intelligent energy services such as DSM services. Both the data producer and the data consumer rely on the cloud service provider to offer the computing resources that include storage, processing, and software. However, this approach is not secure because the cloud service providers who act as trusted third party can access, destroy, alter, or leak users' private data and information.

To address these challenges, various works [9, 59, 60] have proposed the use of blockchain to design data integrity verification systems (DIS), illustrated in Figure 9.10. However, the DIS has limitations at various levels. First, the initial authenticity of data producers cannot be guaranteed. This is because data from IoT systems may be subject to noise, bias, sensor drift, errors, or malicious alteration from compromised devices. Second, while blockchain secures the data verification proof against tempering, distributed denial of service (DDoS), and consistency attacks, the data uploaded to the centralized cloud storage is susceptible to these attacks. Lastly, the applications that relate to data processing and interactions between data consumers (energy service provider) and the service users should ensure the end-to-end integrity of the collected, stored, and processed data. Therefore, DIS requires a privacy-preserving trust mechanism that cuts across these levels to fulfill both transparency and auditability.

9.3.2.1 Threat Model
To illustrate and defend against privacy issues facing citizen-utilities discussed in Section 9.2.3.2, it is useful to have a general model of the operating environment, the different actors, and the assets to protect, as illustrated in Figure 9.11. The actors identified in this threat model are

1. The **data producers/owners** are network entities who produce data and whose data may be sensitive.

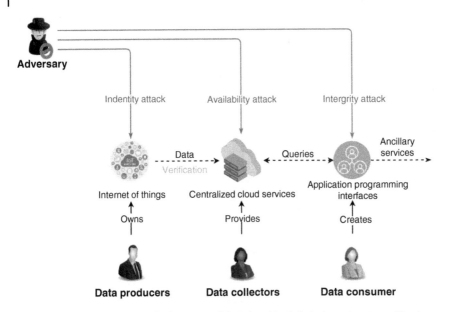

Figure 9.11 Threat model of privacy attacks against blockchain-based systems. The human figures represent actors and the symbols represent the assets. Dashed lines represent data and information flow, while full lines represent possible actions.

2. The **data collectors** are network entities who provide data hosting and other centralized cloud services such as software-as-a-service, platform-as-a-service, or infrastructure-as-a-service.
3. The **data consumers** are entities who utilize data and cloud services hosted by the centralized cloud servers to develop intelligent customer application such as forecasting, control, and monitoring usually delivered via some sort of application programming interface (API) to the end-users.
4. The adversary is a malicious entity within or outside the system whose goal is to compromise the system for malicious activities or selfish gains.

The assets that are sensitive and are potentially under attack include the identity of IoT devices, availability of data and information stored in the cloud, and the integrity of intelligent data analytic models provided by data consumer APIs.

The different privacy attack surfaces against the existing systems can be modeled in terms of **adversarial knowledge**, as illustrated in Figure 9.12. The range of knowledge varies from limited to full knowledge of the centralized cloud services and data consumers' APIs. In between these two extremes, there is a range of possibilities such as partial knowledge of identification, localization/tracking, or profiling [61]. The identification threat lies in associating an identity (user or IoT device) to a specific privacy-violating context. Localization and tracking are the threat of determining and recording a node's location through time and space. Profiling denotes the threat of compiling information dossiers about individual users or devices to infer interests by correlation with other datasets in a centralized cloud server.

In the sensing layer, *data producers* are installing Wi-Fi-enabled smart DERs (e.g. smart meters and EVs) and smart home appliances (e.g. air conditioners, water heater, and space heaters) in their homes. Even older appliances can act as smart devices by adding Wi-Fi-enabled peripherals such as Aquanta[6] and Tando[7]. Aside from producing fine-grained data, these Wi-Fi-enabled IoT devices can also be controlled remotely for energy efficiency, demand response, and spinning reserve DSM services. However, attackers are exploiting the security vulnerabilities of IoT devices connected to the internet to launch botnet attacks [62]. This layer is susceptible to identification threat. For example, if an adversary learns the identity of an IoT device, they could use this information to perform a False Data Injection (FDI) attack, where false data are sent to the cloud, blockchain, or fog computing nodes to generate misleading insights for DSM services.

The cloud computing layer is a centralized system owned by *data collectors*, which consist of data centers and traditional cloud servers with sufficient computing resources, unlike the IoT devices in the sensing layer or the Fog computing nodes. Cloud computing is responsible for delivering data storage and data computing services. Therefore, the cloud layer is susceptible to profiling attacks. For example, profiling may lead to violation of privacy through price discrimination in P2P energy trading or erroneous automatic DSM decisions.

Lastly, the application layer is responsible for processing both real-time and none real-time tasks. Particularly, fog computing also termed as edge computing is suitable for real-time tasks such as demand-supply balancing, frequency control, and surveillance and monitoring. Therefore, this layer is suitable for running DET applications that deliver DSM and other intelligent services provided by *data consumers*. Intelligent home assistance services such as Google Home[8] and Amazon Alexa[9] are examples of data consumer services. This layer is susceptible to localization and tracking attacks. For example, an attacker can use Manipulation of demand via IoT (MadIoT) attacks, first proposed by Soltan et al. [62], to manipulate the total power demand of certain regions using compromised high-energy electric appliances only.

9.3.2.2 PDCH System

In our ongoing study, we are investigating how to preserve data and information privacy in citizen-utility. To address the challenges outlined above, we present the high-level design concepts of the proposed privacy-preserving data clearing house (PDCH) framework. The framework, illustrated in Figure 9.12, is a blockchain-based data management tool for preserving on-ledger and off-ledger transactions data privacy.

To address the identity threat, we propose a blockchain-based scheme that provides batch verification and signature aggregation in the DET network. The scheme should guarantee the traceability of IoT data and the anonymity of the IoT devices in order to prevent false data injection from compromised IoT devices. The scheme should also verify aggregated signatures to reduce the computation cost for application in resource-constrained devices. Lastly, the scheme should ensure the traceability of identities in order to prevent illegal behaviors of IoT devices.

---

6 https://aquanta.io/
7 https://www.tado.com/all-en/
8 https://www.home-assistant.io/integrations/google_assistant/
9 https://www.home-assistant.io/integrations/alexa/

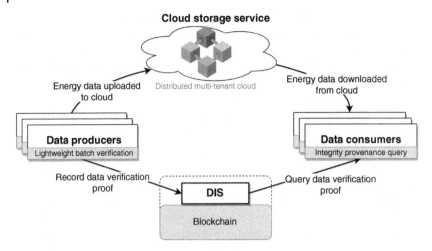

Figure 9.12 Overview of the proposed frameworks.

To address the availability threat, we propose a distributed privacy-preserving multi-tenanted cloud storage mechanism for trust-less and cost-effective data storage. The mechanism addresses the availability issue by replacing the centralized cloud storage with a distributed one. To design a distributed cloud storage, we incentivize the network participants to create a multi-tenant cloud storage service based on P2P personal cloud storage devices. In a P2P cloud storage system, participants exploit the large amount of idle disk space in personal devices within the network, significantly eliminating high data storage costs and availability risks in centralized cloud storage (e.g. the DDoS attack). Moreover, to ensure data privacy in the distributed cloud storage, we use Homomorphic data encryption [60].

Lastly, to address the issue of data integrity, we propose a privacy-preserving integrity provenance query model based on secure multiparty computation technique [63]. The model ensures the efficiency, integrity, and correctness of the query results regardless of the data volume. In our next steps, we plan to design and implement the proposed PDCH system and evaluate its security, privacy, and performance.

9.4 Conclusions

In this chapter, we studied the existing blockchain-based DET solution for P2P energy trading and DSM in smart grids, referred to as citizen-utilities. We categorized the DET market into two main categories, namely, PCG and DSM. Under prosumer communities, we review the application of citizen-utilities in microgrids, VPP, and VEN, illustrated using case studies. Similarly, we review case studies and other proposed blockchain-based DET projects for DSM services including DR, EE, and SR. Further, we reviewed some of the technological challenges hindering the production deployment of citizen-utilities that include lack of scalability, interoperability, and privacy. To provide scalability, interoperability, and privacy in citizen-utilities, we proposed a scalable citizen-utility based on the HARB framework.

The HARB framework provides complex high-order relationships for clustering the physical network for scalability, a distributed digital notary for value transfer between network clusters and dTAM for preserving transaction privacy. Lastly, we proposed the PDCH system, a privacy-preserving system for DSM applications. In future, we aim to complete the design and implementation of the PDCH system and evaluate its privacy feasibility.

References

1 M. Sabounchi and J. Wei. Towards resilient networked microgrids: blockchain-enabled peer-to-peer electricity trading mechanism. *2017 IEEE Conference on Energy Internet and Energy System Integration, EI2 2017 - Proceedings*, 2018, pages 1–5, Jan. 2018.

2 A. Dorri, A. Hill, S. Kanhere, R. Jurdak, F. Luo, and Z. Y. Dong. Peer-to-peer energytrade: a distributed private energy trading platform. *ICBC 2019 - IEEE International Conference on Blockchain and Cryptocurrency*, pages 61–64, 2019.

3 J. Green, D. Martin, and M. Cojocar. The untouched market: distributed renewable energy in multitenanted buildings and communities. In *Urban Energy Transition*, pages 401–418. Elsevier, 2018.

4 J. Green and P. Newman. Citizen utilities: the emerging power paradigm. *Energy Policy*, 105: 283–293, Jan. 2017.

5 J. Abdella and K. Shuaib. Peer to peer distributed energy trading in smart grids: a survey. *Energies*, 11 (6): 1560, 2018.

6 J. Bruinenberg, L. Colton, E. Darmois, J. Dorn, J. Doyle, O. Elloumi, H. Englert, R. Forbes, J. Heiles, P. Hermans, and M. Uslar. CEN -CENELEC - ETSI: Smart Grid Coordination Group - Smart Grid Reference Architecture Report 2.0. November, 2012.

7 E. S. Kang, S. J. Pee, J. G. Song, and J. W. Jang. A blockchain-based energy trading platform for smart homes in a microgrid. In *2018 3rd International Conference on Computer and Communication Systems (ICCCS)*, pages 472–476. IEEE, Volume 4 2018.

8 F. Lombardi, L. Aniello, S. De Angelis, A. Margheri, and V. Sassone. A blockchain-based infrastructure for reliable and cost-effective IoT-aided smart grids. *IET Conference Publications*, 2018 (CP740), pages 1–6, 2018.

9 B. Yu, J. Wright, S. Nepal, L. Zhu, J. Liu, and R. Ranjan. IoTChain: establishing trust in the internet of things ecosystem using blockchain. *IEEE Cloud Computing*, 5 (4): 12–23, 2018.

10 S. Nakamoto. Bitcoin: a peer-to-peer electronic cash system. http://bitcoin.org/bitcoin.pdf.

11 S. Karumba, S. S. Kanhere, R. Jurdak, and S. Sethuvenkatraman. HARB: a hypergraph-based adaptive consortium blockchain for decentralised energy trading. *IEEE Internet of Things Journal*, 4662 (C): 1, 2020.

12 M. Andoni, V. Robu, D. Flynn, S. Abram, D. Geach, D. Jenkins, P. McCallum, and A. Peacock. Blockchain technology in the energy sector: a systematic review of challenges and opportunities. *Renewable and Sustainable Energy Reviews*, 100: 143–174, 2019.

13 E. Espe, V. Potdar, and E. Chang. Prosumer communities and relationships in smart grids: a literature review, evolution and future directions. *Energies*, 11 (10): 2528 2018.

14 A. J. D. Rathnayaka. Development of a community-based framework to manage prosumers in smart grid. July, 2014.

15 R. Belk. You are what you can access: sharing and collaborative consumption online. *Journal of Business Research*, 67 (8): 1595–1600, 2014.

16 A. Goranovic, M. Meisel, L. Fotiadis, S. Wilker, A. Treytl, and T. Sauter. Blockchain applications in microgrids: an overview of current projects and concepts. *Proceedings IECON 2017 - 43rd Annual Conference of the IEEE Industrial Electronics Society*, 2017, pages 6153–6158, Jan. 2017.

17 E. Mengelkamp, J. Gärttner, K. Rock, S. Kessler, L. Orsini, and C. Weinhardt. Designing microgrid energy markets: a case study: the Brooklyn Microgrid. *Applied Energy*, 210: 870–880, 2018.

18 G. Wood. Ethereum: a secure decentralised generalised transaction ledger. *Ethereum Project Yellow Paper*, pages 1–32, 2014.

19 N. Martyniuk and K. Kaspar, Wepower Whitepaper. WePower, 2019. [Online]. https://wepower.network/media/WhitePaper-WePower_v_2.pdf (accessed 08 December 2019).

20 Power Ledger. WHITEPAPER We believe empowering individuals and communities to co-create their energy future will underpin the development of a power system that is resilient, low-cost, zero-carbon and owned by the people of the world.

21 LO3 Energy. Transactive energy: a new approach for future power systems. 2019.

22 M. Mihaylov, S. Jurado, and N.-S. Avellana. NRGcoin: virtual currency for trading of renewable energy in smart grids. 2014.

23 K. Matsumura, M. Marmiroli, and Y. Tsukamoto. Smart grid for distributed resources and demand response integration. *IET Conference Publications*, 2018(CP757), 2018.

24 R. Ma, H. Zhou, W. Qian, C. Zhang, G. Sun, and H. Zang. Study on the transaction management mode of virtual power plants based on blockchain technology. *2020 12th IEEE PES Asia-Pacific Power and Energy Engineering Conference (APPEEC)* 3: 1–5, 2020.

25 AEMO. Virtual Power Plant Demonstration, 172–175, May 2020.

26 S. Seven, G. Yao, A. Soran, A. Onen, and S. M. Muyeen. Peer-to-peer energy trading in virtual power plant based on blockchain smart contracts. *IEEE Access*, 8: 175713–175726, 2020.

27 IEA. Renewables 2019. October 2019, 1–15, 10 2019.

28 Y. Wang, Z. Su, and N. Zhang. BSIS: blockchain-based secure incentive scheme for energy delivery in vehicular energy network. *IEEE Transactions on Industrial Informatics*, 15 (6): 3620–3631, 2019.

29 F. Yucel, K. Akkaya, and E. Bulut. Efficient and privacy preserving supplier matching for electric vehicle charging. *Ad Hoc Networks*, 90: 101730, 2019.

30 I. A. Umoren, S. S. A. Jaffary, M. Z. Shakir, K. Katzis, and H. Ahmadi. Blockchain-based energy trading in electric vehicle enabled microgrids. *IEEE Consumer Electronics Magazine*, 9 (6): 66–71, 2020.

31 I. S. F. Gomes, Y. Perez, E. Suomalainen, I. Silvestre, F. Gomes, Y. Perez, and E. Suomalainen. Coupling small batteries and PV generation: a review. *Renewable and Sustainable Energy Reviews*, 126: 109835, Mar. 2020.

32 P. Palensky and D. Dietrich. Demand side management: demand response, intelligent energy systems, and smart loads. *IEEE Transactions on Industrial Informatics*, 7 (3): 381–388, 2011.

33 S. M. S. Hussain, S. M. Farooq, and T. S. Ustun. Implementation of blockchain technology for energy trading with smart meters. *2019 Innovations in Power and Advanced Computing Technologies, I-PACT 2019*, 2019.

34 M. R. M. Cruz, D. Z. Fitiwi, S. F. Santos, and J. P. S. Catal ao. A comprehensive survey of flexibility options for supporting the low-carbon energy future. *Renewable and Sustainable Energy Reviews*, 97: 338–353, Aug. 2018.

35 A. Khatoon, P. Verma, J. Southernwood, B. Massey, and P. Corcoran. Blockchain in energy efficiency: potential applications and benefits. *Energies*, 12 (17): 1–14, 2019.

36 C. Pop, T. Cioara, M. Antal, I. Anghel, I. Salomie, and M. Bertoncini. Blockchain based decentralized management of demand response programs in smart energy grids. *Sensors*, 18 (2): 162 2018.

37 Z. Zhou, B. Wang, Y. Guo, and Y. Zhang. Blockchain and computational intelligence inspired incentive-compatible demand response in internet of electric vehicles. *IEEE Transactions on Emerging Topics in Computational Intelligence*, 3 (3): 205–216, 2019.

38 J. C. Ferreira and A. L. Martins. Building a community of users for open market energy. *Energies*, 11 (9), 9 2018.

39 A. K. Erenoğlu, İ. Şengör, O. Erdinç, and J. P. S. Catal ao. Blockchain and its application fields in both power economy and demand side management. *Blockchain-Based Smart Grids*. Elsevier, 103–130, 2020.

40 G. M. Madhu, C. Vyjayanthi, and C. N. Modi. A novel framework for monitoring solar PV based electric vehicle community charging station and grid frequency regulation using blockchain. *2019 10th International Conference on Computing, Communication and Networking Technologies, ICCCNT 2019*, pages 1–7, Jan. 2019.

41 R. J. Bessa and M. A. Matos. Medpower2010_Paper_126. Nov. 1–9, 2010.

42 C. Wu, H. Mohsenian-Rad, and J. Huang. Vehicle-to-aggregator interaction game. *IEEE Transactions on Smart Grid*, 3 (1): 434–442, 2012.

43 S. Jeong, N.-N. Dao, Y. Lee, C. Lee, and S. Cho. Blockchain based billing system for electric vehicle and charging station. In *2018 Tenth International Conference on Ubiquitous and Future Networks (ICUFN)*, pages 308–310. IEEE, 2018.

44 V. Dedeoglu, A. Dorri, R. Jurdak, R. A. Michelin, R. C. Lunardi, S. S. Kanhere, and A. F. Zorzo. A journey in applying blockchain for cyberphysical systems. *2020 International Conference on COMmunication Systems and NETworkS, COMSNETS 2020*, pages 383–390, 2020.

45 S. Omaji. Privacy Aware Energy Management in Smart Communities by Exploiting Privacy Aware Energy Management in Smart Communities by Exploiting Blockchain By CIIT / FA17-PCS-013 / ISB PhD Thesis In Computer Science COMSATS University Islamabad Islamabad - Pakistan. Sept. 2020.

46 A. Dorri, S. S. Kanhere, R. Jurdak, and P. Gauravaram. Blockchain for IoT security and privacy: the case study of a smart home. *2017 IEEE International Conference on Pervasive Computing and Communications Workshops, PerCom Workshops 2017*, pages 618–623, 2017.

47 A. Panarello, N. Tapas, G. Merlino, F. Longo, and A. Puliafito. Blockchain and IoT integration: a systematic survey. *Sensors (Switzerland)*, 18 (8): 2575, 2018.

48 E. Androulaki, A. Barger, V. Bortnikov, C. Cachin, K. Christidis, A. De Caro, D. Enyeart, C. Ferris, G. Laventman, Y. Manevich, S. Muralidharan, C. Murthy, B. Nguyen,

M. Sethi, G. Singh, K. Smith, A. Sorniotti, C. Stathakopoulou, M. Vukolić, S. W. Cocco, and J. Yellick. Hyperledger fabric. In *Proceedings of the 13th EuroSys Conference*, pages 1–15, New York, NY, USA, 4 2018. ACM.

49 E. Abebe, D. Behl, C. Govindarajan, Y. Hu, D. Karunamoorthy, P. Novotny, V. Pandit, V. Ramakrishnan, and C. Vecchiola. Enabling enterprise blockchain interoperability with trusted data transfer (industry track). *Middleware Industry 2019 - Proceedings of the 2019 20th International Middleware Conference Industrial Track, Part of Middleware 2019*, pages 29–35, 2019.

50 V. Buterin. Chain Interoperability [white paper]. 2016.

51 S. Thompson. The preservation of digital signatures on the blockchain. *See Also: the University of British Columbia iSchool Student Journal*, 3(Spring), 2017.

52 A. Shrestha, R. Bishwokarma, A. Chapagain, S. Banjara, S. Aryal, B. Mali, R. Thapa, D. Bista, B. P. Hayes, A. Papadakis, and P. Korba. Peer-to-peer energy trading in micro/mini-grids for local energy communities: a review and case study of Nepal. *IEEE Access*, 7: 131911–131928, 2019.

53 S. M. Sajjadi, P. Mandal, T. L. Tseng, and M. Velez-Reyes. Transactive energy market in distribution systems: a case study of energy trading between transactive nodes. *NAPS 2016 - 48th North American Power Symposium, Proceedings*, pages 1–6, 2016.

54 K. Gai, Y. Wu, L. Zhu, M. Qiu, and M. Shen. Privacy-preserving energy trading using consortium blockchain in smart grid. *IEEE Transactions on Industrial Informatics*, 15 (6): 3548–3558, 2019.

55 J. Jung, S. Chun, and K. H. Lee. Hypergraph-based overlay network model for the Internet of Things. *IEEE World Forum on Internet of Things, WF-IoT 2015 - Proceedings*, pages 104–109, 2015.

56 R. Zhang, R. Xue, and L. Liu. Security and privacy on blockchain. *ACM Computing Surveys*, 52 (3): 1–34, 2019.

57 R. Mercer. Privacy on the blockchain: unique ring signatures. 2016.

58 K. Soska, A. Kwon, N. C. Cmu, and S. Devadas. Beaver: a decentralized anonymous marketplace with secure reputation. *IACR Cryptology ePrint Archive*, page 464, 2016.

59 S. Feng, W. Wang, D. Niyato, D. I. Kim, and P. Wang. Competitive data trading in wireless-powered internet of things (IoT) crowdsensing systems with blockchain. *2018 IEEE International Conference on Communication Systems, ICCS 2018*, pages 389–394, 2019.

60 Y. Wang, F. Luo, Z. Dong, Z. Tong, and Y. Qiao. Distributed meter data aggregation framework based on blockchain and homomorphic encryption. *IET Cyber-Physical Systems: Theory and Applications*, 4 (1): 30–37, 2019.

61 J. H. Ziegeldorf, O. G. Morchon, and K. Wehrle. Privacy in the internet of things: threats and challenges. *Security and Communication Networks*, 7 (12): 2728–2742, 2014.

62 S. Soltan, P. Mittal, and H. V. Poor. BlackIoT: IoT botnet of high wattage devices can disrupt the power grid. *Proceedings of the 27th USENIX Security Symposium*, pages 15–32, 2018.

63 Q. Xia, E. B. Sifah, A. Smahi, S. Amofa, and X. Zhang. BBDS: blockchain-based data sharing for electronic medical records in cloud environments. *Information (Switzerland)*, 8 (2), 2017.

10

Blockchain-enabled COVID-19 Contact Tracing Solutions

Hong Kang, Zaixin Zhang, Junyi Dong, Hao Xu, Paulo Valente Klaine, and Lei Zhang

James Watt School of Engineering, University of Glasgow, G12 8QQ, Glasgow, United Kingdom

10.1 Introduction

The coronavirus disease 2019 (COVID-19) is an infectious disease that is caused by the severe acute respiratory syndrome coronavirus 2 (SARS-CoV-2) [1]. In 2020, the world completely stopped as all countries across the globe were affected by the pandemic, with billions of people trapped into quarantine, self-isolation, and full lockdown [2, 3]. Before the vaccine is given to people, non-pharmaceutical interventions (NPIs), which aim at slowing down the transmission of the disease by reducing the contact rate of people [4], are being implemented by various countries across the globe. NPIs largely target social distancing (also known as physical distancing) by keeping a certain distance from others and avoiding large gatherings, in combination with the utilization of face covers and hand washing [5]. These measures have proved themselves highly effective, as seen in [5], because there is an appreciable decline in the number of cases in communities and cities that have implemented NPIs early in the COVID-19 pandemic.

However, despite the benefits of NPIs, several other strict measures were also adopted in order to reduce the transmissibility of the virus, including the closure of workplaces and travel restrictions, which incurred in an immediate threat to the global economy. In a study by the Goldman Sachs, it was predicted that the United States' economy would shrink by 24% in the second quarter of 2020, more than twice as much as any decline ever recorded [6]. Under these circumstances, developing balanced strategies to take both the economy and a rebound of COVID-19 cases into consideration have become imperative. One such strategy adopted by countries such as South Korea and Singapore is the utilization of contact tracing [7, 8].

Contact tracing is the process of identifying people who may have come into contact with an infected person and collecting information about recent close contacts that this person had [9]. A risk score can be given and evaluated based on the duration, distance, and other relevant information (such as ventilation). Contact tracing has proven to be a pillar of communicable disease control in public health for decades, such as in the control of sexually transmissible diseases as well as in the Ebola crisis in Africa [10]. Recently, it also showed its effectiveness on the control of the COVID-19 pandemic in some countries, such as Singapore and South Korea [10]. Thus, it is clear that the utilization of contact tracing can help

Wireless Blockchain: Principles, Technologies and Applications, First Edition.
Edited by Bin Cao, Lei Zhang, Mugen Peng and Muhammad Ali Imran.

governments ease out the social distancing restrictions, while focusing on a more track and trace approach, allowing business to open in a controlled manner, further helping the economy while also saving lives.

Despite its benefits, traditional contact tracing approaches have their limitations. The main issue with this approach is that it highly depends on the accuracy of the infected persons' memory; thus, its efficiency is limited. As a result, the workload of the follow-up interviews as well as from the healthcare workers are immense. In addition, recent studies show that current healthcare infrastructures are not prepared to perform contact tracing in such a large scale, such as the COVID-19 pandemic. In the United States, for example, an additional 50 000 people would need to be hired and trained, while around US$3.6 billion of funding would be needed [11]. In this context, other approaches might be more viable. One such approach that has gained popularity in recent years is the utilization of digital contact tracing (DCT), which has been developed and tested in some countries to identify contacts between infected people and potential spreaders in a faster and more reliable manner. By utilizing mobile devices (such as smart phones) to record close contacts between people the drawbacks in terms of having to remember all the visited locations and contacts of an infected person is removed.

In the design of the DCT applications, the location where people encounter an infected person is not relevant, but rather the contact event is of significant importance; therefore, sensitive information, including the user's detailed location information, is not required. In DCT, the Bluetooth protocol is often used as a positioning technology. Most of the time, Bluetooth can only be used over short distances, but this feature enables the discovery and record of close contacts nearby. In Singapore, their DCT application, TraceTogether, powered by the Bluetrace protocol, makes use of Bluetooth Low Energy. and it has helped the local government control the spread of the disease by adopting a strategy of test, trace, and contain [12]. However, Bluetooth has security concerns on its vulnerable wireless interface. Threats including bugging, sniffing, and jamming are prominent to all Bluetooth-based contact tracing solutions; thus, there is a high risk of replay attacks to the contact tracing network, which can cause a massive scale of panic to the public. Meanwhile, the Bluetooth protocol is potentially liable to be used against user security because the identification of hardware on the Bluetooth physical layer may not be concealed, which brings the exposure of the physical hardware [13].

Another application that relies on Bluetooth is the Google Apple Contact Tracing. Unlike TraceTogether, the service provider does not get hold of the user's real identity; instead, a random identifier is assigned to different users. Thus, users' privacy is protected. However, the user is required to use their central server for contact matching and generating notifications, which brings the concern of trajectory attack on user privacy and enables the reconstruction of the user's profile using access information in the server. Similarly, in the United Kingdom, a DCT application jointly developed by the government with the National Health Service (NHS) also showed issues in terms of potential exposures of user privacy. Besides, fake contacts can be generated by relay-based wormhole attacks, which both the NHS and the Google Apple Contact Tracing are vulnerable to, which may cause confusion among citizens and high pressure on public healthcare systems.

In addition to Bluetooth, the Global Position System (GPS) was also used in some applications for proximity estimation. AAROGYA Setu, for example, collects personally identifiable information, contact information, GPS coordinates, self-assessment data, and analyzes how many confirmed COVID cases are there in a user's vicinity. However, the utilization of GPS signals for user tracking has some drawbacks. Firstly, the high power consumption caused by the GPS positioning may affect users' daily mobile phone use experience. Secondly, sensitive privacy issues may lead to users' distrust of service providers, as whoever stores a user's GPS data can easily reconstruct a person's routine. In addition, location information is sometimes inaccurate indoors, such as in a shopping mall or restaurant, which might hinder the potential benefits of GPS signals, or even cause incorrect estimations of cases to be recorded in inaccurate locations.

Another promising technique that can be used for user tracking is the Health Code System. It is different from the above methods, as it does not use Bluetooth nor proximity detection, but it is rather based on relational cross-matching by scanning QR codes associated with users. In this system, user privacy is not respected because of centralization, and the identity of the user is not hidden to the system's authority. However, the health code is only scanned at the time of passing checkpoints; hence, it saves the user battery and does not consume any data. Additionally, thanks to its highly centralized hierarchy, the coverage can be extended easily. Many other protocols and solutions are emerging to deal with contact tracing for the current pandemic, such as COVIDSafe, Decentralized Privacy-Preserving Proximity Tracing, Pan-European Privacy-Preserving Proximity Tracing, etc. They are similar to the solutions described above, albeit each with their own tweaks on certain features. Thus, for the sake of brevity, we will not dive into too many details about them but interested readers can refer to [14–17] for more details.

Considering the problems of the existing contact tracing methods, it is clear that a novel technology that can improve security and provide greater transparency to the public should be added to contact tracing solutions. One potential alternative to solve such issues is the utilization of blockchain, which has already shown great value in areas such as finance and the Internet of Things [18, 19]. Blockchains are distributed databases organized using a hash tree, which is naturally tamper-proof and irreversible [20]. Each block has a header with a hash value associated with the previous block's content, establishing a retroactive connection from the latest block to the genesis block, the first block in the chain. It provides an unbreakable linkage to the fully traceable records in the order of blocks. As such, users can verify the integrity and authenticity of any known block by calculating the hashed value and comparing it with the next block. Thus, any changes to the previous block will tamper its integrity, leading to a verification failure. Moreover, every participant in the network holds a copy of the data, in order to locate the latest block on the longest chain, which makes the network distributed.

Based on the aforementioned advantages of blockchain, as well as the issues present in current DCT applications, it can be seen that integrating blockchain with contact tracing can bring lots of benefits. Based on that, we propose a framework that integrates blockchain as the underlying infrastructure of DCT and risk assessment in public locations. In DCT, messages between users can be exchanged and confirmed cases can be stored in the blockchain, whereas in risk assessment, public locations can advertise their risk levels to the chain, so that it is visible for all participants. In this process, Blockchain can play a

Table 10.1 Comparison between two modes of BeepTrace.

Modes	Relative	Relative
BeepTrace-active	Bluetooth	Single chain
BeepTrace-passive	GPS	Gemini chain

neutral role in a distributed manner to bridge the user/patient and the authorized solvers to desensitize the user identity and location information.

Recently, we also introduced a novel integration of DCT (Bluetooth and GPS service) and blockchain technology named BeepTrace [21, 22], which inherits the advantages of these two technologies, ensuring the privacy of users and eliminating the concerns about the third-party trust while protecting the population's health. BeepTrace is an open initiative framework involving governments, authorities, public places, and software developers and researchers committed to developing an open and efficient platform for solving the contact tracing problem of COVID-19, minimizing the damage of public health and the economy from this pandemic, as well as maintaining user privacy. Furthermore, Beep-Trace is designed into two different modes based on different sensing technologies used and different blockchain architectures, namely, BeepTrace-active and BeepTrace-passive, with active utilizing Bluetooth, while passive utilizes GPS. As we mentioned before, these sensing technologies have shown their strength in contact tracing; thus, both modes of Beep-Trace are equally effective. Because of differences shown in Table 10.1, BeepTrace-active and BeepTrace-passive are responsible for micro and macro contact tracing, in that they trace and match cases based on different views, individual view and general view, respectively. At the same time, the designed data encryption and blockchain technology in both modes preserve all users' privacy.

In this chapter, we summarize and compare both modes of BeepTrace and indicate their working principles and privacy preservation mechanisms in detail. After that, we demonstrate a preliminary approach of BeepTrace to prove the feasibility of the scheme. Lastly, we look forward to the development prospects of BeepTrace or other decentralized contact tracing applications and point out some development directions.

10.2 Preliminaries of BeepTrace

10.2.1 Motivation

According to the data provided by the World Health Organization (WHO) by the end of January 2021, there have been more than one hundred million confirmed cases of COVID-19, including 2.2 million deaths [23]. This has caused great damage to the global economy and, more importantly, the health of the global population. Despite traditional DCT being implemented in some countries to identify potentially infected people and control the spread of the virus, as we previously discussed, traditional methods pose a concern about people's privacy. Even though traditional centralized contact tracing solutions provide security from a health perspective, the risk of data corruption and manipulation raises concerns about the privacy of its users. Therefore, when BeepTrace was conceived, we regarded the privacy protection as an uncompromising demand. As such, blockchain technology was chosen because of its anonymous network, which completely removes the privacy concerns. Based on that, some aspects of BeepTrace are discussed in more detail below:

10.2.1.1 Comprehensive Privacy Protection

Under the conception of BeepTrace, users' privacy should be protected in a full life cycle through generation, sharing, and storage. At the beginning of the service, any sensitive information shared on the blockchain is encrypted by either symmetric-key or asymmetric encryption algorithms in both modes of BeepTrace. Meanwhile, the trusted government or third-party agencies are limited to obtain and share users' sensitive information with the help of Public Key Infrastructure(PKI) based key distribution. Moreover, the blockchain-enabled platform minimizes the risk of disclosure of users' personal information based on the credential management functionality with cryptography. Lastly, the life cycle of the tracing information on users' storage is also regulated under General Data Protection Regulation (GDPR) [24], and health agencies' recommendations, for example, WHO recommends that 14 days is the minimum length of a tracing cycle. For this reason, all trace and match data stored on the user side is recommended to be stored for a period of 14 days in BeepTrace, which greatly reduces the storage pressure on a user's device.

10.2.1.2 Performance is Uncompromising

Besides protecting the privacy of users, the tracing performance is also an important consideration, which demonstrates the effectiveness of infection prevention and the coverage of its services. Firstly, the more sources of tracing information, the more reliable the tracing results. Therefore, both Bluetooth and GPS data can be used in different modes of BeepTrace to provide duplicate protection. As for the sphere of application, contact tracing projects initiated by governments or Apple Google joint efforts [12, 25, 26] are difficult to be widely promoted worldwide for political or technical reasons. In this case, BeepTrace is proposed, allowing users to have global access to the tracing network, thanks to the accessibility of blockchain. While blockchain is not as responsive as today's centralized databases, contact tracing is not a service that requires real-time feedback, and the performance of the blockchain network can be sufficient to meet the demand.

10.2.1.3 Broad Community Participation

The vision of BeepTrace is that all aspects of society should participate and help in the contact tracing system. In addition to calling on citizens to participate in contact tracing, we hope that data from some public areas, such as shops, malls, or pharmacies, will also be involved in our BeepTrace application. Because these are areas of high population density and human mobility, where the virus can spread easily, it is significant for users to know the risk level of such places. In such cases, these places can record and show the current number of people at that location, their average proximity, and whether there have been recently confirmed infections around these places or not. Nevertheless, because of economic and political interests, the feasibility of these third parties is a question worth discussing. Thus, blockchain can be a perfect medium to solve such issues according to its characteristics of decentralization and openness. A metric about the risk level of exposure to COVID-19 in public areas should be published on the blockchain; in this way, users can make plans to go out based on the risk index and their situation.

10.2.1.4 Inclusiveness and Openness

It is worth noting that the BeepTrace framework for both active and passive modes is compatible, and in the long run, BeepTrace will evolve into a synthesis that encourages everyone to share any types of contact tracing information in any cryptography with different authorities or individuals. In this scenario, it can be regarded as an interface of information tracing center open to the world, providing privacy-preserving contact tracing services globally. Apart from that, the consensus mechanism (CM) of the proposed blockchain framework is not limited to a specific type, Proof of Work (PoW), Proof of Stake (PoS), or Direct Acyclic Graph (DAG) can all be utilized, as any CM that meets network performance requirements and guarantees security can be considered for insertion into the network.

10.2.2 Two Implementations are Based on Different Matching Protocols

Based on a sharing concern about user privacy, there are two different modes of Beep-Trace, BeepTrace-active, and BeepTrace-passive. The backbone of both approaches is the blockchain, and their main difference is in the method of generating contact information and the positive cases matching protocols, as seen in Figure 10.1.

For BeepTrace-active, thanks to the real-time information from Bluetooth or WiFi-direct, the application on users' smartphones could generate a contact list for users whenever they have close contact in a crowded public area. Specifically, contact information is actively exchanged by Bluetooth, which represents the connectivity status of users' devices. In general, the sensing range of Bluetooth is about 10 m. Therefore, users in the same

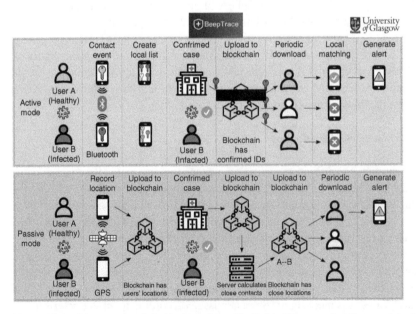

Figure 10.1 Diagram showing the operation of both BeepTrace modes. Source: Adapted from [21].

Bluetooth area can be regarded as in direct geographical proximity. Apart from that, users can inquire about positive cases uploaded on the blockchain and check the potential matches between these cases and the contact data on users' local storage. As a result, we conclude this implementation as an active mode of BeepTrace because contact information is actively exchanged among users, and the matching of positive cases is initiated and finished on users' terminal.

However, the contact data generation and positive cases matching protocol for BeepTrace-passive is distinct from the former. Firstly, a TraceCode with a prefix containing the user's pseudonym and suffix containing encrypted GPS data replaces Bluetooth data as the data circulating on the blockchain and serves as the basis for contact matching of positive cases. GPS provides location estimates, which can be translated to contact information, such as when two people come into close contact with each other. Secondly, a notification chain, separate from the blockchain on which the user-uploaded TraceCode, is set up to notify the user if they have been in contact with a positive case. In this case, both proximity estimation and case matching are done by a Geo Solver, and the user is passively notified if they have been in close contact with a positive case. Therefore, BeepTrace-passive greatly alleviates client-side computing and storage overhead.

To summarize, the key differences between the two modes rely on the sensing technology, with BeepTrace-active utilizing Bluetooth, while BeepTrace-passive utilizes GPS. Moreover, the information stored in the blockchain is also different. On the one hand, the active mode stores at the blockchain the pseudo-IDs of confirmed cases of COVID-19; after that, users that have downloaded the application can periodically query the blockchain for its data and match the downloaded pseudo-IDs with the ones stored locally in their phones. In case there is a match, then the person is alerted and should follow specific self-isolation guidelines. On the other hand, in the passive mode, users periodically upload their GPS positions to the blockchain. Thus, whenever a person is tested positive for COVID-19, this person's state in the blockchain is updated by the endorsement of a health staff member, and third-party servers utilize this information in order to calculate close contacts that this person had with other participant users. Then, risk levels are uploaded to the blockchain, which can be downloaded by its users. In case a person's GPS data matches the one in the blockchain, they are passively notified of a potential exposure and will have to follow the self-isolation guidelines.

10.3 Modes of BeepTrace

In this section, we give a detailed workflow description and an explanation of key concepts of both modes of BeepTrace. In the following parts, we firstly introduce the workflow of the contact tracing framework and then demonstrate how the privacy of users is ensured in each mode of BeepTrace.

10.3.1 BeepTrace-Active

10.3.1.1 Active Mode Workflow

For BeepTrace-active, the first step is to generate temporary pseudo-IDs consisting of a series of random letters (identification of users) and the created timestamp, as indicated by Figure 10.2. Furthermore, the Bluetooth names of users' devices can periodically change

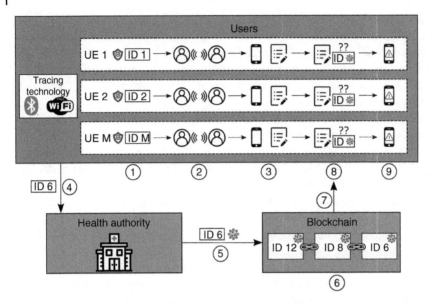

Figure 10.2 Framework of BeepTrace-active. Source: Klaine et al. [21] / with permission of IEEE.

to avoid the leakage of users' privacy, as in [12]. Because the pseudo-IDs are also periodically changing, the positions of users are impossible to be tracked. Apart from that, only the confirmed cases' pseudo-IDs are accessible on the blockchain; thus, the privacy of users is well protected. In step 2, the users' devices broadcast the local pseudo-IDs to devices in the same Bluetooth area. In step 3, a local list of contacts is generated, including all pseudo-IDs received through the broadcasting of other Bluetooth devices. At any point in time, a person might need to go to a hospital in order to perform a COVID-19 nucleic acid testing (NAT), as shown in step 4.

In case a person is tested positive, either the health authorities or the user himself should take the responsibility to upload the IDs stored on the users' mobile phone (from the last 14 days) in step 5. Then, these IDs are published as a new block on the blockchain, which is accessible for all users (step 6). There is no concern about privacy, in that the pseudo-IDs do not contain any personal information of confirmed patients. The application using BeepTrace-active mode downloads and update local pseudo-IDs of positive cases from blockchain periodically, as in step 7. The application compares the risky pseudo-IDs from the blockchain and the local contact list locally on their mobile phones in step 8. In the last step, the user is then alerted if the local contact list contains any risky pseudo-IDs from the blockchain, notifying the user of a possible contact with an infected person.

10.3.1.2 Privacy Protection of BeepTrace-Active

In BeepTrace-Active, we use two mechanisms to protect the privacy of users. First, the contact data generated by Bluetooth is hash-encrypted before being uploaded to the blockchain. The local data are also encrypted by a hash algorithm, so it does not affect the matching process. The second is the anonymity of the blockchain. In the active mode of BeepTrace, we use blockchain as the medium for information broadcasting and storing the pseudo-IDs supplied by confirmed patients. The local storage identified in Figure 10.3 is a list of all

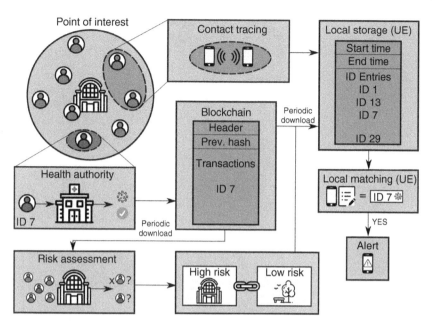

Figure 10.3 BeepTrace-active framework for blockchain. Source: Klaine et al. [21] / with permission of IEEE.

pseudonyms by users. It is worth noting that the user personal information is never shared in any case, avoiding privacy leakage. By publishing these pseudo-IDs into the blockchain, the data are protected by it with ultimate privacy preservation, as the access information is not trackable if a user matches the record in locally. This leaves the user with more security and privacy and eliminates the possibility of malicious manipulations of test results. In addition, Figure 10.3 also shows how public locations can advertise information in the blockchain. By estimating the number of people in a determined location, or by periodically downloading data from the blockchain, the risk of public locations can be determined and uploaded to a separate chain. These data can also be downloaded by users, so they can assess the risks of going outside or visiting certain places. Despite blockchain being widely used as a distributed ledger for recording information agreed by different parties that perform transactions, in this chapter, the application scenarios of blockchain are further extended. By integrating blockchain into DCT, confirmed COVID-19 cases can be published in the chain, allowing a tamper-free and public access to this data in order to perform local matching. This enhances user privacy, minimizes misinformation, and can also lead to an increase in trust and, potentially, to the adoption contact tracing, benefiting millions of people and mitigating the impact of future pandemics.

10.3.2 BeepTrace-Passive

10.3.2.1 Two-Chain Architecture and Workflow

BeepTrace-passive mode is designed to use PKI-distributed keys to users, diagnosticians, and geodata solvers, which are represented in Figure 10.4. The authorities are regarded as Certification Authority (CA) to run this process in step 1. BeepTrace application of user

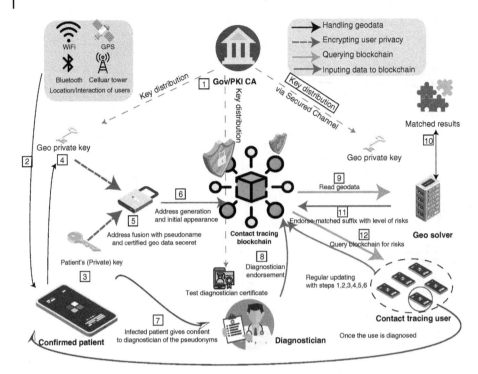

Figure 10.4 Workflow and framework of BeepTrace-passive. Source: Onireti et al. [22] / with permission of IEEE.

terminal collects the raw geodata using GPS services, as shown in step 2. Then, users are expected to generate private keys locally in step 3, normally, one for each day. These private keys can be 256 bits pseudo-random numbers in an encrypted chip, such as Apple T2 Security Chip [27]. Like Ethereum [28], a public address using the ECDSA-secp256k1 algorithm based on the private key can be generated. Then, the hash value of this public address is obtained using keccak-256 while a certain number of bytes is kept in order to generate a pseudonym as a prefix of the TraceCode. In step 4, an encryption of the raw geographic information with a timestamp is generated by using the public key from CA to form the suffix of the TraceCode. An address fusion pseudonym and geodata encryption are performed in step 5. As previously mentioned, both symmetric-key and asymmetric encryption algorithms are used to make sure the privacy of users is well protected. The TraceCode can be regarded as a connection of the pseudo-identity and their geodata. Moreover, all users are expected to upload this indexable TraceCode to the blockchain (step 6) and repeat steps 1 to 6 until they are confirmed as a positive case by the diagnostician.

As for confirmed patients, they have choices to exchange pseudonyms in step 7. After getting the consent of patients, the specific diagnostician is responsible for assisting users in finishing this process. The diagnostician replaces the prefix (pseudonyms) with a new pseudonym section to generate a new TraceCode and then endorses it on the blockchain network, as shown in step 8. This process can eliminate all users' concerns about their privacy because the patient-related pseudonym information is completely separated from the geographic information. Once TraceCode is uploaded to the blockchain, the processing of TraceCode is handled by a third party, a publicly trusted entity with a geodata private key issued by the government, known as a geo solver. This agency is expected to access the blockchain and decrypt the geodata and timestamp of TraceCode in step 9. Next, as in step 10, the geo solver authorized by the CA performs contact tracing by finding out all pseudonyms whose geodata can match the geodata with the endorsement of the diagnostician. A risk-level announcement is also introduced to show the degree of cross-infection between several TraceCodes. The geo solver endorses the risk level and related pseudonyms on a separate notification blockchain in step 11.

When users choose BeepTrace-passive mode, they can download the information about users' risk level of current local situation from the notification blockchain in step 12. In the case that their geodata matches the geodata with an endorsement of the diagnostician, they are notified passively about their risk level. In this process, because the pseudonyms are generated by the encryption algorithm based on the pseudo-random number, it does not contain any user-related privacy information, so the user's privacy is well protected.

10.3.2.2   Privacy Protection in BeepTrace-Passive

The working principle of TraceCode and how the TraceCode can be decoupled from the users' privacy with the signed pseudonym by diagnostician and geodata sharing are illustrated in Figure 10.5. Users' privacy is not compromised by sharing pseudonyms, which are generated by their private keys. Several cryptography algorithms can be applied in this process; a possible solution is shown in the last section using both symmetrical and asymmetrical encryption. According to the figure, both pseudonym prefix and geodata suffix are encrypted information. As shown in Figure 10.4 steps 3 and 4, users generate pseudonyms based on their local private keys as the head of the TraceCode and use a public key distributed by CA to encrypt raw geographic data generated by their devices as the tail of TraceCode. Thus, the complete TraceCode connects the users' pseudo-identity and their geodata. The TraceCode of confirmed patients should be verified by the diagnostician to ensure the reliability of the data on the blockchain and avoid the social panic caused by untrusted data.

The sharing data on the blockchain network is all cyphertext; only authorized case-matching agencies (geo solvers) with the private key can decrypt the geographic information. However, the pseudonym prefix cannot be cracked by any authorities or individuals; thus, the privacy is well preserved. A notification blockchain is proposed to unload the tracing blockchain, publishing filtered risk exposure information from trusted sources. Alternatively, under the premise of guaranteeing the network performance, this double-link structure can be simplified to two smart contract addresses on a single chain. Most importantly, the users' access traces to the blockchain, such as IP addresses, routing information, and even the ISP records are completely isolated from the blockchain network. In short, information about users other than pseudonyms and geodata is not available to any institution or individual, including miners; thus, the network is privacy preserving in nature.

Another measure is taken to keep the privacy of users in geodata generation. Before the geodata is encrypted by public keys from CA and upload to blockchain, perturbations are required to avoid identical suffix match tracing against the patient's private key, by either

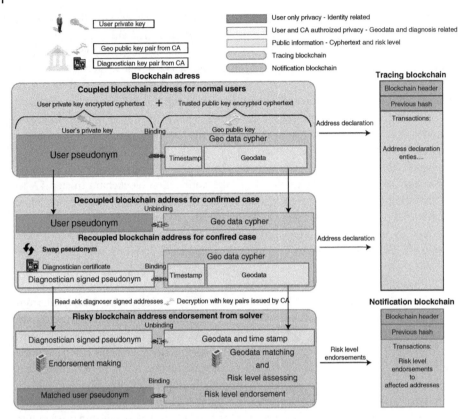

Figure 10.5 TraceCode construction of BeepTrace-passive. Source: Onireti et al. [22] / with permission of IEEE.

adding salt to geodata or transforming the geodata. Three possible transformation methods can be adopted.

1. Use geo datum with an elliptical encrypted system with perturbation, a well-known example of the implementation is GCJ-02 datum reference system.
2. Convert GPS geo datum (WGS84) [29] to the Grid reference system (OSNG: OSGB36, where the accuracy is limited).
3. GIS aggregation/geodata generalization and perturbation to avoid trajectory privacy tracking.

Last but not least, users are empowered to customize the service and ensure their privacy. For example, users have complete freedom of choosing which level of geographic information they want to share, as long as it is accurate and compliant. The diagnostician is responsible to use the above methods to fine-tune geodata into a coarser or finer grain to avoid trajectory tracing by malicious users or even completely reconstructing the geodata with a dedicated key for secured geodata sharing, if required. Also, users have the right to change their pseudonyms at any time to eliminate possible leakage of privacy.

10.4   Future Opportunity and Conclusions

10.4.1   Preliminary Approach

Based on the technologies and concepts discussed above, an Android platform-oriented COVID-19 close contact detection application has been developed. Android platform is a good starting point for application development and test because it occupies more than 70 percentage of the smartphone market [30]. The preliminary design achieves basic ideas of BeepTrace-active, including helping users know whether they are close contacts, while keeping users' real identities confidential. At the same, it can operate normally on devices with minimum Android 7.0 (API 24), which means more than 73.7% Android users can access this application.

The Bluetooth protocol is based on Bluetooth Low Energy (BLE) technology. BLE remains in sleep mode constantly except for when a connection is initiated. Meanwhile, its high data rate results in a much shorter connection time compared with classic Bluetooth. Therefore, usage of BLE can reduce power consumption while maintaining a similar communication range and quality. This makes BLE work better in applications requiring high real-time performance, including BeepTrace. Although BLE uses the same 2.4 GHz ISM frequency band as classic Bluetooth, it only allows the Android application to communicate with other BLE applications instead of all Bluetooth applications nearby. This feature reduces the workload of information filtering.

In BeepTrace-active, pseudo-IDs are generated by combining a Universally Unique Identifier (UUID) and the manufacturer ID. A UUID is a 128-bit number used to identify information, while the manufacturer ID consists of unique numbers assigned by the Bluetooth SIG to member companies requesting one. Several bits of the manufacturer ID are modified as checking code to ensure the accuracy of information. The randomly generated UUID will be directly spliced with the modified manufacture ID to generate pseudo-code. If the ID collected fails to have the correct checking code in these digits, it will not be taken down. For information storage, three databases, namely, personal-ID database, ID-collected database, and ID-downloaded database, are created locally via SQLite. All information is stored in string form, and data written for more than 14 days will be erased. In this way, BeepTrace will not take up much storage space. For each user, his/her pseudo-ID will be regenerated periodically and every time it changes, the new ID will be stored in a personal-ID database. Personal pseudo-ID is advertised continuously via BLE. At the same time, the BeepTrace application continuously searches for BLE devices nearby, while collecting and storing pseudo-codes with correct format in the ID-collected database.

Ethereum is used as the blockchain platform; the process of uploading pseudo-ID is implemented in the form of smart contract functions. Once a positive pseudo-ID is uploaded, an Ethereum event is emitted to broadcast this pseudo-ID on the blockchain network. All users who are in this network and listen to this event will receive the updated data. Users download all confirmed pseudo-IDs in the recent 14 days from the blockchain to the local ID-downloaded database when they use the application for the first time. Also, users are free to run the matching process locally without the interference of any organizations. If any pseudo-IDs from the ID-collected database and ID-downloaded database are matching, the application will notify them they are COVID-19 close contacts.

In our test, both the uploading and downloading rates are near 100 transactions per second (TPS). Considering the throughput requirements, this result demonstrates the feasibility of the scheme of BeepTrace.

10.4.2 Future Directions

The destructive threat of the COVID-19 virus comes from its significant infectiousness. Therefore, timely detection of close contacts and potential patients is of great help to the prevention and containment of the epidemic. A trend of promoting a virus tracking application has been started by the COVID-19 crisis. Although many policies against COVID-19 have been introduced, the epidemic is still raging in many countries. Therefore, an effective virus tracking software is still worth waiting. At the same time, if the BeepTrace application is deployed in advance, it will fight other viruses or epidemics that may break out in the future. Therefore, continuous optimization of BeepTrace is necessary. We identify the following as some important topics. Those problems are what we are facing and trying to overcome.

10.4.2.1 Network Throughput and Scalability

Some unignorable issues with our proposed contact tracing scheme are the massive traffic caused by a large amount of addresses declaration because of frequent (global) geodata update; hence, great computing resources are required for geodata matching. Meanwhile, we face a great challenge of blockchain processing throughput for single-chain operation. It is a great challenge running the desired hundreds of millions of transactions per second on any existing blockchain solution. Luckily, the needs of such high TPS are rare in the real world. For example, it is reasonable to assume that a user does not travel internationally often; therefore, the needs of the user data are completely met in the domestic blockchain network. In addition, all parameters are selected at typical maximum values to see the peak requirement. For instance, it is not reasonable to assume all people (in all ages) are active 14 hours every day. We also encourage the use of multiple blockchains by regionally grouping the users via PKI and public keys management. By dividing users into smaller groups, the network capacity can be easily scaled up. Besides, the emerging high-throughput ready blockchain can be introduced to the deployment of BeepTrace, for instance, directed acyclic graph (DAG) in theory has no throughput limit thanks to its intentionally designed forking schemes [31]. In addition, when the technology is ready for high-throughput performance, we can easily migrate two or more regional chains together and speed up the sharing of the information.

The computing resources are limited from time to time; however, the geodata complexity can be dramatically reduced if the user's quantity on a single chain is below the thresholds. In the case of international passengers, the country can employ the server to look up the data in both regional networks and hence reduces the needs of massive networks all the time. We have made a comparison of simulations based on an assumption of different size networks. For a medium-sized country with 70 million population, the required computing resource is as little as dozens of cloud server instances; however, for the large population bases, such as the combined population match of the top seven most populated countries (a sum of 4 billion people), it takes tremendous computing resources equaling to

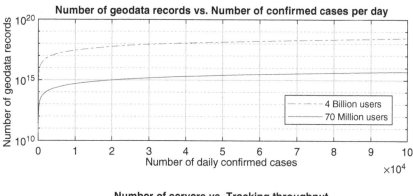

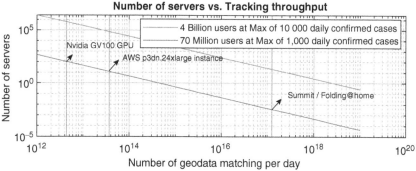

Figure 10.6 Computing resource requirement for contact tracing geodata. Source: Onireti et al. [22] / with permission of IEEE.

23 of Summit [32] (the fastest supercomputer in 2019) and hardly achievable using current technology, although it might be possible in the near future. A comparison of simulation results is shown in Figure 10.6.

10.4.2.2 Technology for Elders and Minors

It is believed that some usage barriers exist when some people, especially elders and minors, try to access technology. However, they might have even greater risk of getting infected. Therefore, they should never be left out. The limits of technology reach to certain groups of people may become a major issue at rolling out digital contact tracing. However, it is not completely impossible to include them. An idea is to establish an in-time information network between these people and their guardians. Wearable technology and wireless IoT [33] can be used by the elders and minors to enable them to join the contact tracing program. The private keys can be stored locally but also be transferable to guardians and carers. By transferring the private keys in a secured channel, the parents and carers can take responsibility to keep their beloved under protection, without giving up on their privacy.

At the same time, it is very likely that elders and minors will not be putting enough effort to receive the notifications. In this case, a trusted third party is needed. By giving consent of privacy to some other users or third-party service providers, they can start sending push messages to the vulnerable once there is a risk. People naturally do give their privacy consent to the above parties, for example, care homes, online health companies, parents, and adult children of elders. By combining these avenues, we believe no one should be left out in this crisis.

10.4.2.3 Battery Drainage and Storage Optimization

For recently proposed contact tracing programs, the challenge of battery drainage and storage optimization are unavoidable because of the requirements in terms of active broadcast and record of GPS coordinates. However, our scheme can be more energy efficient by conducting recording and uploading periodically instead of accessing a cloud database continuously. Further improvements can be achieved by making the phone wait until it is plugged in and within the Wi-Fi coverage. By sending the data only when the mobile device is being charged, our scheme becomes more battery friendly. Delaying the information upstream can induce lower performance in the contact tracing network, but it is completely acceptable to be notified a few hours later rather than immediate response because of the nature of tracing lag. From the solver side, if the duplicates of endorsement are made, the solver will only upload the unique address to the blockchain, which reduces the pressure on the user.

10.4.2.4 Social and Economic Aspects

It is well known that centralized systems are more efficient and economical than decentralized systems in most cases. Blockchain is a representative of distributed systems and deploying such a system in a nationwide manner may cost taxpayers more. However, from another side, the decentralized blockchain system is also well recognized among citizens as a non-governmental solution that can preserve privacy in a much better way than a centralized system. Such a consensus can effectively minimize the resistance from human rights organizations and fear of citizens of infringing rights or other fundamental civil liberties. This can also increase the uptake of DCT among citizens and is thus of paramount importance to winning the battle with COVID-19 as early as possible and to save billions of lives each day.

From the blockchain mining perspective, attracting sufficient independent miners to contribute to the blockchain construction is key to maintaining its distributed nature. In the most successful blockchains such as Bitcoin, the rewards to the miners come from the transaction fees and/or the creation of a new block. In BeepTrace, it could be difficult to build such an ecosystem in a short time frame and there are no real transactions (thus no transaction fees) in such a system. As possible solutions, the rewards can come from the government, by paying the miners who created and maintained the blockchain, or in the case of sharing some existing blockchains, transaction fees can be claimed back from the government. Of course, conquering COVID-19 is the common mission of all mankind; thus, each user could be part of the miners to voluntarily support, legitimize, and monitor the blockchain network.

10.4.3 Concluding Remarks

BeepTrace has explored a blockchain-based methodology to track patients and close contacts while protecting people's personal information to fight the COVID-19 pandemic, which also has great potential value against other epidemic diseases. Two main frameworks have been proposed. BeepTrace-active mode, based on Bluetooth or WiFi-direct, relies on users to actively upload and download patients' information via blockchains and contact matching of positive cases is conducted on users' terminals, whereas BeepTrace-passive mode works on GPS. Authorized geodata solvers warn users if they have been in close contact with a positive case. These two frameworks are compatible, and throughout the whole process, users will appear under pseudonyms and pseudo-IDs to protect privacy. Detailed procedures and functions of each entity are presented and compared with existing solutions to show the advantages. Challenges are also discussed from blockchain performance, solvers complexity, user's battery and storage, and economic and social aspects. BeepTrace application has been proved to work well in most Android-based mobile phones. Its universality and safety make it suitable for wide promotion among the people to fight the COVID-19 epidemic.

References

1 A.E Gorbalenya, SC Baker, RS Baric, RJ de Groot, C Drosten, AA Gulyaeva, BL Haagmans, C Lauber, AM Leontovich, BW Neuman, D. Penzar The species severe acute respiratory syndrome-related coronavirus: classifying 2019-nCoV and naming it SARS-CoV-2. *Nature Microbiology*, 5: 536–544, 2020. ISSN 2058-5276. info:doi/https://doi.org/10.1038/s41564-020-0695-z.

2 World Health Organization, Coronavirus disease 2019 (covid-19) situation report-85. 2020. https://www.who.int/docs/default-source/coronaviruse/situation-reports/20200414-sitrep-85-covid-19.pdf.

3 G. Ramachandran A. Hekmati, and B. Krishnamachari. Contain: privacy-oriented contact tracing protocols for epidemics. 2020.

4 M. C. J. Bootsma and N. M. Ferguson. The effect of public health measures on the 1918 influenza pandemic in U.S. cities. 104: 7588–7593, 2007. ISSN 0027-8424. info:doi/https://doi.org/10.1073/pnas.0611071104.

5 C. Corinne, M. Harris, G. Adhanom, L. Tedros, R. Tu, "Mike" J. Michael, V. K. Vadia, D. Maria, F. Diego, O. Imogen, and G. Charles. WHO audio emergencies coronavirus press conference, 2020. https://www.who.int/docs/default-source/coronaviruse/transcripts/who-audio-emergencies-coronavirus-press-conference-full-20mar2020.pdf.

6 M. McKee and D. Stuckler. If the world fails to protect the economy, COVID-19 will damage health not just now but also in the future. *Nature Medicine*, 26 (5): 640–642, 2020.

7 S. Brack, L. Reichert, and B. Scheuermann. Privacy-preserving contact tracing of COVID-19 patients. 2020.

8 D. Normile. Coronavirus cases have dropped sharply in South Korea. Whats the secret to its success? 2020. https://www.sciencemag.org/news/2020/03/coronavirus-cases-have-dropped-sharply-southkorea-whats-secret-its-success.

9 L. Ferretti, C. Wymant, M. Kendall, L. Zhao, A. Nurtay, L. Abeler-Dörner, M. Parker, D. Bonsall, and C. Fraser. Quantifying SARS-CoV-2 transmission suggests epidemic control with digital contact tracing. *Science*, 368 (6491), 2020. ISSN 0036-8075. info:doi/https://doi.org/10.1126/science.abb6936.

10 A. M. McCollum, K. Mirkovic, R. Arthur, A. L. Greiner, K. M. Angelo, and F. J. Angulo. Addressing contact tracing challengescritical to halting Ebola virus disease transmission. *International Journal of Infectious Diseases*, 41: 53–55, 2015.

11 J. B. Watson, A. Cicero, and M. Fraser. A national plan to enable comprehensive COVID-19 case finding and contact tracing in the US. 2020.

12 J. Bay, J. Kek, A. Tan, C. S. Hau, L. Yongquan, J. Tan, and T. A. Quy. BlueTrace: a privacy-preserving protocol for community-driven contact tracing across borders. page 9, 2020.

13 D. L. Becker, K. Johannes, and D. Starobinski. Tracking anonymized bluetooth devices. In *Proceedings on Privacy Enhancing Technologies*, pages 50–65.

14 India Government. Aarogya Setu Mobile App, 2020. https://www.mygov.in/aarogya-setu-app/.

15 Department of Health Australia. The COVIDSafe Application, 2020. https://www.health.gov.au/sites/default/files/documents/2020/04/covidsafe-application-privacy-impact-assessment-agency-response.pdf.

16 DP-3T Task Group. Decentralized Privacy-Preserving Proximity Tracing. April, 33, 2020.

17 PePP-PT e.V. i.Gr. Pan-European Privacy-Preserving Proximity Tracing, 2020. https://www.pepp-pt.org/content.

18 F. Tian. An agri-food supply chain traceability system for China based on RFID & blockchain technology. In *Proceedings of the 13th International Conference on Service Systems Service Management (ICSSSM), Kunming, China*, pages 2–5.

19 A. Tapscott and D. Tapscott. How blockchain is changing finance. *Harvard Business Review 1.9*, pages 2–5.

20 S. Underwood. Blockchain Beyond Bitcoin. Technical Report. Tech. Rep. 11, Sutardja Center for Entrepreneurship & Technology Technical Report. UC Berkeley, June 2016.

21 P. V. Klaine, L. Zhang, B. Zhou, Y. Sun, H. Xu, and M. Imran. Privacy-preserving contact tracing and public risk assessment using blockchain for COVID-19 pandemic. *IEEE Internet of Things Magazine*, 3 (3): 1–8, 2020.

22 H. Xu, L. Zhang, O. Onireti, Y. Fang, W.J. Buchanan, and M.A. Imran. Beeptrace: Blockchain-enabled privacy-preserving contact tracing for covid-19 pandemic and beyond. *IEEE Internet of Things Journal*, 8(5), 3915–3929, 2020.

23 World Health Organization. WHO Coronavirus Disease (COVID-19) Dashboard. https://covid19.who.int/.

24 P. Voigt and A. von dem Bussche. *The EU General Data Protection Regulation (GDPR): A Practical Guide*. Springer Publishing Company, Incorporated, 1st edition, 2017. ISBN 3319579584.

25 Apple Inc. and Google LLC. Exposure Notification. May 2020.

26 J. Snow and M. Mallon. The security behind the NHS contact tracing app, pages 1–14, 2020.

27 Apple. Apple T2 Security Chip. Technical Report October, 2018. https://www.apple.com/euro/mac/shared/docs/Apple_T2_Security_Chip_Overview.pdf.

28 Ethereum.org. Ethereum accounts, ethereum.org. October, 2021. https://ethereum.org/en/developers/docs/accounts/.

29 U.S.A Department Of Defense. Global Positioning System Standard Positioning Service. Www.Gps.Gov, September, 1–160, 2008. http://www.gps.gov/technical/ps/2008-SPS-performance-standard.pdf.

30 Statcounter Global Stats. Mobile Operating System Market Share Worldwide. 2020. URL https://gs.statcounter.com/os-market-share/mobile/worldwide.

31 B. Cao, Y. Li, L. Zhang, L. Zhang, S. Mumtaz, Z. Zhou, and M. Peng. When internet of things meets blockchain: challenges in distributed consensus. *IEEE Network*, 33 (6): 133–139, 2019.

32 Oak Ridge National Laboratory. ORNL Launches Summit Supercomputer. Technical report, Oak Ridge National Laboratory, 2018. https://www.ornl.gov/news/ornl-launches-summit-supercomputer.

33 Y. Sun, L. Zhang, G. Feng, B. Yang, B. Cao, and M. A. Imran. Blockchain-enabled wireless internet of things: performance analysis and optimal communication node deployment. *IEEE Internet of Things Journal*, 2019. ISSN 23274662. info:doi/https://doi.org/10.1109/JIOT.2019.2905743.

11

Blockchain Medical Data Sharing

Qi Xia, Jianbin Gao, and Sandro Amofa

University of Electronic Science and Technology of China, Chengdu, Sichuan Province, 610054, Peoples Republic of China

11.1 Introduction

From the earliest times, the ability of a population to achieve the desired goals of the family or the state has depended on factors such as its general well-being. Medical health therefore is a key priority of any state that takes its own longevity seriously. To guarantee adequate healthcare for the population, societies have developed hospitals and other health institutions to manage the collection, analysis, use, and security of health data. These factors, although varied in their complexity, are further compounded when the collected data has to be transmitted from one institution to the other [1]. The difficulty of protecting the data whether at rest, in transit, or enforcing permissible behavior with the data upon reception by other users is a challenge with limited, sufficiently secure candidate solutions. While it is generally perceived that sharing the data maximizes its utility through analyses by multiple parties, the development of varying perspectives on the data, and provable objective interpretation, sharing also introduces several security and privacy risks that require significant financial, technical, operational, and managerial resources to address. The sharing arrangement must be conducted carefully by all participants and evaluated periodically and systematically to ensure the safety and security of the systems and data (see Figure 11.1). As alluded to earlier, medical data sharing increases the cumulative usage, value, and utility of a given medical dataset. However, while the increase in these three factors is important and welcome by all beneficiaries, the process to deliver more value from data through sharing also poses severe risks to individuals and institutions engaged in the sharing arrangement [2]. To begin, sharing was naturally restricted when patients' records, including transcripts of patient–doctor interactions, medical histories, test results, etc., were stored as hard copies or handwritten notes in file cabinets in the respective hospitals where the doctors worked. To initiate sharing, patients officially request for their data from the primary hospital to a secondary one such as a specialist practice. If it was deemed appropriate to respond to the written query, the data were duplicated in the requisite format and physically transported to the next institution. Till the requested data was received, it was subject to all the variables that affected the delivery of ordinary mail. Aside the obvious problems of storage space for

Wireless Blockchain: Principles, Technologies and Applications, First Edition.
Edited by Bin Cao, Lei Zhang, Mugen Peng and Muhammad Ali Imran.
© 2022 John Wiley & Sons Ltd. Published 2022 by John Wiley & Sons Ltd.

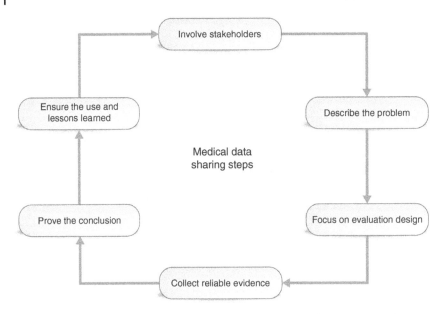

Figure 11.1 Medical data sharing strategic steps (*which ensures feasibility, security, propriety, and accuracy*).

the documents, inappropriate access to sensitive patient records [3], and physical tampering with the data, they could also be rendered unusable or even destroyed in disasters such as fires or floods. Thus, the advent of electronic medical records provided an avenue to store patient data in a format that was conveniently accessible, transmissible, and with relatively small input could be kept sufficiently secure. Hence, electronic medical records also ushered in a paradigm of sharing that boosted patient mobility and afforded more options with respect to patients' choice of caregivers.

However, the mobility of patient data in medical data networks creates a critical vulnerability as increased digital access and advancing technical implementations present more people with greater access to their data. Therefore, while electronic medical records advantageously facilitate larger volumes of data and improved data access in the healthcare space, security challenges centering on legitimate access, data confidentiality, and integrity also emerge on a corresponding scale [4]. While a database may be used to grant and manage permissions to patient data, problems of scalability make this solution impractical with escalating number of users. An effective solution to mitigate the security challenges inherent in the sharing arrangement regardless of the participants' particulars is the incorporation of blockchain technology. The blockchain is a distributed digital infrastructure that combines game-theoretic behavior control and cryptographic data protection techniques to facilitate the exchange of digital assets with the assurance of immutable, auditable transactions. The blockchain was first employed as the underlying technology of the Bitcoin cryptocurrency which was created and popularized by Satoshi Nakamoto and Bitcoin [5].

Fundamentally, the Bitcoin blockchain is a clever coupling of a linked-list data structure with a cryptographically secure consensus mechanism that renders received financial transactions on the blockchain network immutable. As such, transactions from any and all

Table 11.1 Some differences between the cryptocurrency and healthcare applications of blockchain technology.

Cryptocurrency uses	Healthcare uses
1. Focus on currency creation as a reward for successful transaction validation.	1. Focus on patient records management concerning access control, data transfer and security.
2. Automatic regulation of network transaction speed through target difficulty adjustment.	2. No need for target difficulty regulation as it does not profit the network.
3. Uses open, permissionless blockchain.	3. Uses closed, permissioned blockchain.
4. Miners behavior maximizes reward earned.	4. Validators' behavior maximizes the trust and security of the blockchain.
5. Transaction privacy contingent on pseudo-anonymity.	5. Transaction privacy must be designed into the network.
6. Data transfer implies a change of ownership.	6. Data transfer confers access rights and permissions for specified operations.

participants on the network are welcome and included in the next block of published transactions so long as they are valid and also have the cryptographically secure proof-of-work. However, researchers soon recognized that the properties of the blockchain could be leveraged to securely exchange other digital assets instead of restricting it to financial uses only. Thus, the Office of the National Coordinator, ONC of the United States Department of Health and Human services launched a competition that required researchers to adapt the blockchain to assist the management of healthcare and healthcare data [6]. As the research showed, even though the blockchain can be adapted to facilitate transactions in healthcare, fundamental differences arise as a result of the varying application. Table 11.1 presents some differences between the blockchain as used in support of cryptocurrency and as compared to blockchain as a tool for healthcare data management. As is evident in the sheer volume of publications globally, the infusion of blockchain technology into healthcare and particularly into medical data sharing presents several advantages, some of which have been systematically investigated and published widely by both industry and academia. The goal of this chapter is to provide new perspectives on some known methods while throwing more light on new ones. While we strive to present new material on the subject matter, it may be necessary to revisit some already established angles of blockchain medical data sharing in order to properly contextualize it and to highlight new perspectives on the logical outworking of blockchain-enabled sharing arrangements. We list the contribution of the chapter as follows:

- We highlight three paradigms of medical data sharing that can benefit from blockchain technology,
- We present three special cases that are especially suited to blockchain medical data sharing,
- We present an architecture to support each paradigm presented, and
- We analyze medical data sharing to highlight privacy and security benefits to data owners.

11.1.1 General Overview

Recently, providing healthcare is progressively dependent on the collaboration of multiple institutions of varying specialties. Medical data therefore moves from one platform to the other in support of healthcare outcomes. The need to combine the medical data with other datasets emphasizes the role of interoperability in the success of the sharing enterprise. Again, current methods of data access, querying, and filtering responses endanger the privacy and security of patients. The inability to guarantee the security of data therefore necessitates a critical look at the issues defining medical data sharing while emphasizing the role of the blockchain in solving these challenges.

11.1.2 Defining Challenges

The difficulties that result from medical data sharing cluster around data security, data privacy, and the use researchers put the data to. While the blockchain does not negate the requirement for vigilance in sharing arrangements, it can compensate for vulnerabilities that may serve as vectors for malicious attack depending on the architecture employed. We list these factors and their components and briefly outline how the blockchain addresses each of them.

11.1.2.1 Data Security

The data must be secure from access by unauthorized entities, systems, and processes [7]. To achieve this, it is critical to ensure that all access by users occurs within the context of a network. As an additional measure, an underlying peer-to-peer network can be employed. This provides two benefits: it ensures that users on the network can independently verify one another's identity before transactions. It also provides a basis for a global view of network transactions. The blockchain provides the peer-to-peer network environment, a shared global history of transactions and where required can provide the identities of network actors in a closed blockchain.

11.1.2.2 Data Privacy

The data shared must be restricted to only those systems, entities, processes, and procedures that legitimately require it as an input [8]. Because of the time and entity restrictions that often define privacy, it is necessary to generate a log of the entities that access the data. This can effectively provide the provenance required to securely track access patterns and better protect data privacy. The blockchain can contribute to the data privacy effort because by default, it can facilitate the use of a different identity/pseudonym for every new transaction. Thus, user privacy is protected by the difficulty of linking multiple transactions to one particular individual.

11.1.2.3 Source Identity

A critical factor in assessing the utility of shared data is the origin of the said data. Integrity attacks such as data swapping, data replay, or unauthorized modifications of the data in transit or at rest in storage can have fatal consequences because of its role in decision support. As such, data source validation is a vital step in medical data sharing [9].

A blockchain middle-ware service can effectively validate a dataset against a known list of contributing devices, systems, agents, and processes. Thus, with source identity validation, the blockchain can supply the provenance required to monitor sensitive health data along its life cycle.

11.1.2.4 Data Utility

The value of the shared data resides in the utility to which data analysts put it [10]. Hence, a dataset may be analyzed multiple times by different teams to derive the maximum benefit from it. This can support abuse in situations where the data is used to correlate trends other than the ones for which it was obtained. With the inclusion of smart contracts in the sharing, specific conditions of acceptable use may be embedded in the data to enforce desirable behavior on the part of recipients. The Hyperledger platform provides tools and services to create smart contracts that can be deployed on networks to provide access to whole dataset or to specific portions, monitor its use, and guarantee expected provenance.

11.1.2.5 Data Interoperability

The health industry has several competing standards and formats that create complex, unstructured heterogeneous data, all of which are equally important in their support for clinical decision making as required for diagnosis. Thus, it can be challenging to combine it with data obtained from other sources [11]. Considering the often-occurring necessity of combining images (scans), textual and multimedia data, it is important to create an abstracting layer that extracts crucial data for use by health personnel regardless of the format of the input or output data. The resulting abstraction layer has to match the input data with outputs in spite of the probability of integrity attacks. The blockchain can play a watchdog role by keeping a hash-table of input formats and outputs that can be referenced by agents and processes to facilitate interoperability of various data formats and sources.

11.1.2.6 Trust

Anonymized datasets acquired through data harvesting and processing may still retain sensitive elements that attackers may explore and exploit. Data such as address, date of birth, and contact details can be used to uniquely identify a person and therefore considered as sensitive. This can undermine trust in data source, it's uses, and the ultimate results of analyses conducted using the said data. Technical guarantees to restore trust in the data must combine provable anonymity with guaranteed utility of the shared data [12]. As a remediation tool, the blockchain can guarantee trust on multiple levels from data source validation, data transfer channel protection, and secure data storage. In all of these arrangements, the blockchain serves as disinterested intermediary that limits access to only pre-authenticated users and provides a timestamped log of network transactions.

11.1.2.7 Data Provenance

This refers to the metadata charged with the establishment of causal relationships between objects and the determination of attribution regarding object creation, mobility, and conditions of current status. This is critical in healthcare where provenance of records can establish their authenticity as well as facilitate transparency and auditability [13]. As a means of objectively determining the integrity of systems or data, provenance can effectively answer questions as to how and when data objects were created, who has accessed them and what operations have been performed using the data in question. The blockchain provides effective provenance through the use of timestamps and Merkle trees, which together guarantee the originating node and time a digital record was created. The Bitcoin blockchain provides pseudonymity for transactions, which may prove a challenge in healthcare. Therefore, a closed permissioned blockchain can be used as is appropriate in the implementation context.

11.1.2.8 Authenticity

Data authentication is the two-step process of confirming the origin of data as well as its integrity. Its importance in healthcare cannot be overstated because of the increasing reliance of healthcare systems on digital platforms for the collection, processing, transmission, and long-term storage of sensitive healthcare records. For the data to pass authenticity tests, it must be the result of acquisition from a verified entity and demonstrate it has not undergone any corrupting process. Thus, to achieve effective provenance of electronic medical records, data must remain authentic or continually pass authenticity checks throughout its life cycle [14]. The blockchain can assist in data authenticity protection with its emphasis on timestamped transactions and Merkle trees that pledge with certainty that a network of users can conveniently and independently verify the metadata on stored records.

11.1.3 Sharing Paradigms

Effective record management is vital in healthcare for apparent reasons. At the point of care, doctors require the information they receive to be as complete and as accurate as possible to ensure that any diagnosis they make and their subsequent decisions for disease management is factually grounded. Patients, on the other hand, require trust in the established processes to ascertain the symptoms of their condition to aid prescription of appropriate treatment while respecting their privacy to all extents guaranteed by doctor–patient confidentiality regulations. Thus, depending on the extent of care required, sharing occurs on three fundamental levels: These are

- Institution-to-institution data sharing,
- Patient-to-institution data sharing, and
- Patient-to-patient data sharing.

It is these foundational sharing paradigms that enable and underlie well-defined Big Data technologies such as data mining, machine learning, and the Internet of Things in healthcare. They are listed in the order given above for systematic presentation. More realistically, patient-to-institution data sharing occurs first and is usually followed by either of the remaining two. Patient-to-patient sharing was introduced and popularized through patient online social networks. As the data proceeds from one point to the other, properties such as its origin and the correctness of algorithms that analyze it must engender trust. The more the number of entities involved in the sharing, the greater the tendency for dubious operations and hence data abuse [15]. The blockchain, as an impartial algorithmic judge, can mediate the aspects of sharing that are prone to abuse. Using a secure sharing architecture (as shown in Figure 11.2), challenges inherent in sharing such as access control,

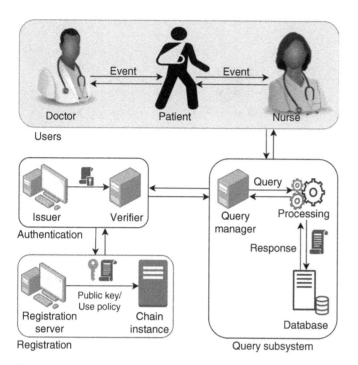

Figure 11.2 An overview of secure architecture for exchange of sensitive medical data.

data ownership, acceptable use, interoperability, privacy, and security can be minimized or outright eliminated by the incorporation of blockchain medical data sharing in healthcare. In the architecture, separate modules for the administration of users, query management, registration, and authentication cooperate to facilitate data sharing in a secure manner. Completed transaction data are sent to the blockchain and can be accessed when required by authorized agents.

11.1.3.1 Institution-to-Institution Data Sharing

Confronted by the requirement of managing population health with dwindling budgets and decreasing resources, researchers increasingly have to rely on data from sources other than their own. While electronic medical records and the ubiquity of high-speed data networks in hospitals and other healthcare provision institutions present several advantages for the exchange of patient data, it is fraught with risks that are known to undermine patients' confidence in data sharing arrangements. For instance, the sharing of data at some point in time necessitates a regime of control to ensure effective access to datasets for research. However, this very often leads to centralization of the data, or centralization of control over the data, both of which have serious drawbacks for privacy [16].

Data centralization creates a single point of failure in the data flow with its attendant problems. Hardware failure, breach by hackers, or tampering with the data can render the data inaccessible, unusable, or undermine the validity of research conducted using the said data. To achieve secure data sharing requires a systematic recognition of challenges posed by data exchanges between cooperating institutions. While the data may be shared freely, it

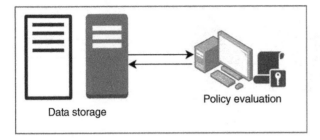

Data storage

Policy evaluation

Figure 11.3 Data retrieval from storage must satisfy specified policies for added security.

is important for the sake of continued collaboration to respect the limits imposed by ethical, regulatory, and legal frameworks (see Figure 11.3). For instance, ethical considerations may limit the operations that can be executed with a given dataset even though there are no legal or technical challenges. Again, healthcare industry regulations may require specific procedures to protect data while in the custody of an institution [17]. The challenge inherent in multi-institution medical data sharing therefore is how to control the data after sharing to prevent unauthorized processing or operations. In this instance, control refers the capacity to determine and enforce the following:

- *who* else can access the data,
- what *operations* are permissible on data and
- *how long* the data should be available to the receiving participant.

This is a critical consideration because of different interests, requirements, and motivations that each research institution is subject to. Traditionally, controlling access to digital infrastructure is achieved using the three steps of identification, authentication, and authorization. When properly implemented, these steps provide sufficient basis for accounting for user actions on the network while guaranteeing that privacy intrusions will be prevented or detected when it occurs. This therefore warrants the creation of a network of sufficient resources and features that is capable of accepting users on an ad hoc basis and can still provide a collective history of network actions that participants find computationally satisfactory. Consequently, a peer-to-peer network with participants' nodes can be employed in the solution. However, the nodes exiting and rejoining the network will have difficulties with network views of events that occurred in their absence. It is to overcome these challenges that the blockchain includes a consensus mechanism and so it can be used as an intermediary for controlling access to the sensitive data and providing a tamper-proof log of network events (see Figure 11.4).

Medical data exchange platforms can therefore be "updated" to source their data from blockchain middleware [18] that provides data management and security services that ameliorates the security of shared data (see Figure 11.5). The next paragraph presents a description of the blockchain as applied in healthcare data sharing as well as its procedures.

Blockchain in healthcare The blockchain is a term used to describe distributed ledger technologies that are employed to facilitate and manage multiple cooperating institutions, especially where there is the need to jointly administer the shared infrastructure and its data. However, the blockchain is specifically the technology that underpins the Bitcoin

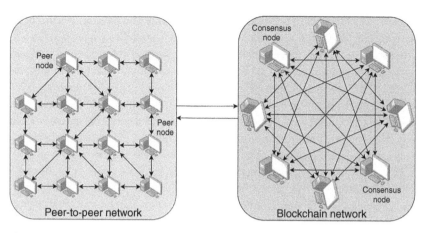

Figure 11.4 Conceptually the blockchain is constructed as a peer to peer network as with a consensus mechanism for agreeing on the state of submitted transactions.

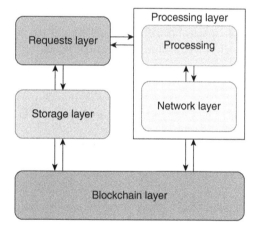

Figure 11.5 A secure blockchain-based architecture for the sharing medical data.

cryptocurrency and is distinguished by immutability of records (transactions), consensus of participating nodes, a continuously extending list of transactions, competition to validate transactions and claim accompanying reward (mining and mining fees), and a built-in limit to control the rate of issuance of new currency by the Bitcoin network. These features, while necessary for the creation and management of a pecuniary system, provide little advantage in healthcare. To begin, the motivation for action in a financial setup is completely different from motivation for action in healthcare. Outcomes to maximize in healthcare are qualitative and non-financial. These include patient satisfaction, trust, privacy protection, efficiency of data sharing, and data control. In finance, the emphasis is on quantitative measures such as maximization of fees for services rendered. Therefore, blockchain in healthcare must support the previously stated qualitative features using cryptographic techniques and other best practices adapted for use with the blockchain.

Blockchain network The blockchain is a linked-list data structure and a consensus protocol that work together with an underlying peer-to-peer network model to confer validity on transactions executed on the network. Each list of transactions is referred to as a block and is cryptographically chained to the list before it. By design, the transaction details are immutable once placed on the blockchain and rely on pseudonyms instead of identities. The continuously extending list of blocks is distributed to all network participants to ensure a consistent view of data accepted by the network. However, the need to restrict access to data compels a blockchain model that limits transaction authoring to a group of pre-approved nodes. This is known as a permissioned blockchain because of the requirement to permit only known and verified users. For multi-institution data sharing as occurs in healthcare, the transfer of the dataset from "Hospital A" to "Hospital B" must include verification by network nodes. This facilitates the immutable capture of attributes such as ownership and usage permissions as Hospital B becomes the custodian of the dataset.

Cryptographic keys Requests for data and replies are denoted as actions on the network and must be signed and encrypted using appropriate cryptographic keys in order to be valid. Thus, the use of cryptographic primitives guarantees confidentiality of data while also ensuring non-repudiation of action. To participate in the data exchange network, a user requires a membership key and a transaction key pair. These are described below:

1. *Membership key*: This key is generated by the network and sent to the user after receiving a request to participate in the network. It is checked after transactions are initiated.
2. *Transaction private key*: This key is used to sign requests when they are created. It is kept secret by the key owner.
3. *Transaction public key*: This key is used to verify the authenticity of received requests. It is published on the network after it is created.

System components The system to achieve the goals of blockchain medical data sharing needs network, cryptographic, and blockchain components. The next few lines briefly describe the roles of the authenticator, processing nodes, and the smart contract center. These components are required to coordinate user action of the network.

- *Authenticator*: The authenticator node checks the validity of requests by ensuring that the membership key of the requester is valid on the network. It verifies the digital signature on requests to determine the source. It compares the signature to its own list of identities to find a match. When verified, a request can then proceed.
- *Processing nodes*: These nodes perform miscellaneous tasks such as acting as intermediaries between requests and replies, collecting metadata, and preparing reply packages. They select unprocessed requests from a common pool and process them according to user-determined policies (see Figure 11.6).
- *Smart Contract Center*: This oversees the creation of smart contracts that grant permissions and automate the routine aspects of sharing. They can also be used to monitor user behavior with received data, evaluate network conditions, and generate reports for system administration.

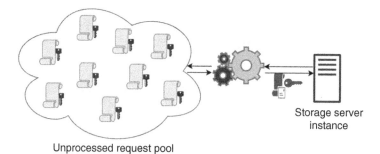

Storage server
instance

Unprocessed request pool

Figure 11.6 Transactions are submitted to a pool of requests where the processing nodes can retrieve them for processing and forward results to a requester or storage server.

Process   To access data, the user sends a request to the system that is received by the authenticator. The authenticator queries the requesting node to determine the validity of its membership key. Once the membership key is validated, the authenticator forwards the request to processing nodes for security checks, i.e. validity of signature on the request. The processing nodes then forward the request to the data store with a package denoting the data required and the processing to be performed. It also sets a time counter so that access to the data does not remain indefinite. To reply to the request, the data store queries the data owner for permission to grant access. The data owner responds to the request by signing a hash of the data store query. It then forwards it to the data store. When received, the data store signs a hash of the initial request and compares it to the one received from the data owner. After this step, the data owner then verifies the signature on the received hash. After successful verification, the data store can then transfer the required access rights to the processing nodes. The processing nodes then prepare the access rights as a package and retrieve the transaction public key of the requester to sign the package. The processing nodes then send the package to the requester. When received, the requester can then sign the package with their own private key to verify the authenticity of the package before accessing the permissions. The parameters of the request-reply process such as the timestamps, specific permissions, user identities, and other metadata are recorded in an immutable log and processed into a block that is transferred to the blockchain.

There are some benefits to the above sharing arrangement: First, and most important for multi-institution sharing, the raw data are not transmitted as in traditional sharing setting (compare Figures 11.7 and 11.8). Only the required access permissions are transmitted to the requester. As such, it is convenient to specify timed access, monitor operations on data, and revoke access in case of violations. Because all actions are network-based, they can be processed after occurrence and transmitted to the blockchain to form part of the provenance record of a given dataset. Second, it can help regulate data access so as to respect owner-specified restrictions on the data even after sharing. This facilitates the derivation of more value from the data while still retaining control over it.

Institutions-to-institution data sharing has as its primary objective the derivation of more value from shared data. Thus, competing interests, different research practices and orientations, and varying levels of skill all drive multi-institution data sharing to assist innovation in healthcare. Its main challenge of data control after sharing can be effectively remedied when the sharing is implemented on a blockchain-based platform.

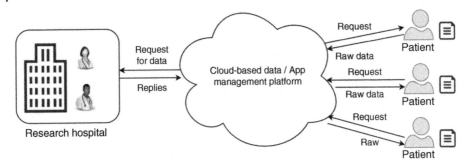

Figure 11.7 Traditional medical data sharing models request and access raw data from patients and their devices, thereby compromising the privacy and security of patients.

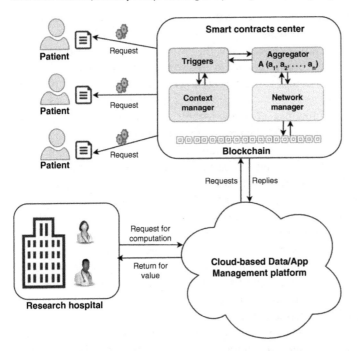

Figure 11.8 Blockchain-based data sharing protects data and privacy through models such as requests for computation that return desired value instead of raw data.

11.1.3.2 Patient-to-Institution Data Sharing

The era of Big Data technologies has sharply focused the attention on data as the primary fuel of innovation. Thus, many new products and services are now designed or augmented with data harvesting modules that collect and transmit data to waiting databases. While this practice is relatively new, the ubiquity of wireless networks, internet-enabled devices and willing participants ensure that new data reach the receiving databases continually. As a result, relatively new companies have accumulated very large databases that can be analyzed to provide insights that can be conveniently and speedily commercialized. This trend of sensitive data collection underlies several of the digital apps that accompany wearable sensing devices.

The popularity of these devices lie in their affordability and the quick analytics they provide through their partner apps. Thus, while they present a convenient avenue to quantify and visualize individual activity levels with convenient metrics, they also access longitudinal data that the patient may not otherwise relinquish [19]. This uneven arrangement in the sharing of the healthcare data from sensing devices implies that companies create multiple revenue streams using patient data in ways the patient does not agree to but cannot effectively revoke or object to. However, suppose a means can be deduced to share the financial profits from the use of such data, the patient does not have an objective method to determine the fairness or lack thereof of the arrangement. Thus, in patient-to-institution data sharing, questions that arise border on privacy of data control over the uses of data, who has permission to perform operations on the data and the equitable distribution of financial proceeds in the event of making money from the said data. The technical methods employed to implement institution-to-institution medical data sharing can be modified to serve here as well. The qualifying features necessary to invoke the use of the blockchain still remain cooperation among mutually distrustful parties and the need to maintain an immutable log of valid network events [20]. As is evident, these features characterize patient-to-institution data sharing.

11.1.3.3 Patient-to-Patient Data Sharing

One challenge of modern healthcare is patient engagement and support to achieve the desired outcomes of improved health and better quality of life. These two goals are intrinsically connected to the patients' own input in their care. In many ways, patient engagement can be facilitated by the quality of information received from trusted sources to aid decisions made for the management of chronic conditions such as diabetes, cardiovascular diseases, and other age-related illnesses. It has long been realized that access to other patients' experience in managing their treatments helps to develop new ways of coping with particular conditions [21]. In time, therefore, the advent of social network platforms permits the digital assembly and grouping of patients of specific disease categories. Thus, while it began as a means of informal communication among professionals, hobbyists, and other well-defined groups who could quickly create and share content with their peers, digital social networks have also found application in organizing individuals to attain specific goals relating to their health. Therefore, there are some digital networks dedicated to serving the needs of particular groups such as patients with diabetes or those who are physically disabled. However, these social networks find it difficult to attract the very people for whom the networks were designed. Especially with disabled patients, the reasons for avoiding such social networking sites are few and well known:

1. *Discrimination*: some insist that accessing the social network site constitutes the creation of an enclave of the disabled. In many countries, this constitutes or reinforces the act of discrimination. Disabled persons therefore avoid it.
2. *Privacy*: signing up for social networking sites involves collection of names, birth dates, gender, and other sensitive personal identifying details that can help attackers focus sharply on how to exploit the data collected by the network.
3. *Stereotype*: this reinforces the mental picture that disabled people or socially challenged patients should only assemble in locations designated for them.

Others argue in favor of digital social networks citing how it impacts and improves patients quality of life through continuous monitoring and group cohesion [22]. These are briefly expanded on below:

1. *Accessible communication method*: Users on digital social networks usually communicate using text and multimedia files. This is especially beneficial because it accommodates a wide variety of disabilities. Hence, communication becomes possible regardless of the particular disability. Those with speech difficulties can focus on text data, while those with visual problems can obtain information from audio. The digital network also permits users to access the experience of particular individuals regardless of their location.

2. *Socialization online*: Ongoing health conditions may compel young patients to be confined in one location such as a home. This can adversely affect their socialization as they are less like to meet and form the bonds that create a satisfactory social life. A social network can facilitate socialization by allowing the confined patient to find and join groups to explore opportunities of mutual interest.

3. *Raise awareness*: Online platforms provide the access, affordability, and anonymity required to create awareness about any particular issue. Many patients may be willing to provide detailed accounts of their conditions to others but do not want to compromise their privacy to do so. The digital platforms allow patients to create blogs, websites, and services that assist patient communities to manage themselves. The challenge is to share useful experience data without compromising the security of the sharing partners.

4. *Similar patients*: In the case of particularly rare diseases, it is helpful for the patient to meet others who have a same condition. Meeting in person may not be possible because of geographic, social, and economic restrictions. On the internet, however, such restrictions are eliminated, thus allowing access to supportive and valuable information from a fellow sufferer.

Thus, while social networks present opportunities to assist some patient groups with timely information on self/condition management, they also threaten the privacy and security of patients on the network. To address the problem at hand, there is the need to create a mechanism to

1. allow patients to access data and experience from other patients while remaining anonymous,
2. guarantee that input data actually comes from patients in the group,
3. maintain a secure log of network actions for audit purposes, and
4. prove identity of users in case of dispute.

Therefore, to facilitate secure exchange of information on such a social network requires a balance of strong privacy controls with guaranteed authenticity of shared data (see Figure 11.9). Blockchain technology can be especially beneficial in remedying the challenges [23] posed here. First, because the technology is network based, it can source authentic data from verified users after blockchain oracles have determined the authenticity of data. Second, the need to control user input can be addressed using a closed permissioned blockchain. Lastly, the blockchain can provide the history of network actions as a secure digital log of network events. Thus, while patient-to-patient medical data

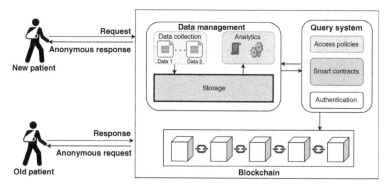

Figure 11.9 Overview of patient-to-patient medical data sharing using blockchain.

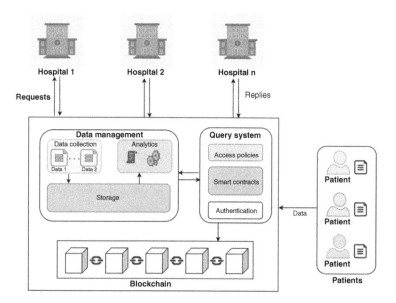

Figure 11.10 Blockchain medical data sharing involving multiple institutions.

sharing is not a fully established paradigm, blockchain technology presents patients with the benefits of personal medical data sharing between patients while keeping its attendant problems of privacy and security at bay.

While some of the difficulties may be answered with solutions prescribed in other sections, it is important to consider how the individual user can opt out of the data sharing while still being protected. With patient-to-institution sharing, the patient must retain the option to quit the sharing of data while being able to use a wearable sensor or digital app in a manner that is still beneficial as opposed to patient-to-patient sharing where personal identifying characteristics are more likely to be abused. Again, the patient should be able to rejoin the sharing program at any time if they so desire. Therefore, as with any secure data sharing solution, an effective option has to be network based to account for every action executed using the data as input. Hence, an architecture such as shown in Figure 11.10

where individuals and institutional users can connect to a data management and query system with a blockchain recording transaction details. With the required parameters, the details on the blockchain can be accessed to support healthcare when necessary. We will proceed to outline and explain the use and benefits of the blockchain in benefiting all sides in the sharing.

1. *Complexity of sharing* Medical data sharing may begin as a simple exchange of data between two entities. However, because of recent developments in data protection challenges and regulations, protecting the data can evolve very quickly to include others who are not part of the initial sharing arrangement and are not legally bound to respect the agreement [24]. Thus, a sharing arrangement that began as a one-to-one relationship can metamorphose into a one-to-many arrangement without the parties renegotiating the terms. This complexity benefits the institutions at the expense of the patient and termination can be difficult or impractical. Even in termination of the sharing agreement, the patient cannot request that his data be removed completely from the database of the hosting institutions. Thus, the implications of complexity affect security and privacy of data. To resolve the issue of variable complexity in data sharing arrangement, the candidate solution must maintain the initial arrangement to bind all subsequent users to respect the terms till its termination.

The threats to this paradigm are largely integrity based and have a high likelihood of occurrence as the data are transmitted from one institution to the other. We assume that the value of the data while it remains with the originating institution or after a recipient receives, it insulates it from unauthorized modifications and tampering. Thus, successful vectors of attack must be from malicious attackers that swap the data requested, replay previously transmitted data, or modify data in transit without proper authorization, all of which are exercised to compromise the integrity of the data.

2. *Swapping*: This is a violation of data integrity where a malicious attacker intercepts the intended response for a request, and a different dataset is forwarded to the requester. The swap attack relies on access privileges on the network where both the requester and respondent are present. In this attack, the privacy of the data is compromised while maintaining the value to both the attacker and the intended recipient. To detect a swapping attack, an authentication check can reveal the data source as accurate while the data itself fail the authentication test.

3. *Replay*: In the replay attack, a malicious user intercepts messages and forwards them to recipients at a time of his choosing. He may also forward several variants of the same message to the unsuspecting recipient. In this case, an authentication test against a skilled attacker will pass the origin of the data and must focus on detection methods that employ secure hashing.

11.1.4 Special Use Cases

In this section, we highlight some new and emerging services that can benefit from the security, privacy, data control, granular data access, and trust blockchain medical data sharing infuses into healthcare. These are precision medicine, monetization of healthcare data, and patient record regeneration. They are described below:

11.1.4.1 Precision Medicine

Precision medicine, also called personalized medicine, is a relatively new method of patient care. It is characterized by minimizing risk factors based on individual attributes and choosing a treatment that may cure or control the patient's condition. To be effective, the caregivers need to have a genetic understanding of a patients' particular disease. This new approach holds several attractions that can optimize patient care from the individual level to the management of population health. The promise of precision medicine is the promotion of health management strategies that leverage an individual's biological composition, lifestyle, environments, and choices that have a direct bearing on health [25]. This facilitates a very high level of customization of patients' treatments so that medical decisions, practices, and products can all be tailored to one patient, a family, or an entire population that is known to have specific biological traits. The current paradigm of treatment focuses on giving patients tests, drugs, and subjecting them to procedures based on previously observed statistical effectiveness. Therefore, in all probability, patients suffering from one disease will be given the same treatment regardless of their lifestyle, ethnicity, diet, and other common factors. Precision medicine, on the other hand, is directly dependent on all observable factors of a patient's life insofar as it impacts their health (see Figure 11.11). Indeed, to be effective, precision medicine requires full access to complete and accurate clinical data. To grasp the full potential of precision medicine, it is important to appreciate and secure the promising but vulnerable components that make it possible. These components are cybersecurity, the quantified self, telehealth, automated data aggregation, privacy protection, and secure data sourcing. As one may discern, all these depend on or are directly affected by medical data sharing. Thus, as has been the trend so far, blockchain technology can serve as a remediating factor that compensates for the security challenges inherent in precision medicine [26] by reducing risks posed to patient data, minimizing the vulnerability of host systems to malicious attack and providing a cryptographically secure service and transaction log that is immutable and can prevent or withstand current modes

Figure 11.11 Precision medicine combines aspects of current healthcare with advanced computational data modeling to take advantage of new discoveries for optimal health management.

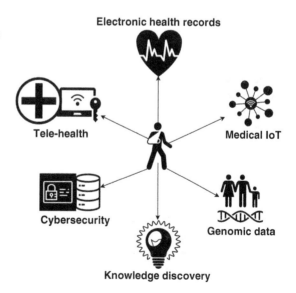

of attack. We will, therefore, examine each factor according to its contribution to and effect on precision medicine.

1. *Cybersecurity*: The driving technologies behind the data security management precision medicine requires are all network dependent for their application services. Hence, while helping to disseminate data beyond the initial collection points, medical data networks can also present potential threats to data privacy and patient confidentiality. Appropriate Cybersecurity frameworks for precision medicine will do well to incorporate layered protections schemes into platforms, devices, and procedures. This can better alleviate the threats that electronic data is vulnerable to. A successful attack on the security infrastructure, applications, or services that support precision medicine can have catastrophic consequences as valid, accurate, and timely health data is a critical component in clinical decision making. Hence, whether implemented for a single patient, a nuclear family, or a large population, precision medicine must be supported by a Cybersecurity framework that is scalable while protecting data and privacy [27]. It must be based on industry standards, integrate precision medicine component factors, and also facilitate speedy recovery of data and service availability in the event of a successful attack. When implemented effectively, Cybersecurity can facilitate the development and operational effectiveness of precision medicine strategies while guaranteeing trust in the system.

2. *Quantified self/Internet of Things*: The proliferation of wearable consumer health devices and the longitudinal data they generate provide multiple opportunities for the advancement of precision medicine. These devices collect data on patients in between hospital visits to give doctors a data-driven basis for evaluating treatment efficacy and patients compliance with given directives. Their service to the precision medicine paradigm lies in the data generated and collected, the apps that analyze and structure the data, and the ability to transmit it to others [28]. First, the apps that accompany these devices facilitate the structured collection, analysis, and presentation of the data. Again, being present on the patient for most of the time, the data generated give valuable quantitative insights into patients lifestyle. Despite the benefits of wearable sensors, challenges with data synchronization, failed transmission of collected data, device failures, and patient attitudes can affect the collected data. With timestamped transactions and immutable record-keeping, the blockchain can secure IoT data for precision medicine.

3. *Tele-health*: Patient-generated health data can furnish caregivers with sufficient information to empirically assess lifestyle as a factor of healthcare. Such data can provide a complete data-driven overview of patients life, including sensitive elements such as activity levels, duration, and location data. Following from the point above, risk-based monitoring can be implemented for patients within specific vulnerability groups by taking advantage of streaming continuous, real-time data. Thus, as a component of precision medicine, tele-health strategies must allow for more efficient transfer and integration of electronic medical data for virtual patient care [29]. In such an arrangement, the role of blockchain as access control mechanism can secure patient data as it traverses multiple platforms to assist in care delivery. The blockchain's immutable record system also serves as a source of metadata that can be interrogated to determine data provenance, access patterns, and sequence of patient-generated data inputs that warranted particular restorative actions by caregivers.

4. *Automated Data Aggregation*: A key feature of the blockchain is the execution of prede-termined actions by the network when required conditions are met. Referred to as smart contracts, this feature can collect, analyze, transmit, or store data. Popular smart con-tract platforms such as Ethereum and Hyperledger can be leveraged to securely aggregate individual patient data by developing unique smart contracts that mine the data for criti-cal insights to improve the patients' quality of life, transfer information to caregivers, and secure the privacy of data and actions. However, because the data provided are external to the blockchain, it is important for it to be verifiable and accurate in order to be allowed on the blockchain. This can improve data accuracy and limit or eliminate falsified data from entering the blockchain. We can achieve through the use of decentralized oracles as input infrastructure [30]. The decentralized oracles can query the data providers and retrieve patient information vital in decision making. Because a central oracle can be eas-ily compromised, we ensure that multiple oracles can submit queries for information to electronic health data platforms. For example, an oracle can query a collection of health data sources collectively named *Src*. We thus make queries to nodes $Src_1, Src_2, ..., Src_k$. Replies $a_1, a_2, ...a_k$ are collated from the responding nodes. The replies from the various oracles are then aggregated into a single answer $A = agg(a_1, a_2, ..., a_k)$. If a majority of sources return the identical value a, the function $agg()$ returns a; otherwise, it returns an error. Thus, provided a majority ($>k=2$) of sources are functioning correctly, the oracle will always return a correct value A. The data are then sent to the blockchain through an API provided in the smart contract.

11.1.4.2 Monetization of Medical Data

This is a revenue generation model where a digital asset that is non-revenue-generating turns into one that now brings financial returns (money). This has been firmly applied to several social media platforms where the demand for videos and other forms of consumer entertainment is so high that content creators are rewarded financially to incentivize their continued contribution to the desired platforms. However, the promise of financial reward can bring with it an infringement of copyright and intellectual property rights. As such, proper attribution of digital files is critical to the success of the sharing platform as well as the harmony of its content contributors. In the medical data sharing space, monetization can be used as an incentive to attract innovation and research [31] by leveraging vast vol-umes of data already available in distributed clinical data networks. As can be expected, the monetization of healthcare data raises fundamental questions (see Figure 11.12) such as

- Ownership: who owns the data?
- Access: Who gets access to the data?
- Processing: What computations can be performed on the data?
- Control: Can the data owner retain control over it even after sharing?
- Equity: How are the benefits of monetization distributed?

The blockchain can provide adequate answers to these questions by using a multilevel smart contract platform to address the following:

1. *Ownership*: Data contributors can create a smart contract that takes three inputs: a pseudonymous identity, a policy for acceptable use, and a reporting module.

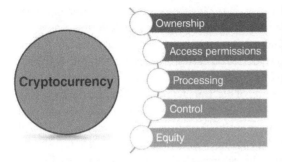

Figure 11.12 By addressing difficulties posed by ownership of data, access permissions, processing, and data control, the blockchain can assist researchers develop equitable models for the monetization of health data.

The contract tags all data from the contributor with unique identifiers before processing nodes can process and validate it. Hence, ownership can, therefore, be determined by the address accessed by the reporting module.

2. *Access*: To control access to the data, a contributor can be required to create a public and private key pair and publish the public key on the blockchain. Requests for data can therefore begin by retrieving the public key from the chain and using it to encrypt the request package. The contributor upon receiving the request can decrypt it with his private and respond as appropriate. Otherwise, the request package can be dropped.

3. *Processing*: This can be accomplished using the smart contract created previously. The accompanying acceptable use policy can contain instructions as to the actions that are permissible on the data. Thus, security of every aspect of processing the data can effectively be achieved using the smart contract feature of blockchain technology.

4. *Control*: The contributor must retain control over shared data on the platform while still making it available to interested parties. Control includes access to the data, duration of access, and access revocation. A smart contract included in the data package can monitor for duration and processing patterns and revoke access in case of violation.

5. *Equity*: The smart contract can also include a counter that logs how often the identifier is accessed to determine how much remuneration is due to the contributor. Thus, once deployed on the blockchain network, the smart contract can provide the metadata required to analyze the demand for a dataset or specific portions of it. Greater demand for a given dataset can therefore translate to greater reward for its owner.

11.1.4.3 Patient Record Regeneration

While there are several methods of ensuring access to patients healthcare data, it is demanding to guarantee the accuracy and completeness of patient's healthcare records at the place and time of care. As is evidenced in the abundant literature, one of the enduring challenges of healthcare since the introduction of electronic medical records (EMRs) is the interoperability of records (patient data) vis-a-vis competing standards, proprietary protocols, and formats that govern the exchange of data from one institution to the other [32]. The matter of interoperability and siloing of patient data comes to the fore primarily because of factors such as the interaction of patient populations with disparate healthcare institutions that lack incentives to engage in institution-to-institution sharing of patient data. Part of the problem also rests with regulatory and compliance regimes that hospitals and healthcare personnel have to abide by entirely. Failure to submit to these industry regulations results in harsh penalties in accordance with HIPAA, HITECH, and recently GDPR as well as other applicable laws. Thus, interoperability suffers because of the undeclared but persistent enforced territoriality or perceived economic advantage of competing hospitals: it is advantageous to institutions when patients' movement is restricted so that they do not oscillate freely from one healthcare provider to the next.

Unfortunately, the increasing rate of migration across international borders as well as fast-emerging public health issues (Covid-19, MERS, SARS, etc.) that affect large populations require the reconsideration of hospital-to-hospital sharing of personal health records. To achieve this largely demands a framework of interoperability to manage patient records in already existing patient databases and to facilitate future patient mobility in a manner that is convenient while guaranteeing access to healthcare with security and privacy already accounted for. While there are some schemes to achieve this objective, they are not designed to permit the patient, who in this case is the primary data owner and ultimate beneficiary, to exercise control over his data. The processes, schemes, and software used may also be vendor-specific, compelling the patient to remain with one or a group of hospitals they no longer prefer. Even where these difficulties may be alleviated, desirable features for the management of sensitive data such as the integrity of records, their availability, and confidentiality depend on the goodwill of hosting institutions and their digital storage platforms. This can easily create a single point of failure from which recovery may be difficult or impossible. It is in this regard that the blockchain presents multiple opportunities to facilitate secure access to patient data [33] while ensuring effective controls over its use and the list of entities permitted to use to the said data. The constructive and resourceful utilization of cryptographic techniques, as well as the consensus mechanism of the underlying peer-to-peer network, provides a safe platform for managing private health records.

Here, we propose a blockchain-based digital health information exchange scheme that can securely facilitate the exchange of personal health records among multiple entities (see Figure 11.10). We first create a permissioned blockchain network to receive all transactions from the participating hospitals. In this context, a transaction is data generated as a result of laboratory tests, doctor's diagnosis, prescriptions, or other such documents generated in support of patient care. Users in the scheme create policies that determine what actions they can perform on the network. They are then permitted to contribute data to particular data stores as determined by their policies. Accordingly, doctors can request data, require tests, and create records while patients can read their personal records and set access rights. Consequently, smart contracts are used to automate actions such as data contribution, data access, and requests for computation. Based on the stored policy, a patient's full record can be regenerated at the point of care using patient ids.

To achieve this, we define two smart contracts: *SortContract* and *AggregatorContract*. The SortContract is executed on the network periodically to find transactions and list them according to the identities and access policy responsible for its creation. The AggregatorContract takes the output of the SortContract and appends the indices of the output to the identity assigned to the use policy. Processing of transactions on the network is done by nodes contributed by participants. An appropriately determined reward scheme can be used to ensure that all transactions in the processing queue have an equal probability of being part of the next block.

11.1.5 Conclusion

This chapter looked at the infusion of blockchain technology into medical data sharing. We examined the broad paradigms of institution-to-institution, patient-to-institution, and patient-to-patient data sharing. The benefits of each paradigm while presenting desirable features for all involved stakeholders also create vulnerabilities that threaten patient privacy and security. Again the continuing reliance on electronic automated data storage and processing infrastructure for analytics and the management of healthcare data implies that sensitive data are routinely exposed to malicious attack vectors. To protect the data and system users, the blockchain's fundamental architecture of consensus-based user transactions validation provides digital mechanisms that can facilitate secure exchange.

In using cryptographic primitives and consensus algorithms, blockchain-based medical data sharing can provide services such as the facilitation of precision medicine strategies, accurate and complete regeneration of patient data at the point of care, and the monetization of medical datasets. While each service is developed and advanced by different interest, they all rely on medical data sharing to be effective. The inclusion of the blockchain in the service adds layered security features that supply the confidence that is required to sustain the dissemination of sensitive personal data. While network security challenges evolve and security technology tries to remain effective, the blockchain promises to remain the disinterested algorithmic manager that ensures safety, security, and privacy of all data committed to its care.

Acknowledgments

This work was partially supported by Sichuan Science and Technology Program (2019YFH0014,2020YFH0030,2020YFSY0061).

References

1 Q. Xia, E. B. Sifah, A. Smahi, S. Amofa, and X. Zhang. BBDS: blockchain-based data sharing for electronic medical records in cloud environments. *Information*, 8 (2): 44, 2017.

2 H. Jin, Y. Luo, P. Li, and J. Mathew. A review of secure and privacy-preserving medical data sharing. *IEEE Access*, 7: 61656–61669, 2019.

3 X. Yang, T. Li, X. Pei, L. Wen, and C. Wang. Medical data sharing scheme based on attribute cryptosystem and blockchain technology. *IEEE Access*, 8: 45468–45476, 2020.

4 C. L. Chen, P. T. Huang, Y. Y. Deng, H. C. Chen, and Y. C. Wang. A secure electronic medical record authorization system for smart device application in cloud computing environments. *Human-Centric Computing and Information Sciences*, 10: 1–31, 2020.

5 S. Nakamoto and A. Bitcoin. A peer-to-peer electronic cash system, 2008. Bitcoin–URL: https://bitcoinorg/bitcoinpdf, pages 4.

6 C. Brodersen, B. Kalis, C. Leong, E. Mitchell, E. Pupo, A. Truscott, et al. Blockchain: securing a new health interoperability experience. *Accenture LLP*, 1–11, 2016.

7 D. Shrier, W. Wu, and A. Pentland. Blockchain & infrastructure (identity, data security). *Massachusetts Institute of Technology-Connection Science*, 1 (3): 1–19, 2016.

8 C. Esposito, A. De Santis, G. Tortora, H. Chang, and K. K. R. Choo. Blockchain: a panacea for healthcare cloud-based data security and privacy? *IEEE Cloud Computing*, 5 (1): 31–37, 2018.

9 L. Yue, H. Junqin, Q. Shengzhi, and W. Ruijin. Big data model of security sharing based on blockchain. In *2017 3rd International Conference on Big Data Computing and Communications (BIGCOM) IEEE*, 2017. pages 117–121.

10 M. Rodriguez-Garcia, M. Batet, and D. Sánchez. Utility-preserving privacy protection of nominal data sets via semantic rank swapping. *Information Fusion*, 45: 282–295, 2019.

11 E. Abebe, D. Behl, C. Govindarajan, Y. Hu, D. Karunamoorthy, P. Novotny, et al. Enabling enterprise blockchain interoperability with trusted data transfer (industry track). In *Proceedings of the 20th International Middleware Conference Industrial Track*, 2019, pages 29–35.

12 T. H. Kim, R. Goyat, M. K. Rai, G. Kumar, W. J. Buchanan, R. Saha, et al. A novel trust evaluation process for secure localization using a decentralized blockchain in wireless sensor networks. *IEEE Access*, 7: 184133–184144, 2019.

13 X. Liang, S. Shetty, D. Tosh, C. Kamhoua, K. Kwiat, and L. Njilla. Provchain: a blockchain-based data provenance architecture in cloud environment with enhanced privacy and availability. In *2017 17th IEEE/ACM International Symposium on Cluster, Cloud and Grid Computing (CCGRID) IEEE*, 2017, pages 468–477.

14 I. Zikratov, A. Kuzmin, V. Akimenko, V. Niculichev, and L. Yalansky. Ensuring data integrity using blockchain technology. In *2017 20th Conference of Open Innovations Association (FRUCT) IEEE*, 2017, pages 534–539.

15 B. K. Zheng, L. H. Zhu, M. Shen, F. Gao, C. Zhang, Y. D. Li, et al. Scalable and privacy-preserving data sharing based on blockchain. *Journal of Computer Science and Technology*, 33 (3), 557–567, 2018.

16 G. Zyskind and O. Nathan Decentralizing privacy: using blockchain to protect personal data. In *2015 IEEE Security and Privacy Workshops IEEE*, 2015, pages 180–184.

17 A. Telenti and X. Jiang. Treating medical data as a durable asset. *Nature Genetics*, 52 (10): 1005–1010, 2020.

18 Q. Xia, E. B. Sifah, K. O. Asamoah, J. Gao, X. Du, and M. Guizani. MeDShare: trust-less medical data sharing among cloud service providers via blockchain. *IEEE Access*, 5: 14757–14767, 2017.

19 S. Amofa, E. B. Sifah, O. B. Kwame, S. Abla, Q. Xia, J. C. Gee, et al. A blockchain-based architecture framework for secure sharing of personal health data. In *2018 IEEE 20th International Conference on e-Health Networking, Applications and Services (Healthcom) IEEE*, 2018, pages 1–6.

20 A. Dorri, S. S. Kanhere, and R. Jurdak. Towards an optimized blockchain for IoT. In *2017 IEEE/ACM 2nd International Conference on Internet-of-Things Design and Implementation (IoTDI) IEEE*, 2017, pages 173–178.

21 I. De Martino, R. D'Apolito, A. S. McLawhorn, K. A. Fehring, P. K. Sculco, and G. Gasparini. Social media for patients: benefits and drawbacks. *Current Reviews in Musculoskeletal Medicine*, 10 (1): 141–145, 2017.

22 D. J. Cote, I. Barnett, J. P. Onnela, and T. R. Smith. Digital phenotyping in patients with spine disease: a novel approach to quantifying mobility and quality of life. *World Neurosurgery*, 126: e241–e249, 2019.

23 A. A. Siyal, A. Z. Junejo, M. Zawish, K. Ahmed, A. Khalil, and G. Soursou. Applications of blockchain technology in medicine and healthcare: challenges and future perspectives. *Cryptography*, 3 (1): 3, 2019.

24 W. O. Hackland A. Hoerbst Managing complexity. From documentation to knowledge integration and informed decision findings from the clinical information systems perspective for 2018. *Yearbook of Medical Informatics*, 28 (1): 95, 2019.

25 G. S. Ginsburg and K. A. Phillips. Precision medicine: from science to value. *Health Affairs*, 37 (5): 694–701, 2018.

26 Z. Shae and J. J. Tsai. On the design of a blockchain platform for clinical trial and precision medicine. In *2017 IEEE 37th International Conference on Distributed Computing Systems (ICDCS) IEEE*, 2017, pages 1972–1980.

27 L. Wang and R. Jones. Big data, cybersecurity, and challenges in healthcare. In *2019 SoutheastCon IEEE*, 2019, pages 1–6.

28 M. Afzal, S. R. Islam, M. Hussain, and S. Lee. Precision medicine informatics: principles, prospects, and challenges. *IEEE Access*, 8: 13593–13612, 2020.

29 A. E. Pritchard, K. Sweeney, C. F. Salorio, and L. A. Jacobson. Pediatric neuropsychological evaluation via telehealth: novel models of care. *The Clinical Neuropsychologist*, 34 (7–8): 1367–1379, 2020.

30 M. Moscatelli, A. Manconi, M. Pessina, G. Fellegara, S. Rampoldi, L. Milanesi, et al. An infrastructure for precision medicine through analysis of big data. *BMC bioinformatics*, 19 (10): 351, 2018.

31 M. B. Hoy. An introduction to the blockchain and its implications for libraries and medicine. *Medical Reference Services Quarterly*, 36 (3): 273–279, 2017.

32 G. G. Dagher, J. Mohler, M. Milojkovic, and P. B. Marella. Ancile: privacy-preserving framework for access control and interoperability of electronic health records using blockchain technology. *Sustainable Cities and Society*, 39: 283–297, 2018.

33 X. Zhang, S. Poslad, and Z. Ma. Block-based access control for blockchain-based electronic medical records (EMRs) query in eHealth. In *2018 IEEE Global Communications Conference (GLOBECOM) IEEE*, 2018, pages 1–7.

12

Decentralized Content Vetting in Social Network with Blockchain

Subhasis Thakur and John G. Breslin

National University of Ireland, Galway, Galway H91 TK33, Ireland

12.1 Introduction

Fake news and misinformation in online social network (OSN) such as Facebook and Twitter are a major financial, social, and political risk [1]. Social bots [2, 3] can facilitate the creation and propagation of fake news in social network. All major social network platforms have recognized the presence of numerous social bots. Malicious entities can create problems in financial, social, and political sectors with fake news.

Financial Markets: Fake news can be used to impact trading in stock exchanges. For example, misinformation regarding US politics had caused a major shock in the US stock market [4]. This misinformation caused a loss of US$17 billion per year for the US retirement savings sector.

Healthcare Systems: Fake news can spread misinformation during a pandemic such as Covid-19. Fake news can encourage people not to participate in vaccination. False information on vaccination costs US$9 billion per year [1].

Political Systems: Fake news is now a political tool. There are numerous examples where fake news was used to gain political advantages by shaping public opinion with false information. Political parties are nowadays hiring companies with expertise in spreading false information on the social network. It is estimated that US$400 million are spent every year on generating fake news for political influence maximization.

Detecting rumor and the source of the rumor is well researched, and several algorithms are developed [5, 6]. Rumor detection techniques use machine learning-based solutions. Identifying the source of rumor [7, 8] can also be an effective tool to deter malicious entities from creating misinformation in social network. However, most of these solutions are centralized solutions. OSNs such as Facebook, Twitter, etc., are centralized platforms. Such centralized platform operators can employ algorithms to detect and prevent misinformation. However, being a centralized entity, an OSN platform operator may be biased and selectively remove misinformation. There are several biased content vetting incidents [9, 10]. A biased content vetting may reduce the credibility and revenue of a social network. In this chapter, we develop a blockchain-based decentralized content vetting for centralized social network.

Wireless Blockchain: Principles, Technologies and Applications, First Edition.
Edited by Bin Cao, Lei Zhang, Mugen Peng and Muhammad Ali Imran.

We use proof of work-based public blockchain (Bitcoin) as the underlying blockchain to execute the decentralized vetting procedure. However, public blockchains have scalability problems. Hence, we use the offline channel network to execute the vetting algorithm. In this vetting procedure, all users get a chance to vote for and against content. If content continuously receives more positive votes, then it continues to propagate. We use token transfer methods for the offline channel network to execute such a voting procedure. Our main contributions in this chapter are as follows:

Unidirectional Offline Channel Model: We have developed an unidirectional offline channel for Bitcoin where a peer can send only a finite number of transactions to another user. Such an offline channel allows us to develop a secure voting mechanism where a user cannot manipulate the voting system.

High-Scale Vetting with Offline Channels: We have developed a high-scale vetting procedure using blockchain offline channels that significantly reduces the number of transactions of the blockchain network.

Channel Network Topology: We developed a method to reduce the number of offline channels needed to execute the vetting procedure.

Evaluation: We prove the efficiency of the content vetting solution using experimental evaluation.

The chapter is organized as follows: In Section 12.2, we discuss the related literature, in Section 12.3, we describe the content propagation model, in Section 12.4, we discuss the decentralized content vetting solution, in Section 12.5, we discuss the method to reduce the number of channels needed to execute the vetting procedure, in Section 12.6, we discuss the social network content propagation simulation algorithms, in Section 12.7, we discuss an experimental evaluation of the content vetting solution, and conclude the chapter in Section 12.8.

12.2 Related Literature

Detecting rumor and the source of the rumor is well researched, and several algorithms are developed [5, 6] to detect rumors. Rumor detection techniques have used machine learning-based algorithms. Identifying the source of rumor [7,8] can also be an effective tool to deter malicious entities from creating misinformation in social network. In this chapter, we develop a rumor prevention mechanism using content vetting by the users.

Blockchain is recently applied to design several social media platforms. SteemIt [11] is a blockchain-based online social media platform that rewards its users for creating and rating new content. SteemIt uses Steem [12], which uses delegated proof of work [13] and is more scalable than a proof of work-based blockchains. Lit [14] is a blockchain-based social network platform that is developed using Ethereum. Users are rewarded for creating content in this social media platform, and the amount of reward depends on the popularity of the content. Sapien [15] is a blockchain-based (Ethereum) social network platform designed to deter false news. A user can join the Sapien platform by locking funds into a smart contract and it may lose these funds if it generates a false content. SocialX [16] is a decentralized social media platform designed to deter fake users from a social media platform. SocialX uses Ethereum as the blockchain. Users are rewarded for checking the validity of media content. Foresting [17] is a blockchain-based social media platform where users are rewarded for creating valuable content and the usefulness of content is judged by users of the Foresting network. Minds [18] is an Ethereum-based social media platform that guarantees that there is no censorship of the content created in this social media platform. Decentralization of the social media platform immunes content from censorship. Minds platform uses both on-chain and off-chain transactions. Guidi [19] presented a detailed characterization of these social media platforms. Jiang and Zhang [20] developed a blockchain-based decentralised social network (DSN). In this social network, user data are kept in the blockchain and a user can modify and delete its data. Additionally, this DSN uses attribute-based encryption to preserve the privacy of the users, and as such, encryption allows access to only a subset of the user data. In [21], a blockchain and IPFS-based DSN model was proposed. It uses Ethereum smart contracts to develop DSN functionalities. Ur Rahman et al. [22] developed a blockchain-based DSN with Ethereum as the blockchain. It uses Ethereum smart contracts for access control over the user data in this DSN. Bahri et al. [23] analyses the security and privacy challenges in developing a DSN platform. Fu and Fang [24] used blockchains for privacy-preserving data management in social network. Yang et al. [25] provides a survey on blockchain-based social network and social media. Guidi et al. [26] analyses reward models for users in DSN. Freni et al. [27] discusses how blockchain can solve privacy and security problems with OSN. Yang et al. [28] proposed a blockchain-based secure friend matching algorithm for OSN.

In this chapter, we use proof of work-based blockchains. It was proposed in [29]. There are several variations of blockchains in terms of consensus protocols. Applications of these various types of blockchains are in various application areas such as energy trade [30], IoT service composition [31], etc. Bitcoin lightning network was proposed in [32], which allows peers to create and transfer funds among them without frequently updating the blockchain. Similar networks were proposed for Ethereum [1] and credit networks [33]. A privacy-preserving payment method in the credit network was proposed in [34]. A routing algorithm for the Bitcoin lightning network was proposed in [35].

Our contributions advance the state of the art in securing social network in the following directions:

- We have developed a content vetting method that allows the users of a social network to evaluate the validity of social media content. The proposed method can securely record such evaluation in a blockchain. It ensures that the evaluation of a user cannot be overwritten.
- We advance state of art in designing social network operations with blockchains by executing social network operation in blockchain offline channels. It improves the scalability of the solution.

12.3 Content Propagation Models in Social Network

There are several models of content propagation in social network [36]. In this chapter, we will use the influence maximization model [37] of content propagation. In this model, a user v_i will share content with its neighbors if it has received the content from at least Δ fraction of its neighbors. $\Delta \in [0, 1]$ is chosen by the user.[1] Δ shows the difficulty to influence a user. We will use the following model of content propagation:

1. Let user B is the creator of content in the social network.
2. User A's decision to share or not share content is as follows:
   (a) If the distance (length of the shortest path between A and B) between A and B is less than λ then A, any neighbor of A can send the content to A and A can share the content if A considers the content as correct information.
   (b) Else, with a fixed probability, A is influenced by its neighbors.
   (c) If A is not influenced, then it makes its own decision regarding the validity of the content. Otherwise, if the number of neighbors who had shared this content is more than a threshold, then A will share the content with its neighbors. Otherwise, it will wait until such several neighbors share the same content (Figure 12.1).

In this content propagation model, an adversarial user will create and share content that will be considered rumor or misinformation by other users (Figure 12.2). We will use the following model of an adversarial user:

1. An adversarial user will create a rumor or misinformation.
2. An adversarial user will share a rumor or misinformation.
3. An adversarial user will share a rumor or misinformation irrespective of how many of its neighbors have shared it.

Social bots can be used by the adversarial user to propagate misinformation and prevent the propagation of correct information. Social bots can be part of the social neighborhood of genuine users. Additionally, social bots may use malware to control information to and from a genuine user. We assume that the adversarial user can control only a finite number of users

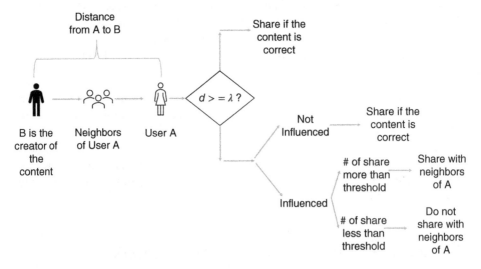

Figure 12.1 Content propagation model.

---

1 This value represents the likelihood that a user can be influenced by its neighbors.

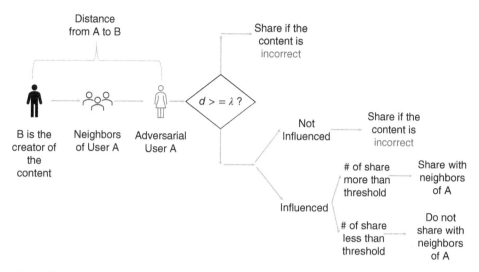

Figure 12.2 Behavior of an adversarial user.

in a social network and there is a cost associated with the number of users the adversarial user can control. $C(k) \in \mathcal{R}^+$ be the function indicating the cost of controlling a fraction k ($k \in [0, 1]$) of social network users. The utility of the adversarial user if misinformation spreads to k fraction of social network users is $U^-(k) \in \mathcal{R}^+$, and utility of the adversarial user if correct information spreads to k fraction of social network users is $U^+(k) \in \mathcal{R}^+$.

12.4 Content Vetting with Blockchains

12.4.1 Overview of the Solution

We use blockchains to develop a decentralized content vetting procedure. It is as follows:

1. Each user of the blockchain network is assumed to be part of a public blockchain network.
2. In the proposed social media sharing procedure, A can share content with B if (a) A can produce a self-attestation that A has examined the content and considered it correct, (b) A can produce a neighborhood content vetting, i.e. similar self-attestation from the neighbors of A.
3. The self-attestation is a random string S for which Hash of S is recorded in the blockchain and anyone can verify the existence of $H(S)$. Self-attestation of a social media content by any user A can be the string (Figure 12.3 S.
4. Blockchains ensure that a user cannot reuse the self-attestation for multiple content i.e. one content and one vote.

12.4.2 Unidirectional Offline Channel

Blockchain offline channels [32] uses multi-signature addresses to open an offline channel among peers of the blockchain. This offline channel [32] is bidirectional and potentially

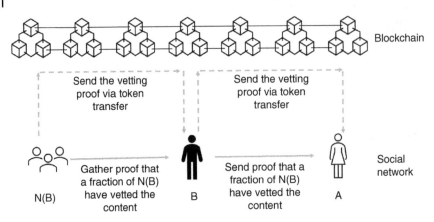

Figure 12.3 Overview of the decentralized vetting procedure.

infinite, i.e. it can execute the infinite number of transfers between two peers provided they do not close the channel and each of them has sufficient funds. We construct an offline channel for proof of work-based public blockchain with the following properties:

- We construct a unidirectional channel between two peers, i.e. only one peer can send funds to another peer of this channel.
- We construct a unidirectional channel that can be used for a finite number of transfers from a designated peer to another peer.

The procedure for creating the unidirectional channel from A to B (A transfers token to B) is as follows: Let A and B are two peers of the channel network H. $M_{A,B}$ is a multi-signature address between A and B. This is a unidirectional channel from A to B.

1. A creates a set of k (k is a positive even integer) random strings $S_A^1, \ldots, S_A^k$. Using these random strings, A creates a set of Hashes $H_H^1 = H(S_B^1), H_B^2 = H(S_B^1), \ldots, H_B^k = H(S_B^k)$, where H is the Hash function (using SHA256). A creates a Merkle tree order λ using these Hashes. Thus, there are k leaf nodes and $k - 1$ non-leaf nodes of this Merkle tree. We denote the non-leaf nodes as $H_A'^1, \ldots, H'(k - 1)_A$.
2. B creates a set of $k1$ random strings $S^1, \ldots, S^k$ and corresponding Hashes $H_B^1, \ldots, H_B^k$.
3. A sends the Merkle tree to B and B sends the set of Hashes $H_B^1, \ldots, H_B^k$ to A.
4. A sends a Hashed time-locked contract $HTLC_A^1$ to B as follows:
   (a) From the multi-signature address $M_{A,B}$, 1 token will be given to A after time T if B does not claim these tokens before time T by producing the key to $H_A'^1$ and 0 token will be given to A if it can produce the key to H_B^1.
   (b) A sends $HTLC_A^1$ to B.
5. Now, A sends 1 token to $M_{A,B}$. A includes the Merkle tree and $H_B^1, \ldots, H_B^k$ in this transaction. This records the Merkle tree and $H_B^1, \ldots, H_B^k$ in the blockchain and any other peer can verify the existence of these Hashes by checking transactions of the public blockchain. Also, at this stage, A's funds are safe as it can get the tokens from $M_{A,B}$ after time T as B does not know $H_A'^1$.

6. Next to send another $(1/k)$ tokens to B, A sends S_A^1 to B and B sends H_B^1 to A. Then, A forms the following HTLC:

   (a) From the multi-signature address $M_{A,B}$, $1 - 1/k$ token will be given to A after time T if B does not claim these tokens before time T by producing the key to $H_A^{\prime 2}$ and $1/k$ token will be given to A if it can produce the key to H_B^2.

   (b) A sends $HTLC_A^2$ to B.

7. This process continues until all keys of the Hashes of non-leaf nodes are revealed by A.

In this model of the unidirectional channel, A is sequentially releasing the keys of the Merkel tree of the HTLCs. Its fund in this channel is decreasing with time. It cannot prevent B from obtaining the tokens as only B can publish the HTLCs. B will publish the HTLC where it gets the maximum value.

12.4.3 Content Vetting with Blockchains

We will use the unidirectional channels described in the previous section to execute the content vetting procedure. It is as follows: We assume that all users of the social network are peers of a blockchain network. We use proof of work-based public blockchains, i.e. Bitcoin. Each user will establish an offline channel with all of its neighbors in the social network. All such channels are unidirectional channels. All channels are marked (such information can be included in the transaction funding the multi-signature address to start the channel) as a positive or negative ballot. Such marking can be verified by any user by checking the transaction record of the blockchain and such marking cannot be changed due to the immutability of blockchain transactions. Briefly, the content vetting procedure is as follows:

1. A user needs to share proof of vetting and proof of neighborhood vetting with its immediate neighbors to share the content.

2. A user's proof of content vetting can be bought by its neighbor by paying the user via a unidirectional channel between them. Such proof is an unknown key of a Hash recorded in the unidirectional channel. For example, (Figure 12.4) A can pay B for content vetting using the channel between them. It will cost A $1/k$ tokens and the most recent key revealed by B for the Hashes $H_B^1, H_B^2, \ldots$, will be regarded as the proof of content vetting.

3. A user's proof of neighborhood vetting can be bought by its neighbor by paying the user via a unidirectional channel between them. Proof of neighborhood vetting is the set of content vetting a user has already bought from its neighbors. A proof of neighborhood vetting can be considered valid by a user if the number of neighbors whose content vetting are included in the neighborhood vetting is more than US50%.

4. Further, proof of content and neighborhood vetting can be categorized as positive and negative voting. Proof of content vetting is regarded as positive voting if the corresponding key belongs to a channel marked as a positive ballot.

5. A rational user will only pay for content vetting if it can sell such vetting information.

We will explain the content vetting procedure with an example. We will use the influence maximization model [37] of content propagation. According to this model, neighbors

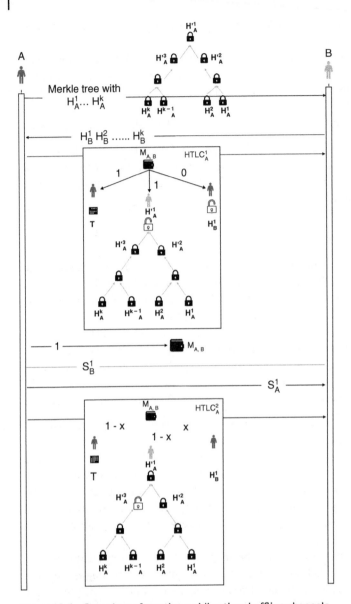

Figure 12.4 Procedure of creating unidirectional offline channels.

of user A can share content with A. If the number of neighbors who had shared the content is more than a fixed threshold, then A may be influenced by its neighbors and may share the content with its other neighbors. The proposed content vetting procedure works similarly. However, a user should produce proof of content vetting and proof of neighborhood vetting while sharing a content with its neighbors. Consider the following scenario (shown in Figure 12.5): Let the creator of content is V_0, v_1 is the neighbor of v_0, v_2 is the neighbor

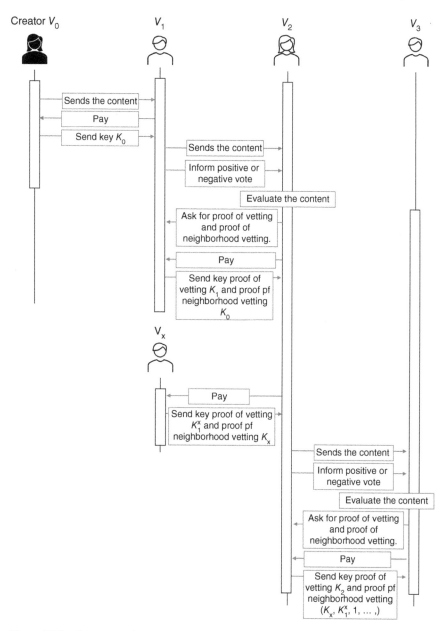

Figure 12.5 Content vetting procedure.

of v_1, and v_3 is the neighbor of v_2. Propagation of the content from v_0 to v_2 with content vetting is as follows:

1. From v_0 to v_1:
   (a) v_0 sends the content to v_1.
   (b) v_1 pays v_0 $1/k$ tokens to get the key k_0 using the unidirectional channel from v_1 to v_0.

2. From v_1 to v_2:
   (a) v_1 sends the content v_2.
   (b) v_1 informs if it will vote positive or negative to v_2.
   (c) v_2 evaluate the content.
   (d) If v_2 evaluate that the content is misinformation (correct) and v_1 informed that it will send a negative (positive) vote then, it asks v_1 to send proof of vetting and proof of neighborhood vetting.
   (e) v_1 sends the proof of vetting by sending the key k_1 in the channel from v_2 to v_1.
   (f) v_2 sends $1/k$ tokens to v_1 using the channel from v_2 to v_1.
   (g) v_1 sends k_0 to v_2 as proof of neighborhood vetting.
3. From v_2 to v_3:
   (a) v_2 sends the content v_3.
   (b) v_2 informs if it will vote positive or negative to v_3.
   (c) v_3 evaluate the content.
   (d) If v_3 evaluate that the content is misinformation (correct) and v_2 informed that it will send a negative (positive) vote then, it asks v_2 to send proof of vetting and proof of neighborhood vetting.
   (e) v_2 sends the proof of vetting by sending the key k_2 in the channel from v_3 to v_2.
   (f) v_3 sends $1/k$ tokens to v_1 using the channel from v_3 to v_2.
   (g) v_3 sends $k_1, k_1^x, \ldots$ (total 0.5λ of such keys from its neighbors) to v_3 as proof of neighborhood vetting.

Note that,

1. Uniqueness of content vetting is guaranteed as an unidirectional channel between two users is updated every time one user asks and pays for content vetting.
2. A user can check the existence of Hashes of the keys presented as neighborhood vetting in the public blockchain. Also, as an unidirectional channel can be used for a finite number of transfers, it ensures that old keys cannot be used as proof of neighborhood vetting.
3. A user pays for the content and neighborhood vetting from its neighbor. A rational user will only do so if it can sell such information to recover such a cost. This means if a user evaluates that content is misinformation and its neighbor is willing to provide a positive vote for it, then it will not buy the content and neighborhood vetting. Similarly, a user will not buy content vetting and neighborhood vetting where the neighbor informed that it is willing to provide a negative vote if it does not consider the content as misinformation.
4. We will show that users who buy negative(positive) vote while the content is correct (incorrect) will have too low funds to buy and share content vetting.

12.5 Optimized Channel Networks

In the above-mentioned content vetting mechanism, we assumed the offline channel network among all users of a social network. It may be difficult to build such a channel network as the number of social neighbors for a user may be too high, and establishing a

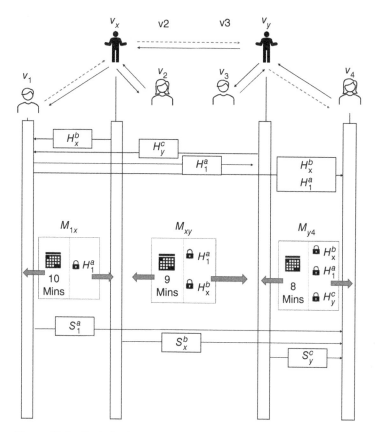

Figure 12.6 Optimized content vetting procedure.

channel will require certain funds in terms of tokens of a blockchain network. We mitigate this problem as follows: We assume that the operator of the social network is represented by multiple nodes of a blockchain network, i.e. the social network operator registers multiple accounts in a blockchain network. A user can establish unidirectional offline channels to and from such peers of the blockchain network representing social network operators. Consider the scenario as shown in the figure in 12.6, v_1 shares unidirectional channels with v_x, v_4 shares unidirectional channels with v_y. v_x and v_y are peers representing the social network operators. v_1 can send proof of content vetting to v_4 as follows (Figure 12.6):

1. $v1$ collects next Hash H_x^b to be used in updating the channel $v_x \rightarrow v_y$.
2. $v1$ collects next Hash H_y^c to be used in updating the channel $v_y \rightarrow v_4$.
3. Let H_1^a be next Hash H_y^c to be used in updating the channel $v_1 \rightarrow v_x$.
4. v_1 informs v_x, v_y, v_4 about all these Hashes.
5. Next, 3 HTLCs are created as follows:
   (a) $HTLC^{1,x}$ states that $1/k$ tokens will be given to v_1 from $M_{1,x}$ (multi-signature address between v_1 and v_x) after 10 minutes unless v_x claims these tokens before 10 minutes by producing the key S_1^a.

(b) $HTLC^{x,y}$ states that $1/k$ tokens will be given to v_x from $M_{x,y}$ (multi-signature address between v_x and v_y) after 9 minutes unless v_y claims these tokens before 9 minutes by producing the keys S_1^a and S_x^b.

(c) $HTLC^{y,4}$ states that $1/k$ tokens will be given to v_y from $M_{y,4}$ (multi-signature address between v_y and v_4) after 8 minutes unless v_4 claims these tokens before 8 minutes by producing the keys S_1^a, S_y^c and S_x^b.

6. After constructing these HTLCs, v_1 sends the key S_1^a to v_4, v_x sends the key S_x^b to v_4, and v_y sends the key S_y^c to v_4.

7. v_4 starts sequential execution of the HTLCs and v_1 gets paid by v_4 via v_x and v_y.

8. In this case, the proof of vetting will be the all keys used in the path from v_1 to v_4, i.e. S_1^a, S_y^c and S_x^b.

The proposed content vetting solution can prevent social bots from spreading rumor or obstructing propagation of correct news:

- A user A (not a social bot or controlled by an adversarial user who wants to spread rumor and prevent propagation of correct information) can verify self-attestation of content vetting and neighborhood vetting by checking if Hash of such proofs exists in the blockchain.
- The same user A can also verify the uniqueness of proof of content vetting and proof of neighborhood vetting, i.e. the same keys are not shared by its neighbors.
- A social bot can bypass the content vetting procedure if A does not follow the content vetting procedure, i.e. does not check for proof of content vetting and proof of neighborhood vetting. However, in this case, A may not be able to spread the rumor as it has not collected proof of neighborhood vetting.

12.6 Simulations of Content Propagation

We will use agent-based modeling of a social network. We will use two sets of simulations, one for modeling the propagation of misinformation and another for the propagation of correct information. The first simulation on the propagation of misinformation (shown in Algorithm 12.1) is as follows:

1. Let $\mathbb{M}$ be the message list of the users, $\mathbb{T}$ be the trust, Δ be the threshold, A be the set nodes controlled by the adversarial node, Frd the number of users who have forwarded the news, and $vlen$ the number of users.

2. At every iteration of the simulation, each user's behavior is as follows:

3. If user i has not sent the social media content, then let d be its distance from the creator of the content. i can check if the creator of the content $\mathbb{S}$ is in its immediate social neighborhood.

4. If $\mathbb{S}$ is not in its immediate social neighborhood, then i will follow the following steps:

   (a) If i is not controlled by the adversarial user then, if the weighted number of neighbors who have sent this content to i (calculated using the trust of i on its neighbors) is more than the threshold $\Delta(i)$ then, with a fixed probability, i will be influenced by its neighbors. In this case, neighbors can influence i to share the content.

   (b) If i is controlled by the adversarial user then, it will share the content with its neighbors.

Algorithm 12.1: Propagation of Misinformation.

Data: $g, \mathbb{T}, \Delta, \mathbb{S}, \mathbb{A}$

Result: Spread = Number of users who had shared the news.

1 **begin**

2    $\mathbb{M} = [0, n \times n]$, $\mathbb{M}[N'(\mathbb{S}), \mathbb{S}] \leftarrow 1$, $Frd \leftarrow [0, 1 \times n]$, $Frd[\mathbb{S}] \leftarrow 1$,
   $sent \leftarrow [0, 1 \times n]$, $sent[\mathbb{S}] \leftarrow 1$, $Spread \leftarrow [1]$

3   **while** *Simulation is not stopped* **do**

4   **for** $i \in [1 : n]$ **do**

5   **if** $sent[i] == 0$ **then**

6   **if** $shortest.paths(\mathbb{S}, i) > 1 \& sent[i] == 0$ **then**

7         $n2 \leftarrow which(\mathbb{M}[i,] > 0), y = \sum(\mathbb{T}[i, n2])$

8   **if** $|n2| > 0$ **then**

9   **if** $i \notin \mathbb{A}$ **then**

10   **if** $y > \Delta[i]$ **then**

11   **if** $Random(1) > .7$ **then**

12             $Frd[i] \leftarrow 1$, $sent[i] \leftarrow 1$, $n3 \leftarrow N'(i)$,
            $\mathbb{M}[n3, i] \leftarrow 1$

13   **else**

14             $sent[i] \leftarrow 1$

15   **else**

16           $sent[i] \leftarrow 1$, $n3 \leftarrow N'(i)$, $\mathbb{M}[n3, i] \leftarrow 1$,
          $Frd[i] \leftarrow 1$

17   **if** $shortest.paths(\mathbb{S}, i) == 1 \& sent[i] == 0$ **then**

18   **if** $|which(\mathbb{M}[i,] > 0)| > 0$ **then**

19   **if** $i \notin \mathbb{A}$ **then**

20           $n1 \leftarrow N(i)$, $n2 \leftarrow which(\mathbb{T}[i,] > 0)$,
          $mean1 \leftarrow mean(\mathbb{T}[i, n2])$
          if $\mathbb{T}[i, start1] > mean1$ **then**

21   **if** $runif(1) > .7$ **then**

22             $sent[i] \leftarrow 1$, $n3 \leftarrow N'(i)$, $\mathbb{M}[n3, i] \leftarrow 1$,

23             $Frd[i] \leftarrow 1$

24   **else**

25             $sent[i] \leftarrow 1$

26   **else**

27           $sent[i] \leftarrow 1$, $n3 \leftarrow N'(i)$, $\mathbb{M}[n3, i] \leftarrow 1$,
          $Frd[i] \leftarrow 1$

28   Add $\sum Frd$ to *Spread*

29   Return(*Spread*)

---

5. If $\mathbb{S}$ is in its immediate social neighborhood, then i will follow these steps:
   (a) If i is not controlled by the adversarial user then, if i's trust in the creator of the content is more than i's average on trust on its neighbors, then with a fixed probability, i will be influenced by the creator of the content and it will share the content.
   (b) If i is controlled by the adversarial user then, it will share the content with its neighbors.
6. Simulation records the number of users who share the content with its neighbors for every iteration.

The second simulation on the propagation of misinformation with content vetting (shown in Algorithm 12.2) is as follows:

1. Let $\mathbb{P}$ be the message list of the users with a positive vote for the social media content, $\mathbb{N}$ be the message list of the users with a positive vote for the social media content, $\mathbb{T}$ be the trust, Δ be the threshold, A be the set nodes controlled by the adversarial node, *Frd* the number of users who have forwarded the news, and n the number of users.
2. At every iteration of the simulation, each user's behavior is as follows:
3. If user i has not sent the social media content, then let d be its distance from the creator of the content. i can check if the creator of the content $\mathbb{S}$ is in its immediate social neighborhood.
4. If $\mathbb{S}$ is not in its immediate social neighborhood, then i will follow these steps:
   (a) If i is not controlled by the adversarial user then, if the weighted ratio between negative and positive votes (weight as trust value) is less than the threshold $\Delta(i)$ then, with a fixed probability i will be influenced by its neighbors. In this case, neighbors can influence i to share the content (and it will send positive votes to such neighbors). Otherwise, it will send negative votes about the content to its neighbors.
   (b) If i is controlled by the adversarial user then, if it will share the content with its neighbors only if the weighted ratio between negative and positive votes is less than the threshold. This is because i needs to prove to its neighbors that it has such a ratio of weighted negative and positive votes.
5. If $\mathbb{S}$ is in its immediate social neighborhood, then i will follow these steps:
   (a) If i is not controlled by the adversarial user then, if i's trust in the creator of the content is more than i's average on trust on its neighbors then with a fixed probability i will be influenced by the creator of the content and it will share the content.
   (b) If i is controlled by the adversarial user then, it will share the content with its neighbors.
6. Simulation records the number of users who share the content with its neighbors for every iteration.

The third simulation on the propagation of correct information (shown in Algorithm 12.3) is as follows:

1. It is similar to the first simulation except:
2. If i is controlled by the adversarial user, then i will not share the content.

The fourth simulation on the propagation of correct information with content vetting (shown in Algorithm 12.4) is as follows:

1. It is similar to the second simulation except:
2. If i is not controlled by the adversarial user and if the weighted ratio between negative and positive votes is less than the threshold, then i will share the content.

Algorithm 12.2: Propagation of Misinformation with Content Vetting.

Data: $g, T, \Delta, \mathbb{S}, \mathbb{A}, \mathbb{P}, \mathbb{N}$

Result: Spread = Number of users who had shared the news.

1 **begin**

2 $\quad\mathbb{P} = [0, n \times n], \mathbb{N} = [0, n \times n] \; \mathbb{P}[N'(\mathbb{S}), \mathbb{S}] \leftarrow 1 Frd \leftarrow [0, 1 \times n) \; Frd[\mathbb{S}] \leftarrow 1$
$\quad sent \leftarrow [0, 1 \times n] \; sent[\mathbb{S}] \leftarrow 1 \; Spread \leftarrow [1]$
$\quad$**while** *Simulation is not stopped* **do**

3 $\quad\quad$**for** $i \in [1 : n]$ **do**

4 $\quad\quad\quad$**if** $sent[i] == 0$ **then**

5 $\quad\quad\quad\quad$**if** $shortest.paths(\mathbb{S}, i) > 1 \& sent[i] == 0$ **then**

6 $\quad\quad\quad\quad\quad n2 \leftarrow which(\mathbb{P}[i,] > 0), y \leftarrow sum(\mathbb{T}[i, n2]),$
$\quad\quad\quad\quad\quad n21 \leftarrow which(\mathbb{N}[i,] > 0), y1 \leftarrow \sum(\mathbb{T}[i, n21])$
$\quad\quad\quad\quad\quad$**if** $|n2| > 0$ **then**

7 $\quad\quad\quad\quad\quad\quad$**if** $i \in \mathbb{A}$ **then**

8 $\quad\quad\quad\quad\quad\quad\quad$**if** $(y1/y) < \Delta[i]$ **then**

9 $\quad\quad\quad\quad\quad\quad\quad\quad$**if** $Random(1) > .7$ **then**

10 $\quad\quad\quad\quad\quad\quad\quad\quad\quad Frd[i] \leftarrow 1, sent[i] \leftarrow 1, n3 \leftarrow N'(i),$
$\quad\quad\quad\quad\quad\quad\quad\quad\quad\mathbb{P}[n3, i] \leftarrow 1$

11 $\quad\quad\quad\quad\quad\quad\quad\quad$**else**

12 $\quad\quad\quad\quad\quad\quad\quad\quad\quad sent[i] \leftarrow 1, n3 \leftarrow N'(i), \mathbb{N}[n3, i] \leftarrow 1$

13 $\quad\quad\quad\quad\quad\quad$**else**

14 $\quad\quad\quad\quad\quad\quad\quad$**if** $(y1/y) < .3$ **then**

15 $\quad\quad\quad\quad\quad\quad\quad\quad sent[i] \leftarrow 1, n3 \leftarrow N'(i), \mathbb{P}[n3, i] \leftarrow 1,$
$\quad\quad\quad\quad\quad\quad\quad\quad Frd[i] \leftarrow 1$

16 $\quad\quad\quad\quad$**if** $d == 1 \& sent[i] == 0$ **then**

17 $\quad\quad\quad\quad\quad$**if** $|(which(\mathbb{P}[i,] > 0))| > 0$ **then**

18 $\quad\quad\quad\quad\quad\quad$**if** $i \notin \mathbb{A}$ **then**

19 $\quad\quad\quad\quad\quad\quad\quad n2 \leftarrow which(\mathbb{T}[i,] > 0)$
$\quad\quad\quad\quad\quad\quad\quad mean1 \leftarrow mean(\mathbb{T}[i, n2])$**if** $\mathbb{T}[i, \mathbb{S}] > mean1$ **then**

20 $\quad\quad\quad\quad\quad\quad\quad\quad$**if** $random(1) > .7$ **then**

21 $\quad\quad\quad\quad\quad\quad\quad\quad\quad sent[i] \leftarrow 1 \; n3 \leftarrow N'(i) \; \mathbb{P}[n3, i] \leftarrow 1$
$\quad\quad\quad\quad\quad\quad\quad\quad\quad Frd[i] \leftarrow 1$

22 $\quad\quad\quad\quad\quad\quad\quad\quad$**else**

23 $\quad\quad\quad\quad\quad\quad\quad\quad\quad sent[i] \leftarrow 1, n3 \leftarrow N'(i), \mathbb{N}[n3, i] \leftarrow 1$

24 $\quad\quad\quad\quad\quad\quad$**else**

25 $\quad\quad\quad\quad\quad\quad\quad sent[i] \leftarrow 1, n3 \leftarrow N'(i), \mathbb{P}[n3, i] \leftarrow 1,$
$\quad\quad\quad\quad\quad\quad\quad Frd[i] \leftarrow 1$

26 $\quad\quad$Add $\sum(Frd)$ to *Spread*.

27 $\quad$Return(*Spread*)

Algorithm 12.3: Propagation of Correct Information.

Data: $g, T, \Delta, \mathbb{S}, \mathbb{A}$

Result: Spread = Number of users who had shared the news.

1 **begin**

2    $\mathbb{M} = [0, n \times n], \mathbb{M}[N'(\mathbb{S}), \mathbb{S}] \leftarrow 1, Frd \leftarrow [0, 1 \times n], Frd[\mathbb{S}] \leftarrow 1,$
   $sent \leftarrow [0, 1 \times n], sent[\mathbb{S}] \leftarrow 1$

3   **while** *Simulation is not stopped* **do**

4   **for** $i \in [1 : n]$ **do**

5   **if** $sent[i] == 0$ **then**

6          $d \leftarrow shortest.paths(g, \mathbb{S}, i)$

7   **if** $d > 1 \& sent[i] == 0$ **then**

8            $n1 \leftarrow N(i) \, n2 \leftarrow which(\mathbb{M}[i,] > 0) \, y = \sum(\mathbb{T}[i, n2])$

9   **if** $|n2| > 0$ **then**

10   **if** $i \in \mathbb{A}$ **then**

11   **if** $y > \Delta[i]$ **then**

12   **if** $runif(1) > .7$ **then**

13                    $Frd[i] \leftarrow 1 \, sent[i] \leftarrow 1 \, n3 \leftarrow N'(i)$
                     $\mathbb{M}[n3, i] \leftarrow 1$

14   **else**

15                    $Frd[i] \leftarrow 1 \, sent[i] \leftarrow 1 \, n3 \leftarrow N'(i)$
                     $\mathbb{M}[n3, i] \leftarrow 1$

16   **else**

17                $sent[i] \leftarrow 1$

18   **else**

19   **if** $|which(\mathbb{M}[i,] > 0)| > 0$ **then**

20   **if** $i \notin \mathbb{A}$ **then**

21                $n1 \leftarrow N(i) \, n2 \leftarrow which(\mathbb{T}[i,] > 0)$
               $mean1 \leftarrow mean(\mathbb{T}[i, n2])$

22   **if** $\mathbb{T}[i, \mathbb{S}] > mean1$ **then**

23   **if** $runif(1) > .7$ **then**

24                    $Frd[i] \leftarrow 1 \, sent[i] \leftarrow 1 \, n3 \leftarrow N'(i)$
                     $\mathbb{M}[n3, i] \leftarrow 1$

25   **else**

26                    $Frd[i] \leftarrow 1 \, sent[i] \leftarrow 1 \, n3 \leftarrow N'(i)$
                     $\mathbb{M}[n3, i] \leftarrow 1$

27   **else**

28                $sent[i] \leftarrow 1$

29   Add $\sum(Frd)$ to *Spread*

30   Return(*Spread*)

Algorithm 12.4: Propagation of Correct Information with Content Vetting.

Data: $g, T, \Delta, \mathbb{S}, \mathbb{A}$

Result: Spread = Number of users who had shared the news.

1 **begin**

2    $\mathbb{P} = [0, n \times n], \mathbb{N} = [0, n \times n], \mathbb{P}[N(\mathbb{S}, \text{``out''}), \mathbb{S}] = 1, F = [0, 1 \times n], F[\mathbb{S}] = 1,$
     $s = [0, 1 \times n], s[\mathbb{S}] = 1$

3   **while** *Simulation is not stopped* **do**

4   **for** $i \in [1 : n]$ **do**

5   **if** $s[i] == 0$ **then**

6   **if** *shortest.paths*$(\mathbb{S}, i) > 1 \& s[i] == 0$ **then**

7            $n2 = which(\mathbb{P}[i,] > 0), y = \sum(\mathbb{T}[i, n2]),$
           $n21 = which(\mathbb{N}[i,] > 0), y1 = \sum(\mathbb{T}[i, n21])$

8   **if** $|n2| > 0$ **then**

9   **if** $i \notin \mathbb{A}$ **then**

10   **if** $\frac{y1}{y} < \Delta[i]$ **then**

11   **if** $runif(1) > .7$ **then**

12                  $F[i] = 1, s[i] = 1 \; n3 = N'(i) \; \mathbb{P}[n3, i] = 1$

13   **else**

14                  $F[i] = 1, s[i] = 1, n3 = N'(i), \mathbb{P}[n3, i] = 1$

15   **else**

16   **if** $y + y1 > .3 \times |N(i)|$ **then**

17                  $s[i] = 1, n3 = N'(i), \mathbb{N}[n3, i] = 1$

18   **else**

19   **if** $\sum(\mathbb{P}[i,]), \sum(\mathbb{N}[i,]) > 0 \& \frac{\sum(\mathbb{P}[i,])}{\sum(\mathbb{N}[i,])} < .14$ **then**

20              $s[i] = 1, n3 = N'(i), \mathbb{N}[n3, i] = 1$

21   **else**

22   **if** $|which(\mathbb{P}[i,] > 0)| > 0$ **then**

23   **if** $i \notin \mathbb{A}$ **then**

24              $n2 \leftarrow which(\mathbb{T}[i,] > 0)$

25   **if** $\mathbb{T}[i, \mathbb{S}] > mean(\mathbb{T}[i, n2])$ **then**

26   **if** $Random(1) > .7$ **then**

27                  $F[i] = 1, s[i] = 1, n3 = N'(i), \mathbb{P}[n3, i] = 1$

28   **else**

29                  $F[i] = 1, s[i] = 1, n3 = N'(i), \mathbb{P}[n3, i] = 1$

30   **else**

31            $s[i] = 1, n3 = N'(i), \mathbb{N}[n3, i] = 1$

32   Add $\sum(F)$ to *Spread*

---

3. If i is controlled by the adversarial user, then i will not share the content (i.e. it will send a negative vote regarding the content) if its fraction of negative and positive votes is more than the threshold and it has received votes from at least a fixed fraction of its neighbors.

Let $\mathbb{M}$ is the message list of the users, $\mathbb{T}$ is trust, Δ is the threshold, A is set nodes controlled by the adversarial node, *vlen* is the number of users, *Frd* is the number of users who have forwarded the news, $\mathbb{P}$ and $\mathbb{N}$ are messages with positive and negative vote, respectively, and *Spread* is the number of users who had shared the content.

12.7 Evaluation with Simulations of Social Network

We use Facebook network data from [38]. This social network is a directed graph with 4039 nodes and 88234 edges. We simulate content propagation in social network using the simulation algorithms shown in Section 12.6. We simulate a blockchain network using agent-based modeling of the blockchain network. We use an asynchronous event simulator (using the SIMPY library of Python). The workflow of each agent (who simulates a peer of the blockchain network) is as follows:

1. Each peer executes four processes in parallel.
2. $Process_3$ receives messages from its neighbors, and if the message is not received before, then it checks if the message contains a new transaction or a new block. If it receives a transaction, then it informs $Process_2$ about the new transaction. If it receives a new block, then it informs $Process_4$ about the new block.
3. $Process_2$ gathers new transactions from $Process_3$, and the new transaction is placed in a queue of undocumented transactions. We assume that the queue model is First In First Out. After adding the new transaction to its queue, a peer forwards the message containing the new transaction to its neighbors.
4. $Process_1$ empties the first k transactions from its queue of undocumented transactions and creates a new block. Then, it solves the puzzle of proof of work protocol and publishes the new block.
5. $Process_4$ examines the new block from $Process_3$, if all transactions of the new block are valid then: if the parent block of the new block is the last blockchain head known to the peer, then it augments its blockchain by placing the new block as child block of its last known blockchain head and recognize the new block as the last know blockchain head. Otherwise, it finds the parent block of the new block in its blockchain and augments the blockchain by adding the new block as its child block.
6. We simulate offline channel network as follows:
   (a) Two multi-signature addresses are needed to create an offline channel between two peers. We simulate such multi-signature address as lists. The initial and final balance of such lists are accessible by all peers, i.e. these lists are shared variables among all instances of the peer class.
   (b) Two peer exchanges messages to exchange HTLCs.

First, we simulate the propagation of incorrect social media content. We execute two sets of experiments. In the first set of experiments, we increase the number of users controlled

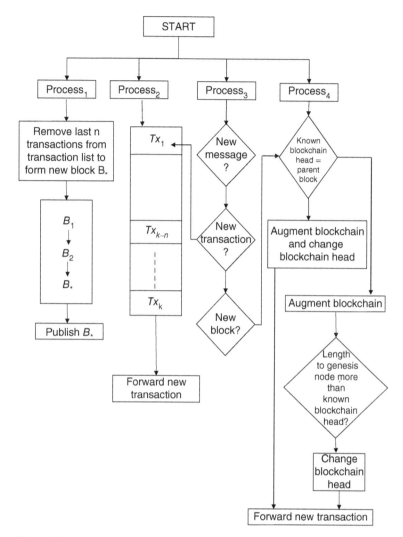

Figure 12.7 Workflow of each miner of the blockchain network.

by the adversarial user from 15% to 22% (1% in each increment) (Figure 12.7). The result of this set of simulations is shown in Figures 12.8 and 12.9. It shows that with content vetting, the incorrect content is shared by a less number of users and more quickly compared with content propagation without vetting. In the second set of experiments, we keep the number of users who are controlled by the adversarial user to 20% but we gradually increase the threshold from 0.3 to 0.65 (increment of 0.5 for each experiment). The result of this experiment is shown in Figures 12.10 and 12.11. It shows that with content vetting, the incorrect content is shared by a fewer number users compared with content propagation without vetting.

Next, we simulate the propagation of correct social media content. We execute two sets of experiments. In the first set of experiments, we keep the number of users who are

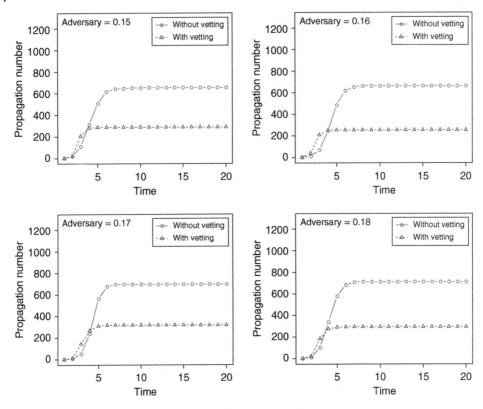

Figure 12.8 Propagation of incorrect information with increasing number of users controlled by the adversarial user. We increase the number of such users from 15% to 18%. It shows that with content vetting, the number of users who shared the incorrect information remains low. Also, it shows that the number of users who shared the incorrect information without content vetting increases as the number of users controlled by the adversarial user is increased.

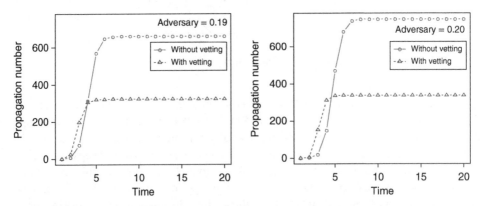

Figure 12.9 Propagation of incorrect information with increasing number of users controlled by the adversarial user. We increase the number of such users from 19% to 22%. It shows that with content vetting, the number of users who shared the incorrect information remains low. Also, it shows that the number of users who shared the incorrect information without content vetting increases as the number of users controlled by the adversarial user is increased.

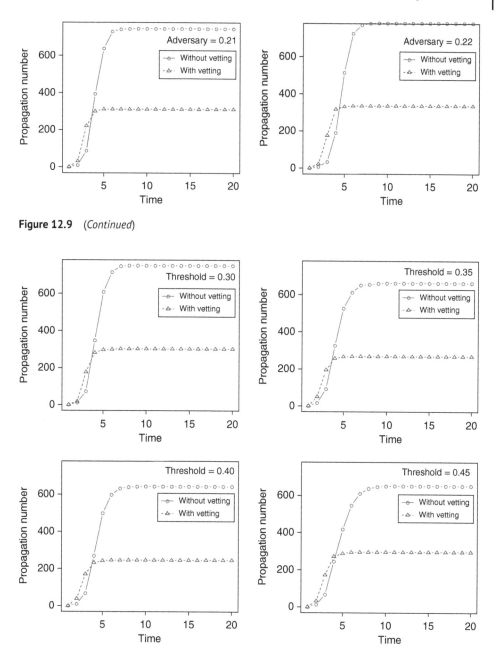

Figure 12.9 (*Continued*)

Figure 12.10 Propagation of incorrect information with increasing threshold of the users. We increase the threshold from 0.3 to 0.45. The threshold is the minimum weighted (calculated with using trust among the users) number of neighbors who had shared the content with the user. This experiment shows that content vetting reduces the number of users who shared incorrect information.

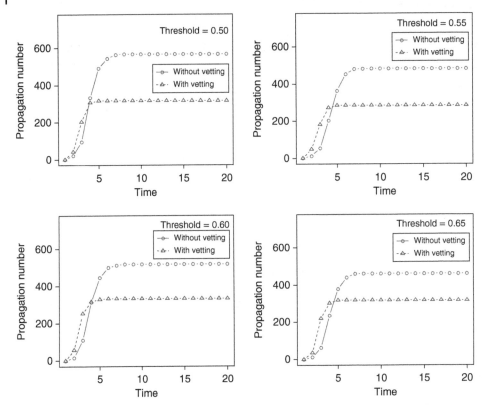

Figure 12.11 Propagation of incorrect information with increasing threshold of the users. We increase the threshold from 0.5 to 0.65. The threshold is the minimum weighted (calculated with using trust among the users) number of neighbors who had shared the content with the user. This experiment shows that content vetting reduces the number of users who shared incorrect information. It also shows that as the threshold is increased, the number of users who share the incorrect content is reduced. This is because as we increase the threshold, it becomes more difficult to influence a user.

controlled by the adversarial user to 20%, but we gradually increase the threshold from 0.3 to 0.55 (increment of 0.5 for each experiment). The result of this experiment is shown in Figures 12.12 and 12.13. It shows that with content vetting, the correct content is shared by more users and more quickly compared with content propagation without vetting.

In the second set of experiments, we keep the threshold level of the users constant (0.3), but we increase the number of users controlled by the adversarial user from 35% to 56% (3% in each increment). The results of this set of simulations are shown in Figures 12.14 and 12.15. It shows that with content vetting, the correct content is shared by more users and more quickly compared with content propagation without vetting. It also shows that an increment of the number of users controlled by the adversarial user significantly reduces the number of users who have shared the correct content.

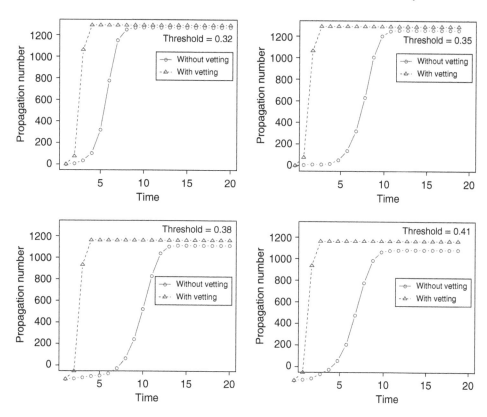

Figure 12.12 Propagation of correct information with increasing threshold of the users. We increase the threshold from 0.32 to 0.41. The threshold is the minimum weighted (calculated with using trust among the users) number of neighbors who had shared the content with the user required to influence a user. This experiment shows that content vetting increases the number of users who shared the correct information.

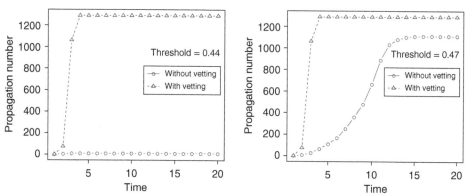

Figure 12.13 Propagation of correct information with increasing threshold of the users. We increase the threshold from 0.44 to 0.53. The threshold is the minimum weighted (calculated using trust among the users) number of neighbors who had shared the content with the user to influence a user. This experiment shows that content vetting increases the number of users who shared incorrect information. It also shows that as the threshold is increased, the number of users who share the correct content without vetting is reduced. This is because as we increase the threshold, it becomes more difficult to influence a user.

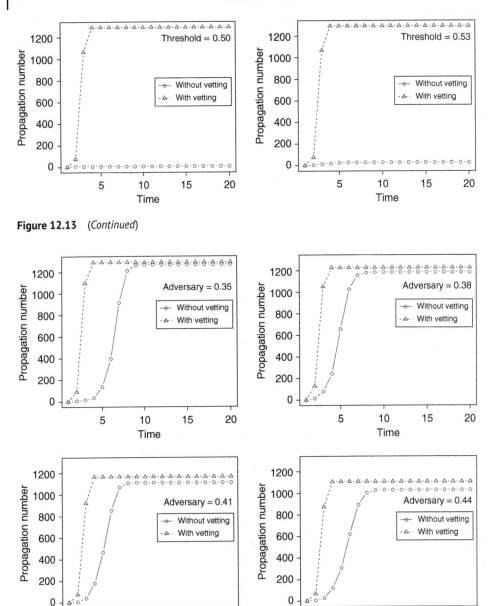

Figure 12.13 (*Continued*)

Figure 12.14 Propagation of correct information with increasing number of users controlled by the adversarial user. We increase the number of such users from 35% to 44%. It shows that with content vetting, the number of users who shared the correct information remains higher. Also, it shows that increasing the number of users controlled by the adversarial user decreases the number of users who shared the correct content.

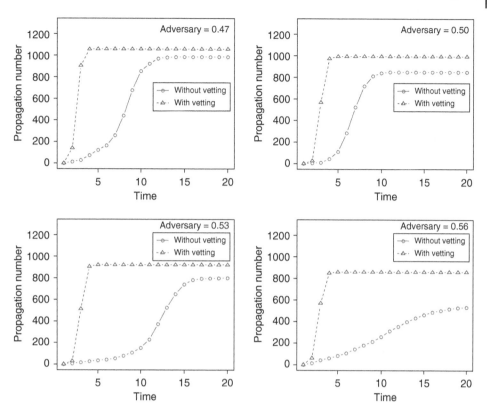

Figure 12.15 Propagation of correct information with increasing number of users controlled by the adversarial user. We increase the number of such users from 47% to 56%. It shows that with content vetting the number of users who shared the correct information remains higher. Also, it shows that increasing the number of users controlled by the adversarial user decreases the number of users who shared the correct content.

12.8 Conclusion

In this chapter, we developed a decentralized content vetting in social network procedure using public proof of work blockchains. We use offline channels to execute the vetting procedure. We showed that the vetting procedure can significantly reduce the propagation of social media content that the users have voted as rumor or misinformation. In the future, we will extend this vetting procedure with functionalities that can protect the privacy of the users who have vetted content.

Acknowledgment

This publication has emanated from research supported in part by a research grant from the Department of Enterprise, Trade and Employment's Disruptive Technologies Innovation Fund (DTIF), managed by Enterprise Ireland on behalf of the Government of Ireland under

Grant Number DT20180040C, and by a research grant from Science Foundation Ireland (SFI) and the Department of Agriculture, Food and the Marine on behalf of the Government of Ireland under Grant Number SFI 16/RC/3835 (VistaMilk) and also by a research grant from SFI under Grant Number SFI 12/RC/2289_P2 (Insight), with the latter two grants co-funded by the European Regional Development Fund.

References

1 University of Baltimore The economic cost of bad actors on the internet: fake news in 2019. https://www.cheq.ai/fakenews. Accessed: 2021-02-17.

2 Y. Boshmaf, I. Muslukhov, K. Beznosov, and M. Ripeanu. Design and analysis of a social botnet. *Computer Networks*, 57 (2): 556–578, Feb. 2013. ISSN 1389-1286. https://doi.org/10.1016/j.comnet.2012.06.006.

3 E. Ferrara, O. Varol, C. Davis, F. Menczer, and A. Flammini. The rise of social bots. *Communications of the ACM*, 59 (7): 96–104, June 2016. ISSN 0001-0782. https://doi.org/10.1145/2818717.

4 Ironman at Political Calculations Follow The cost of fake news for the S&P 500. https://seekingalpha.com/article/4129355-cost-of-fake-news-for-s-and-p-500. Accessed: 2021-02-17.

5 A. R. Pathak, A. Mahajan, K. Singh, A. Patil, and A. Nair. Analysis of techniques for rumor detection in social media. *Procedia Computer Science*, 167: 2286–2296, 2020. ISSN 1877-0509. https://doi.org/https://doi.org/10.1016/j.procs.2020.03.281. International Conference on Computational Intelligence and Data Science.

6 A. Zubiaga, A. Aker, K. Bontcheva, M. Liakata, and R. Procter. Detection and resolution of Rumours in social media: a survey. *ACM Computing Surveys*, 51 (2), Feb. 2018. ISSN 0360-0300. https://doi.org/10.1145/3161603.

7 J. Jiang, S. Wen, S. Yu, Y. Xiang, and W. Zhou. Rumor source identification in social networks with time-varying topology. *IEEE Transactions on Dependable and Secure Computing*, 15 (1): 166–179, 2018. https://doi.org/10.1109/TDSC.2016.2522436.

8 M. Farajtabar, M. G. Rodriguez, M. Zamani, N. Du, H. Zha, and L. Song. Back to the Past: Source Identification in Diffusion Networks from Partially Observed Cascades. In G. Lebanon and S. V. N. Vishwanathan, editors, *Proceedings of the 18th International Conference on Artificial Intelligence and Statistics*, volume 38 of *Proceedings of Machine Learning Research*, pages 232–240, San Diego, California, USA, 09–12 May 2015. PMLR. http://proceedings.mlr.press/v38/farajtabar15.html.

9 E. A. Vogels, A. Perrin, and M. Anderson. Most Americans think social media sites censor political viewpoints. https://www.pewresearch.org/internet/2020/08/19/most-americans-think-social-media-sites-censor-political-viewpoints/. Accessed: 2021-02-17.

10 Social media: is it really biased against us republicans? https://www.bbc.com/news/technology-54698186. Accessed: 2021-02-17.

11 A. Kiayias, B. Livshits, A. M. Mosteiro, and O. S. T. Litos. A puff of steem: security analysis of decentralized content curation. *CoRR*, abs/1810.01719, 2018. http://arxiv.org/abs/1810.01719.

12 C. Li and B. Palanisamy. Incentivized blockchain-based social media platforms: a case study of steemit. In *Proceedings of the 10th ACM Conference on Web Science*, WebSci '19, pages 145–154, New York, NY, USA, 2019. Association for Computing Machinery. ISBN 9781450362023. https://doi.org/10.1145/3292522.3326041.

13 DPOS consensus algorithm - the missing white paper. https://steemit.com/dpos/ @dantheman/dpos-consensus-algorithm-this-missing-white-paper. Accessed: 2021-02-17.

14 Mith. https://mith.io/en-US/. Accessed: 2021-02-17.

15 N. Aurélien B. Ankit, and G. Robert. Sapien. Decentralized social news platform. 2018. https://www.sapien.network/static/pdf/SPNv1_4.pdf.

16 S.P. Ltd. The socialx ecosystem takes the social media experience to the next level. 2018. https://socialx.network/wp-content/uploads/2018/09/Whitepaper-SocialX-v1.1.pdf.

17 Foresting. Rewarding lifestyle social media. 2019. https://cdn.foresting.io/pdf/ whitepaper/FORESTING_Whitepaper_Eng_Ver.1.0.pdf?ver0.2.

18 Minds. The crypto social network. 2019. https://cdn-assets.minds.com/front/dist/en/ assets/documents/Whitepaper-v0.5.pdf.

19 B. Guidi. When blockchain meets online social networks. *Pervasive and Mobile Computing*, 62: 101131, 2020. ISSN 1574-1192. https://doi.org/https://doi.org/10.1016/ j.pmcj.2020.101131.

20 L. Jiang and X. Zhang. BCOSN: a blockchain-based decentralized online social network. *IEEE Transactions on Computational Social Systems*, 6 (6): 1454–1466, 2019.

21 Q. Xu, Z. Song, R. S. Mong Goh, and Y. Li. Building an ethereum and IPFS-based decentralized social network system. In *2018 IEEE 24th International Conference on Parallel and Distributed Systems (ICPADS)*, pages 1–6, 2018.

22 M. Ur Rahman, B. Guidi, and F. Baiardi. Blockchain-based access control management for decentralized online social networks. *Journal of Parallel and Distributed Computing*, 144: 41–54, 2020. ISSN 0743-7315. https://doi.org/https://doi.org/10.1016/ j.jpdc.2020.05.011.

23 L. Bahri, B. Carminati, and E. Ferrari. Decentralized privacy preserving services for online social networks. *Online Social Networks and Media*, 6: 18–25, 2018. ISSN 2468-6964. https://doi.org/https://doi.org/10.1016/j.osnem.2018.02.001.

24 D. Fu and L. Fang. Blockchain-based trusted computing in social network. In *2016 2nd IEEE International Conference on Computer and Communications (ICCC)*, pages 19–22, 2016.

25 F. Yang, Y. Pu, C. Hu, and Y. Zhou. A blockchain-based privacy-preserving mechanism for attribute matching in social networks. In D. Yu, F. Dressler, and J. Yu, editors, *Wireless Algorithms, Systems, and Applications*, pages 627–639, Cham, 2020. Springer International Publishing. ISBN 978-3-030-59016-1.

26 B. Guidi, V. Clemente, T. García, and L. Ricci. A rewarding model for the next generation social media. In *Proceedings of the 6th EAI International Conference on Smart Objects and Technologies for Social Good*, GoodTechs '20, pages 169–174, New York, NY, USA, 2020. Association for Computing Machinery. ISBN 9781450375597. https://doi.org/10.1145/3411170.3411247.

27 P. Freni, E. Ferro, and G. Ceci. Fixing social media with the blockchain. In *Proceedings of the 6th EAI International Conference on Smart Objects and Technologies for*

Social Good, GoodTechs '20, page 175–180, New York, NY, USA, 2020. Association for Computing Machinery. ISBN 9781450375597. https://doi.org/10.1145/3411170.3411246.

28 F. Yang, Y. Wang, C. Fu, C. Hu, and A. Alrawais. An efficient blockchain-based bidirectional friends matching scheme in social networks. *IEEE Access*, 8: 150902–150913, 2020.

29 S. Nakamoto. Bitcoin: a peer-to-peer electronic cash system. www.bitcoin.org, 2008.

30 B. P. Hayes, S. Thakur, and J. G. Breslin. Co-simulation of electricity distribution networks and peer to peer energy trading platforms. *International Journal of Electrical Power & Energy Systems*, 115: 105419, 2020. ISSN 0142-0615. https://doi.org/https://doi.org/10.1016/j.ijepes.2019.105419.

31 I. A. Ridhawi, M. Aloqaily, A. Boukerche, and Y. Jaraweh. A blockchain-based decentralized composition solution for IoT services. In *ICC 2020 - 2020 IEEE International Conference on Communications (ICC)*, pages 1–6, 2020.

32 J. Poon and T. Dryja. The Bitcoin Lightning Network: Scalable Off-Chain Instant Payments. https://lightning.network/lightning-network-paper.pdf.

33 G. Malavolta, P. Moreno-Sanchez, A. Kate, and M. Maffei. SilentWhispers: enforcing security and privacy in decentralized credit networks. *IACR Cryptology ePrint Archive*, 1054, 2016.

34 P. Moreno-Sanchez, A. Kate, M. Maffei, and K. Pecina. Privacy preserving payments in credit networks: enabling trust with privacy in online marketplaces. In *NDSS*, 2015.

35 P. Prihodko, S. Zhigulin, M. Sahno, A. Ostrovskiy, and O. Osuntokun. Flare: an approach to routing in lightning network white paper. 2016.

36 A. Guille, H. Hacid, C. Favre, and D. A. Zighed. Information diffusion in online social networks: a survey. *SIGMOD Record*, 42 (2): 17–28, July 2013. ISSN 0163-5808. https://doi.org/10.1145/2503792.2503797.

37 D. Kempe, J. Kleinberg, and É. Tardos. Maximizing the spread of influence through a social network. In *Proceedings of the 9th ACM SIGKDD International Conference on Knowledge Discovery and Data Mining*, KDD '03, pages 137–146, New York, NY, USA, 2003. Association for Computing Machinery. ISBN 1581137370. https://doi.org/10.1145/956750.956769.

38 J. Leskovec and J. Mcauley. Learning to discover social circles in ego networks. In F. Pereira, C. J. C. Burges, L. Bottou, and K. Q. Weinberger, editors, *Advances in Neural Information Processing Systems*, volume 25, pages 539–547. Curran Associates, Inc., 2012. https://proceedings.neurips.cc/paper/2012/file/7a614fd06c325499f1680b9896beedeb-Paper.pdf.

Index

Wireless Blockchain: Principles, Technologies and Applications, First Edition.
Edited by Bin Cao, Lei Zhang, Mugen Peng and Muhammad Ali Imran.
© 2022 John Wiley & Sons Ltd. Published 2022 by John Wiley & Sons Ltd.